Bones

BONES

STRUCTURE AND MECHANICS

JOHN D. CURREY

PRINCETON UNIVERSITY PRESS

PRINCETON AND OXFORD

Published by Princeton University Press, 41 William Street, Princeton, New Jersey 08540
In the United Kingdom: Princeton University Press, 3 Market Place, Woodstock, Oxfordshire OX20 1SY

Second printing, and first paperback printing, 2006
Paperback ISBN-13: 978-0-691-12804-7
Paperback ISBN-10: 0-691-12804-9

The Library of Congress has cataloged the cloth edition of this book as follows

Currey, John D., 1932–
Bones : structure and mechanics / John D. Currey.
p. cm.
Includes bibliographical references and index.
ISBN 0-691-09096-3 (cl. : alk. paper)
1. Bones. 2. Biomechanics. I. Title.

QP88.2 .C867 2002
573.7′6—dc21 2001043148

British Library Cataloging-in-Publication Data is available

This book has been composed in Sabon

Printed on acid-free paper. ∞

pup.princeton.edu

Printed in the United States of America

10 9 8 7 6 5 4 3 2

CONTENTS

PREFACE TO THE SECOND EDITION

I MENTIONED in the first edition people who have influenced my thinking about bone. In the nearly two decades that have passed since then many more people have made me glad that I chose the field I did. It is invidious to mention names, there are so many, but here are a few. Neill Alexander continues to astonish me with his productivity and clear thinking; Steve Cowin always makes me feel like a student starting graduate school, full of dubious certainties, but he is very kind about it; Rik Huiskes has had an immense influence on the field of biomechanics; David Burr, Alan Boyde, Dennis Carter, Dwight Davy, Lorna Gibson, Tom McMahon, Anders Odgaard, Clint Rubin, Steve Weiner, Peter Zioupos—I could go on and on. I find it interesting, but, I suppose, unsurprising, that most of the people who have influenced me are those whose ideas I often disagree with, but whose experiments are hard to quarrel with. Jillian continues to connect me to the real world.

In writing this book I became increasingly aware of the truth of Tintoretto's statement so applicable to academic study: "The study of painting is exhausting, and the more one gets into it the more difficulties arise and the sea always gets wider." This was a favorite quotation of John Ruskin, and I found it in John Dixon Hunt's splendid study of Ruskin (Hunt 1982). At times I felt that the results coming from the study of bone were accumulating faster than I, in my snail-paced way, could get to grips with and incorporate. In some areas this is certainly true; I could not now pretend to deal with the *biology* of adaptive remodeling, for instance. However, in many areas, although the study of bone has progressed, it has progressed at a rather steady pace, and, indeed, some of the book is not very different from the first edition, although I hope its organization is better. The improvement of the organization owes a debt to the anonymous readers who reviewed the draft of this edition. I have kept an anxious watch on the number of references that postdate 1983 (when the first edition went to bed) and am amused to find that this edition includes a number of new references that predate the first edition. Being old does not always mean being passé!

Steve Bowman cleverly tracked down Ralph Mack and obtained a copy of his 1964 paper. I had been cavalierly quoting it for years without having actually read it.

Katrina Attwood kindly translated the passage from Egils Saga.

Wighart von Koenigswald gave me the stunning photograph of the enamel of *Dicrostonyx torquatus.*

Anusuya Chinsamay-Turan kindly sent me photographs of the remarkable teeth of *Pterodaustro guiñazui.*

Chris Rees and Jillian read parts of the manuscript, in Jillian's case "checking for pomposity."

PREFACE TO THE FIRST EDITION

COMMENTING on an excellent book on vertebrates, a colleague of mine remarked, unfairly, that it gave him the feeling that "animals were just a collection of levers held together by mathematical formulae." There is another class of book that leaves one with the feeling that animals are just "oohs" and "aahs" separated by a great deal of hand-waving. This book will occasionally lurch to one extreme or the other, but I have tried to make it travel down the middle.

As the title implies, I have written this book from the point of view of adaptation, and mechanical adaptation at that. [The title has changed from the first edition, but the philosophy is the same.] As the great population geneticist, Theodosius Dobzhansky said, "Nothing in biology makes sense except in the light of evolution," and our skeletons have been as much part of biology as the rest of us. I regard bones as the best possible solutions to selective problems and, even if they are not, it is better to think of them as such rather than as having particular attributes through historical accident.

I hope this book will be intelligible and interesting to people who have little mechanics, or who have little biology, or whose thoughts about animals have been restricted to mammals. All these people will have to take some things on trust.

Many people have influenced my thinking about bones, but there are a few I wish particularly to mention. Harold Pusey taught me vertebrates at Oxford, and imbued me with the feeling that everything, in the end, was explicable. Arthur Cain, companion on many field trips, convinced me of the optimality of animals practically before the word had been invented. Al Burstein, a generous friend, taught me what little formal biomechanics I know, but failed to engender in me his love of skeet shooting. Neill Alexander, the master biomechanic, has the unsettling tendency to think of my bright ideas two years before I do. Both David White and Julian Bryant critically read much of the manuscript of the first edition, and those equations that are dimensionally correct are probably the result of Julian's disciplining. Margaret Britton introduced me to the dubious pleasures of word processing. Finally, my wife Jillian has taught me little about bones but, by telling me occasionally, in the evenings, about her job as a social worker, has reminded me that there is a real world outside.

Bones

INTRODUCTION

THIS BOOK deals with two complementary aspects of bone and bones: their structure and their mechanical properties. The design of the book is, I hope, more or less evident from the titles of the chapters. I discuss the structure of bone and bones in the first chapter and the mechanics of stiff materials in general in the second chapter, then in the long third chapter I describe the mechanical properties of what one may call "standard" bone tissue. This includes various attempts that have been made to model the mechanical behavior in terms of what we know about its structure and materials science. Bone tissue meets different mechanical problems in different situations, and chapter 4 deals with how the properties of bone vary with the demands placed on it. Cancellous bone is so unlike compact bone in its structure and mechanics that it merits a chapter on its own (chapter 5). Chapter 6 deals with tissues that share with bone the fact that they are stiff and have calcium phosphate as their main stiffening element. However, they are not, conventionally, called bone and, particularly in the case of enamel, have rather different mechanical properties. After chapter 6, I move up from the tissue level to the whole bone level, and chapters 7, 8, and 9 deal with whole bones, their articulations with other bones, and their interactions with the softer tissues that are attached to them. Chapter 10 deals with the general problem of safety factors and how bone shape and mechanical properties change with size. The final chapter deals with the fascinating problem of how, during a single lifetime, bones become adapted to the kinds of loads that are put on them.

When one is familiar with a topic it is difficult to remember how difficult it is to grapple with unfamiliar concepts. I suspect that the majority of the readers of this book will find a large number of unfamiliar concepts here, either on the mechanical or on the "bony" side. All I can say is that I have tried my best to start most topics from the ground up. Probably people will think that there are either too many numbers and equations in this book or not enough. I would have preferred more numbers, though perhaps not more equations, but it is surprisingly hard to find respectable numbers for many properties of bones. The level of physics and mathematical expertise required to enjoy this book (I hope it will be enjoyable as well as instructive) is not great. However, I would urge people not to ignore the structural side of the book; one cannot really understand bone's mechanical properties without having a good feel for its structure. People with a strongly clinical outlook may find some parts rather peripheral to what they consider to be their concerns. I would urge them to plug on and read the whole book!

This book has the advantage, such as it is, of being the product of one mind, and therefore having a coherence of approach and style. I have tried not to obtrude my own prejudices in controversial matters excessively, but a careful reading of many passages will show that I do have views about places where some authors are mistaken. People wishing for a more detailed view of bone mechanics, with a wide variety of expertly written chapters, may like to move on to the huge *Bone Mechanics Handbook* edited by Steve Cowin (2001a). Of course, there is nothing like going to the original literature. Articles on bone mechanics appear frequently in journals such as *The journal of biomechanics*, *The journal of orthopaedic research*, *Bone*, *The journal of bone and mineral research*, *The journal of experimental biology*, *The journal of biomechanical engineering*, *Calcified tissue international*, and *The journal of zoology*. Articles on the structure of bone, as it relates to mechanics, are so widely scattered that it is not possible to give any sensible guidance.

There is an enormous explosion of interest and research at the moment into the cell biology of bone. It is an inchoate and somewhat incoherent field, and I shall deal with it hardly at all. Mainly this is because of ignorance on my part, and the desire to keep this book within sensible size limits—indeed, to finish it at all. However, the approach I have adopted makes sense because, for many purposes, it is possible to treat the way in which bone achieves the structure it does as a black box, knowing that, in the next few decades, the box will be prized open and we shall see how all the marvellous mechanical adaptations of bone are brought about. On the other hand, it would make no sense at all to treat the mechanical properties of bone and bones without a consideration of the structures that produce the mechanical properties. (In this book I try to use the word *bone* to mean the tissue, and *bones* to refer to whole anatomical bones, such as the femur. The context should make it clear whether the tissue or the anatomical bone is being referred to.)

I hope that some people who are new to the subject will be enthused by reading this book to think that the study of bone mechanics may be worthwhile. I certainly have over the last forty years or so, and there is still so much to find out.

Chapter One

THE STRUCTURE OF BONE TISSUE

THROUGHOUT this book I shall be suggesting that the structure of bone tissue, and of whole bones, makes sense only if its function, particularly its mechanical function, is known or guessed. (As Rik Huiskes of Eindhoven is fond of saying [2000]: "If bone is the answer, then what is the question?") However, in this first chapter I shall deal only with the structure of bone, leaving almost all discussion of function until later. Of course, the mechanical properties of bone and bones are determined by their structure, and we cannot begin to understand the function without having a good idea of the structure. Much of the subject matter will be familiar to some readers, but not all to everyone. Indeed, some readers may be coming to bone for the first time, from say materials science, so I shall start with a single paragraph overview of bone structure.

Bone of present-day mammals and birds is a stiff skeletal material made principally of the fibrous protein collagen, impregnated with a mineral closely resembling calcium phosphate. Bone also contains water, which is very important mechanically. Bone is produced inside the body and is usually covered with cells throughout life, though in fish scales, for instance, the external lining of cells may be rubbed off. Most bone not only is covered by cells but has living cells and blood vessels within it. Bone, being hard, cannot swell or shrink; all changes in shape must take place at surfaces. Most bones are hollow and contain hematopoietic or fatty marrow. Marrow probably has little mechanical significance. Tendons and ligaments insert into the bone substance, and the ends of bones are often covered by a thin layer of cartilage for lubrication. Some tissues, such as antler and dentin, are not called bone but are actually bone, or extremely like it. Horn, such as is found in cattle, is a completely different material, usually unmineralized, though the horn core, which supports the horn, is made of bone.

To start straight off talking about the structure of bone begs the question. It is not really at all clear what bone is. Consideration of a present-day mammal or bird would allow a clear distinction to be made, because bone is the only structure that is essentially collagen mineralized with calcium phosphate and containing cell bodies, though in antler bone the cells are all dead by the time the antler comes to be used. Dentin is collagen mineralized with calcium phosphate but it does not contain cell bodies, only tubular extensions of cells. The other significant tissue mineralized with calcium phosphate (as opposed to calcium carbonate) is enamel, and this is very different in that it has virtually no cells or cell processes or, indeed, much organic matrix. However, as soon as one looks outside the mammals and birds the situation becomes much more complex. Bone is found only in vertebrates. Many teleost fish have bone without bone cells, and the range of structures seen in scales of different fish species forms an almost complete spectrum from what is obviously "typical" bone or "'typical" dentin to what is obviously "typical" enamel. This situation is often found in biology, since nature is concerned not with categorization, but with producing effective results. (The great evolutionary biologist John Maynard Smith likes saying that "all biology is false," meaning, of course, that there are very few absolutes in biology.)

For the purposes of this book, the fact that there are so many different types of scales does not matter greatly because virtually nothing is known about the mechanical properties of scales. I shall be concerned almost entirely with the mechanics of typical bone, as found in mammals, but I shall devote some space to tissues such as dentin. Antler, which is dead when functioning, will figure prominently. A good account of the variation of structure in vertebrates is found in Francillon-Vieillot et al. (1990), which, though not an enthralling read, is very clear, comprehensive, and well illustrated.

Even "typical" bone is such a complex structure that there is no level of organization at which one can truly be said to be looking at bone as such. I shall start at the lowest level and work up to a brief description of the variety of shapes one sees in whole bones.

1.1 Bone at the Molecular Level

At the lowest level bone can be considered to be a composite material consisting of a fibrous protein, collagen, stiffened by an extremely dense filling and surrounding of calcium phosphate crystals. There are other constituents, notably water, some ill-understood proteins and polysaccharides, and, in many types of bone, living cells and blood vessels. The amount of water present in bone is an important determinant of its mechanical behavior, and I shall say more about it in chapter 4.

A word about terminology. In this book I use the word *matrix* to mean the water and the soft organic material, mostly collagen, in which the mineral crystals are deposited. This accords with what materials scientists would consider to be the matrix (though some might consider the mineral to be the matrix). However, bone biologists, who are focused almost entirely on the cells of bone, use the word *matrix* to mean the bone tissue itself, that is, the water, the organic material, and the mineral. There is no way round this possible source of confusion; one simply has to be aware of it.

Collagen is a structural protein found in probably all metazoan animal phyla. It is the most abundant protein found in animals, but only in the vertebrates does it undergo a wholehearted transformation into a mineralized skeletal structure, although some soft corals have traveled some way along the road. A classified bibliography of more than 3400 references to collagen, comprehensive up to that time, is given in Kadler (1994).

Unmineralized collagen is also found in the vertebrates, and in many invertebrates, in skin, tendon, ligament, blood vessel walls, cartilage, basement membrane, and in connective tissue generally, in those circumstances where the material is required to be flexible but not very extensible. Collagen makes up more than half the protein in the human body (Miller 1984). Collagen from different sites often has different amino acid compositions; in the mid 1990s 19 types of collagen were known throughout the animal kingdom, and the known number increases relentlessly (Prockop and Kivirikko 1995). The collagens of skin, tendon, dentin, and bone share the same type of composition, and are called *type* 1 collagen. The protein molecule *tropocollagen*, which aggregates to form the microfibrils of collagen, consists of three polypeptides of the same length—two have the same amino acid composition, one a different one. These form on ribosomes, are connected by means of disulfide cysteine links, and leave the cell. Outside the cell the ends of the joined polypeptides are snipped off, the lost part containing the disulfide bonds. The three chains are by now held together by hydrogen bonds in a characteristic left-handed triple helix.

The primary structure of the polypeptides in the tropocollagen molecule is unusual, great stretches of it being repeats of glycine–X–Y, with X often being proline and Y sometimes hydroxyproline. The imino acids proline and hydroxyproline are unlike amino acids in that the nitrogen atom is included in the side chain as part of a five-membered ring. The effect of this is to reduce the amount of rotation possible between units of the polypeptide. It also prevents α-helix formation and limits hydrogen-bond formation. These constraints result in a rather inflexible polypeptide, 300 nm long (Olsen and Ninomiya 1993).

The tropocollagen molecules line up in files and bond, not with molecules in the same file, but with molecules in neighboring files, to form *microfibrils*. The tropocollagen molecules alongside each other are staggered by about one-fourth of their length. There is a gap between the head of one molecule and the tail of the next in the file, the *hole region*, and, because many tropocollagen molecules are stacked side by side, these gaps and other features of the molecules produce a characteristic periodicity, 67 nm long. The whole microfibril becomes stabilized by intermolecular cross-links. Microfibrils aggregate to form *fibrils*. Although the longitudinal arrangement of the tropocollagen molecules in the microfibrils is fairly well understood, the way in which the microfibrils themselves are arranged laterally to form fibrils is much less well understood. A clear introduction to the subject is provided by Prockop and Fertala (1998). Hulmes et al. (1995) produce evidence that the fibrils are arranged in concentric rings. Wess et al. (1998a,b) produce a rather different model that they claim will explain the way in which mineral is able to pack in bone. This is a difficult subject, with the majority view changing often as experimentation becomes ever more sophisticated. It could well be that when a stable view is formed the results will be useful in helping to model the mechanical behavior of bone, but at the moment this is not really the case.

Collagen comprises about 85 to 90% of the protein in bone. The proteins that are not collagen are called, negatively, *noncollagenous proteins* (NCPs). The literature on them is vast and expanding rapidly (Ganss et al. 1999; Gerstenfield 1999; Gorski 1998; Nanci 1999). Some NCPs are restricted to bone, and some are also found in other places in the body. Some of these proteins almost certainly have a role in the initiation and control of mineralization or reconstruction, and some may have a role in binding the collagen and mineral together (Roach 1994). However, we are almost completely in the dark at the moment about any quantitative effect NCPs may have on the mechanical properties of bone.

Impregnating and surrounding the collagen is the bone mineral, which is some variety of calcium phosphate. The precise nature of the mineral of bone, both its chemistry and its morphology, is still a matter of some dispute. The problem is that the mineral in bone comes in very small crystals that have a very high surface-area-to-volume ratio. The size of the crystal is such that in one dimension it is only about 10 atomic layers thick (Lowenstam and Weiner 1989). This makes it reactive, and so most preparative techniques used for investigating it, such as, drying under vacuum for electron microscopy, may cause alterations from the living state. There is agreement that some of the bone mineral is the version of calcium phosphate called hydroxyapatite, whose unit cell (the smallest part of a crystal that is repeated uniformly throughout a crystal) contains $Ca_{10}(PO_4)_6(OH)_2$. The crystals are impure. In particular, there is about 4–6% of carbonate replacing the phosphate groups, making the mineral more truly a carbonate apatite (dahllite). This carbonate substitution takes place more near edges of the bone, close to vascular and marrow spaces and tends to reduce the crystallinity of the crystals (Ou-Yang et al. 2001). Various other substitutions may take place (Boyde and Jones 1998; McConnell 1962).

At the moment, we are ignorant of the mechanical properties of the mineral itself, and all modeling, such as that of Wagner and Weiner (1992) and Sasaki et al. (1991), which I discuss in section 3.7, makes use of somewhat insecurely based (though not necessarily far wrong!) estimates. The mineral is certainly stiff, but its strength, in such small blocks, is unknown.

The positioning of the mineral relative to the collagen fibrils, as well as its shape, is becoming clearer, though there is still controversy. There is some argument as to whether the crystalline mineral, which can be seen in electron micrographs, is needle-shaped or plate-shaped. Ascenzi et al. (1978) claimed that the mineralization process starts off with small granules, about 4.5 nm across, which coalesce or grow into needles about 40 nm long. However, the observations of Landis and his coworkers make it almost certain that in mineralized tendon (Landis et al. 1993) and in embryonic chick bone (Landis et al. 1996) the crystals are platelet-shaped. They have used the technique of taking multiple views of bone using high-voltage electron microscopy to produce a tomographic image. This method shows very clearly the three-dimensional shape of the crystals and to some extent their spatial relationship to the collagen (fig. 1.1). These visualizations show that the crystals' thickness is rather unvarying at about 4–6 nm, their width is about 30–45 nm, and their length is typically 100 nm. Later, these mineral platelets seem to fuse sideways, and lengthways, producing at times sword-shaped blades that are quite long and broad. However, they do not seem to grow in the depth direction, remaining about 5 nm deep. Erts et al. (1994), using scanning probe microscopy, found similar values for turkey tendon.

Reports of the visualization of the crystals directly overwhelmingly supports this view that the crystals in all bone examined are platelet-shaped. Weiner and Price (1986) examined the size of bone mineral crystals, extracting them from the bone by a gentle procedure, and proposed values of about 50 × 20 × 2 nm. Kim et al. (1995) report platelet-shaped crystals from tissues of a taxonomically satisfyingly varied group of species: chickens, bovines, mice, and herring. The average length and breadth, in nanometers, for the four species are given in Table 1.1. Kim et al. did not measure the thickness, but suggested it was about 2 nm. Ziv and Weiner (1994) suggest that most estimates of the size of crystals are underestimates, because the plates are so fragile, and that crystals may be often hundreds of nanometers long in untreated bone.

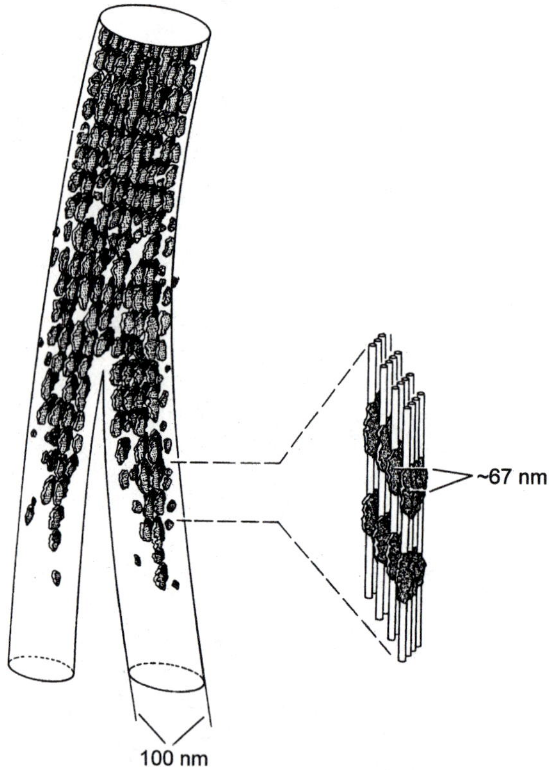

FIG. 1.1 Diagram of mineralizing turkey leg tendon according to the microtomographic investigations of Landis. Mineralization is proceeding from the top down. The crystals are platelet-shaped, and are initially registered in line with the hole region of the fibrils (~67 nm apart). This initial relationship between the platelets and the collagen fibrils is shown in the right-hand part of the diagram. Toward the top of the diagram some platelets are shown as having fused longitudinally. (From Landis, W. J., Hodgens, K. J., Arena, J., Song, M. J., and McEwen, B. F. 1996. Structural relations between collagen and mineral in bone as determined by high voltage electron microscopic tomography. *Microscopy research and technique* 33:192–202. Reprinted by permission of Wiley–Liss, Inc., a subsidiary of John Wiley & Sons, Inc.)

Fratzl et al. (1992) have produced indirect evidence, using small-angle X-ray scattering, that the crystals in ossified tendon are indeed platelet-shaped but that in ordinary compact bone they are more likely to be needle-shaped. On the other hand, Wachtel and Weiner (1994) show that the small-angle X-ray scattering picture from crystals from rat bone is very similar to that from mineralizing turkey tendon, and suggest that it is probably reasonably safe to generalize about the crystal morphology from mineralizing turkey tendon.

TABLE 1.1
Length and Breadth of Mineral Crystals (in nm) in the Bone of Four Species

Species	*Length*	*Breadth*
Chicken	23.3	12.2
Bovine	27.3	15.8
Mouse	21.2	12.0
Herring	37.3	15.4

Source. From Kim et al. (1995).

More contentious is where the mineral is in relation to the collagen fibrils. For years, following a suggestion by Hodge and Petruska (1963), it was thought that the mineral is initially deposited in the holes between the heads and the tails of the tropocollagen molecules (the gap zones). This results in the initial mineralization having a 67-nm periodicity (Berthet-Colominas et al. 1979). Many studies seemed to confirm this, but nearly all were carried out on mineralizing turkey leg tendon, which, although very convenient to study because the collagen fibrils are so well arranged, is not typical bone, particularly in relation to the arrangement of the crystals (Wenk and Heidelbach 1999). It is probable that in some way the particular conformation of the collagen molecule allows it to act as a nucleation site, permitting the precipitation of lumps of mineral that, without the presence of the energetically favorable sites, could not come out of solution. There is some evidence that the mineral deposits preferentially in parts of the fibril that are high in hydrophilic residues (Maitland and Arsenault 1991). Later, the mineral is deposited all over the collagen fibrils, and also within them. Weiner and Traub (1986) have published stereopairs of mineralizing turkey leg tendon, showing how the crystals lie within the fibers. Landis et al. (1993, 1996) show similar pictures (fig. 1.1), and point out that the individual platelets seem to remain separated by a space in the depth direction of about 5 nm. This would, of course, allow collagen microfibrils to exist between the platelets. Jäger and Fratzl (2000) suggest, though with no observations to back the suggestion up, that the crystals may be arranged circumferentially round the center of the fibril. This would accord with the radial fibril model of Hulmes et al. (1995).

All these observations relate to early stages of mineralization, and it is much less obvious what happens when mineralization has proceeded to its full extent. There must be considerable derangement of the initially very uniform collagen structure. It is strange that some quite modern works (e.g., An [2000], table 3.1) still hold to the idea that the only place that the mineral resides in mature bone is in the gap region between the ends of adjoining collagen molecules. This idea is quite wrong.

Much work has been done on the mineralizing turkey leg tendon, because, although it is called tendon, it is actually proper bone and mineralizes in a particularly regular way, so that there is a close relationship along the tendon between distance and the progression of mineralization. This makes it very convenient for studies of mineralization. However, most bone mineralizes on surfaces, rather than at the end of tendons. On surfaces what happens is that a matrix of collagen, plus a few other organic components, is laid down first. This organic material is called *osteoid*. Mineral is deposited in the collagen, initially, it seems, in the gap zones, but then, unlike the situation in tendon, all along the length of the collagen fibrils. The plate like mineral crystals as they grow tend to form quite large lumps, and the individual mineral crystals tend to be oriented in the direction of the collagen fibrils. Within the collagen fibrils the precipitation is not random; one of the long axes of the mineral plates is always fairly well aligned with the collagen fibrils. Also, later, mineral is deposited between the fibrils, in the amorphous and rather tenuous ground substance. The relative amount deposited within and between the collagen is quite hotly argued even now, with some people proposing that more is between than within (Bonar et al. 1985; Lees et al. 1990; Pidaparti et al. 1996). Furthermore, it is likely, certainly in mineralized tendon and possibly ordinary bone, that the mineral outside the fibrils may have a different orientation from that within the fibrils (Lees et al. 1994; Pidaparti et al. 1996). In some fish bone, however, the vast majority of the mineral is within the fibrils, and not between them (Lee and Glimcher 1991). Knowing the truth of these matters is important for understanding why bone behaves mechanically as it does.

Added to our ignorance of the disposition of the mineral in bone is our ignorance of how, and the extent to which, the collagen and the mineral are bound together. The relationship between those mineral crystals that are inside the collagen fibrils and the collagen is extremely intimate, and such short-range forces as van der Waals forces may well be important. Also, ionic bonding probably occurs, which I discuss in section 3.7.

I am conscious, in reviewing the last few pages, of how often I have said that we are ignorant of the true situation. I suspect that much of our ignorance will be removed in the next decade. It will then become possible to try to understand in detail how collagen/mineral interactions determine the mechanical properties of bone at the molecular level, in the same way that metallurgists have a good idea of how, for instance, steel behaves at this level. However, bone has several levels of structural hierarchy above the molecular level, and these all have important effects on the mechanics of bone, as we shall see.

1.2 The Cells of Bone

Bone is permeated by and lined by various kinds of specialized cells, which will be introduced later. I here list them and briefly describe their properties.

Bone-lining cells cover all surfaces of bones, including the blood channels, forming a thin continuous sheet that controls the movement of ions between the body and the bone (Miller et al. 1988). The layer of cells on the outside of the bone is called the *periosteum*, although this word is often used to include the strong collagenous sheet covering the outer surface. The layer of cells on the inside of the bone is called the *endosteum*. The bone-lining cells, which are often considered to be quiescent osteoblasts, are derived, via complex series of changes, from osteoprogenitor cells. These stages are described by Lian and Stein (1996).

Osteoblasts derive from bone-lining cells and are responsible for the formation of bone. They initially lay down the collagenous matrix, *osteoid*, in which mineral is later deposited, and they probably also have a role in its mineralization.

Osteocytes are the cells in the body of the bone. In cancellous bone the density of osteocytes varies from about 90,000 mm^{-3} in rats to about 30,000 mm^{-3} in cows. In general, the larger the animal the lower the density of osteocytes (Mullender et al. 1996). They derive from osteoblasts. They are imprisoned in the hard bone tissue and connect with neighboring osteocytes and with bone-lining cells by means of processes that are housed in little channels (canaliculi), of about 0.2–.03 μm diameter (Cooper et al. 1966). The actual connections with neighboring cells are by means of *gap junctions* that allow small molecules through easily.

Osteoclasts are bone-destroying cells. They are large, multinucleated cells derived from precursor cells circulating in the blood. In time-lapse photography they give the appearance of being extremely aggressive, clamping themselves to the bone's surface and leaving a space underneath a *ruffled border* that is very mobile and beneath which the bone can be seen dissolving. Debris, both organic and mineral, are packed into little vesicles and pass through the cell body of the osteoclast and are dumped into the space above (Nesbitt and Morton 1997; Salo et al. 1997). When osteoclasts have done their job they disappear and presumably die. (A colleague of mine, who initially studied osteoclasts, said that for some strange reason most people who studied them were as aggressive as their subjects. My colleague eventually gave up, and turned to studying benign osteoblasts!)

1.3 Woven and Lamellar Bone

Above the level of the collagen fibril and its associated mineral, mammalian bone exists in two usually fairly distinct forms: woven bone and lamellar bone. Parallel-fibered bone is intermediate.

Woven bone is usually laid down very quickly, more than 4 μm a day and often much more, most characteristically in the fetus and in the callus that is produced during fracture repair. The collagen in woven bone is variable, the fibrils being 0.1–3 μm or so in diameter and oriented almost randomly, so it is difficult to make out any preferred direction over distances greater than about a millimeter (Boyde 1980; Boyde and Jones 1998). The mineralization process involves roughly spherical centers, impregnating both the collagen and ground substance at the same time, in which the crystals seem to be randomly arranged. As these mineralization centers spread they abut and often leave mineral-free spaces (Boyde 1980). As a result, woven bone, though highly mineralized, is often quite porous at the micron level. As in most bone, woven bone contains cells (osteocytes) and blood vessels. Rather frequently, the spaces surrounding the osteocytes are extensive and differ in this way from those in lamellar bone. There are of the order of 60 canaliculi per osteocyte (Boyde 1972), though no doubt this number varies greatly between osteocytes and between species. "Woven" bone is a misnomer, because there are very few examples of weaving, that is, true interlacing, in biology. (It would be a trick almost impossible to bring off, though, surprisingly, the enamel of some rodents has something almost as good—structures arranged in three orthogonal directions, see Section 6.3.)

Lamellar bone is more precisely arranged, and is laid down much more slowly than woven bone, less than 1 μm a day (Boyde 1980). The collagen fibrils and their associated mineral are arranged in sheets (lamellae), which often appear to alternate in thickness. The final degree of mineralization of lamellar bone is less than that of woven bone. The classical view is that the fibrils lie within the plane of the lamella, rarely passing from one to the next and that the fibrils tend to be oriented in one direction within the lamella. Indeed, some workers suggest that the collagen fibrils in a particular lamella are all oriented in the same direction (Ascenzi et al. 1978). However, this is probably not the case; in many lamellae the fibrils are in small *domains* about 30–100 μm across. Within a domain the fibril orientation is constant, but it changes, within one lamella, from one domain to the next (Boyde and Hobdell 1969; Frasca et al. 1977). The collagen fibrils in lamellar bone form branching bundles, 2–3 μm in diameter (Boyde 1980), thicker than in most woven bone. The osteocyte lacunae in lamellar bone are oblate spheroids, the equatorial diameters being about five times longer than the polar axis. The shorter axis of each lacuna is oriented parallel to the direction of the thickness of the lamella.

The division between one lamella and the next looks abrupt under the light microscope, particularly under polarized light. However, scanning electron microscope pictures show a much messier situation. Frequently, lamellae *seem* to come in alternating thicknesses. There is a relatively thick one, about 5 μm thick, whose mineral crystals are roughly arranged all in one direction, over small distances at least, at an angle both to the long axis of the bone and to the circumferential extension of the lamella itself The thin lamella is about 1 μm thick, and has mineral plates oriented in the plane of the circumferential extension of the lamellae, with their smallest dimension normal to the long axis of the bone. This kind of lamellar organization is held to occur in mice by Weiner et al. (1991), and from my observations of fracture surfaces of bone of a large number of species I think this arrangement is widespread. If the mineral is arranged like this, presumably the collagen fibrils also lie in the same direction. As we shall see in the next chapter, such changes in direction have mechanical consequences. There is disagreement at the moment as to whether there is a difference in the density of mineral in the thick and thin lamellae. Marotti (1993) argues, giving evidence from the appearance of the collagen fibrils, that the thicker lamellae are looser and have more mineral, the thinner lamellae are somewhat more collagen-rich, and there is not a great deal of difference in the collagen orientation between the different layers. Yamomoto et al. (2000) argue convicingly that, at least in human dentin, the appearance found by Marotti is an artifact and that the differently appearing lamellae are only different in respect of their orientation. The uncertainty in the literature may be partially because some people concentrate on the mineral and some on the collagen.

Giraud-Guille (1988) produces strong evidence that in some lamellae the collagen fibrils, and presumably, therefore, their associated mineral crystals, are arranged in a “twisted plywood” or helicoidal structure. In a helicoidal structure there are many layers, and within a layer the fibers all point in the same direction. There is a change of angle between the layers, usually quite small, and this is constant in size and sense. The result is a multilayer composite with no preferred orientation. Ziv et al. (1996), Weiner et al. (1997), and Weiner and Wagner (1998) produce further evidence that such continuous transitions between, and even through, apparently discontinuous lamellae are common. The model of Weiner and his co-workers suggests that collagen fibrils, only 80 nm or so in diameter, are the basic unit. These may lie parallel to each other through the thickness of the lamella, or be at an angle, usually about 30°, to their neighbors. Helicoidal arrangements of fibers are found frequently in biological tissues, and Neville (1993) has written a whole book about them. They occur in all sorts of structures, such as insect cuticle and plant cell walls, and are a prime example of the ability of organisms to produce complex structures by self-assembly, that is to say, without cells being involved directly in the placing of the succeeding layers. Helicoidal structures give rise to a confusing artifact if cut and examined at an angle to the plane of the sheets: The structure appears to consist of a series of arcs (Kingsmill et al. 1998, fig. 5f), but, in fact, there are no arcs.

Parallel-fibered bone (described by Ascenzi et al. [1967] and by Enlow [1969]) is structurally intermediate between woven bone and lamellar bone. It is quite highly calcified, but the collagen fiber bundles are much more parallel than those in woven bone.

1.4 Fibrolamellar and Haversian Bone

In mammals there are, at higher levels of structure, four main types of bone. Woven bone can extend uniformly for many millimeters in all directions. Such a large block is found only in very young bone (of rather large mammals) and in large fracture calluses. Lamellar bone may also occupy quite large volumes. Usually, in mammals it does so in circumferential lamellae, initially wrapped around the outside or the inner cavity of bones. There are blood channels in such bone, but they do not much disturb the general arrangement of the lamellae.

Lamellar bone also exists in a quite separate form: Haversian systems, or secondary osteons. The British and other Europeans are inclined to use "Haversian systems," whereas Americans prefer "secondary osteons." It does not matter which is used, except that it is critical to distinguish primary osteons, which I shall describe in a moment, from secondary osteons. Haversian systems form like this: many bone-destroying cells, called osteoclasts, move forward in a concerted attack on the bone tissue. They form a so-called *cutting cone*, shaped, as Martin and Burr (1989) say in an excellent account of the process, like "half an eggshell which is about 200 μm in diameter 300 μm long." Osteoclasts are not derived from cells that occur locally, but instead come from cells circulating in the blood. As the cutting cone advances it leaves a cylindrical cavity of diameter about 200 μm behind. Almost as

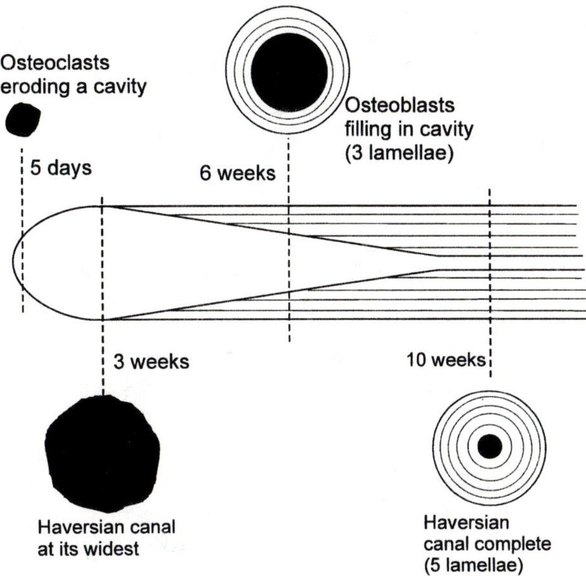

FIG. 1.2 Diagram of a forming secondary osteon (Haversian system) in longitudinal and cross-sectional views. The system is extending toward the left. The times give, very roughly, the time course of the process in humans. At 5 days the osteoclasts are still widening the cavity in the bone. At 3 weeks the cavity it at its widest. By 6 weeks the cavity is half filled in by osteoblasts, and by 10 weeks or so the process is completed, although it will take a long time for the bone to become completely mineralized. In the cross sections the central cavity is shown black.

soon as the cavity forms, it begins to fill in (Parfitt 1994). The walls of the cavity are made smooth, and bone is deposited on the internal surface in concentric lamellae (fig. 1.2). The end result is arranged like a leek *Allium porrum*, with usually clearly distinguishable cylindrical layers, except that there is a central cavity in the Haversian system, which contains one, or sometimes two, blood vessels, and nerves (Marotti and Zallone 1980). In humans the whole process from initiation of the osteoclastic activity to the completion of the filling in takes about 2–4 months.

The Haversian system is the classic result of the process of remodeling. There are two ways in which new bone may appear. In *modeling* the gross shape of the bone may be altered; that is, bone may be added to the periosteal or endosteal surfaces or it may be taken away from these surfaces. In *remodeling* all surfaces of the bone may be affected, including the internal body of the bone, by the formation of Haversian systems. In remodeling the bone involved is usually a small individual packet called a basic nulticellular unit (BMU), and typically the amount of bone remaining after the remodeling cycle is little changed; new bone has more or less replaced old bone. This distinction will be seen to be important in chapter 11.

There is a type of secondary osteon that is different from the type described above; this is called the drifting osteon. It is found in younger animals, in general. After the cavity is formed by the cutting cone, it is not filled in at once; instead, one side continues to be eroded by osteoclasts, while the opposite side is filled in by osteoblasts. As a result the cylinder of the osteon drifts *sideways* through the tissue, and in so doing erodes and replaces a great deal of preexisting bony tissue (Robling and Stout 1999). Nothing is known about the mechanical consequences of the presence of such secondary osteons.

There is an outer sheath to the Haversian system, called the cement sheath (or line, because it is usually viewed in cross section). This is formed when the cutting cone stops its erosional activity and just before new lamellar bone is laid down on the raw surface so formed. The composition of the cement sheath is still controversial. Some, for instance, Frasca (1981), propose that it is more highly mineralized than surrounding bone; others, for instance, Schaffler et al. (1987), propose that it is less highly mineralized. There is agreement that there is very little collagen in it. Schaffler et al. produce evidence that it has more sulfur and less calcium and phosphorus than the neighboring bone lamellae, and is therefore probably less mineralized than them. It may be that the cement line is simply a very thin layer of osteoid that remains when the process of bone erosion is replaced by bone deposition (Martin and Burr 1989). Zhou et al. (1994) suggest that there are globular accretions (of unknown composition) on the ends of the degraded collagen fibrils that have partially eroded away, and that these accretions act as a bridge between the old degraded collagen fibrils, and the newly deposited ones. The cement line is also the site of a concentration of osteopontin, a bone protein that probably has some role in bone remodeling (Gerstenfield 1999; McKee and Nanci 1996; Terai et al. 1999).

It is a pity that this matter is not settled, because it is very important to know, for the purpose of understanding bone mechanics, whether cement sheaths are likely to be more or less compliant, and more or less strong, than the surrounding lamellae. Some canaliculi cross the cement sheath, so cells outside do have some metabolic connection with the blood vessel in the middle of the Haversian system. This is shown in figure 1.3 in which the blood vessels, osteocytes, and canaliculi are impregnated with resin and stand proud when the outer few microns of a specimen of bone are etched away. The density of canaliculi in bone is

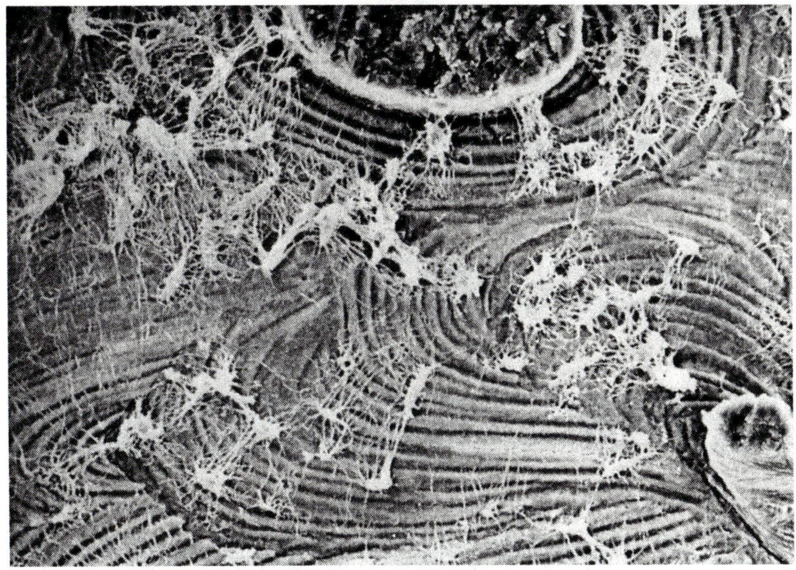

FIG. 1.3 Scanning electron micrograph of a resin-impregnated etched piece of human cortical bone. The resin stands proud of the bone. Two vascular channels can be seen (*top center, bottom right*). The lamellation of the bone is made obvious by the differential erosion produced by the etching acid. Osteocyte lacunae have radiating canaliculi, which can be seen passing over places where the lamellae obviously belong to two generations of osteogenesis, and which therefore have a cement sheath between them (*upper left*). However, the density of canaliculi crossing the sheath is generally less than the density nearby (bottom left, top right). The resorption and redeposition associated with the bottom right vascular channel starts at the bottom left of the picture. Width of image ~500 μm. (From a Christmas card kindly sent by Dr. Peter Atkinson.)

high, averaging about 5 canaliculi per hundred square microns in the plane normal to the predominant direction of the canaliculi (Marotti et al. 1995). The canaliculi account for a porosity of about 1.5% (Frost 1960). The cell biology of osteoclasis, the process of destruction of bone tissue, was reviewed by Zaidi et al. (1993).

A fourth characteristic type of mammalian bone is known as plexiform, or laminar, or fibrolamellar bone (Enlow and Brown 1956, 1957, 1958; Currey 1960; de Ricqlès 1977; Francillon-Vieillot et al. 1990; Stover et al. 1992). It is found particularly in large mammals, whose bones have to grow in diameter rather quickly. Lamellar bone cannot be laid down as fast as woven bone (as I said above, lamellar bone is laid down at a rate of less than 1 μm a day, while woven bone is laid down at more than 4 μm and, indeed, often much more, a day). If a bone has to grow in diameter faster than lamellar bone can be laid down, woven bone must be laid down instead. For reasons that will be discussed in

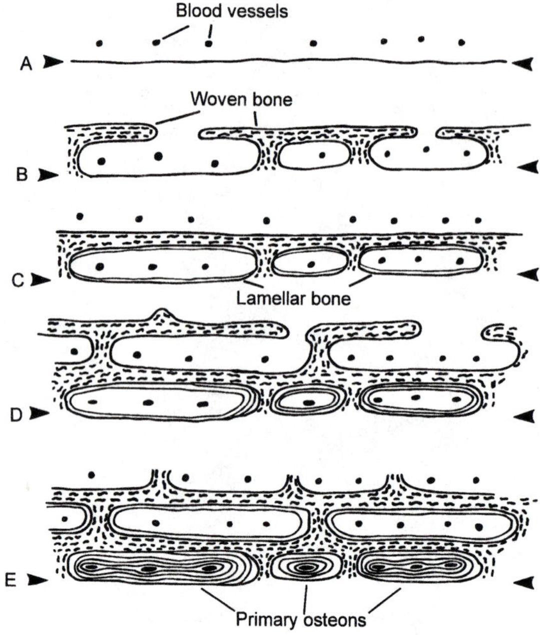

FIG. 1.4 The formation of fibrolamellar bone. These are cross sections of the outer surface of a rapidly growing bone. The arrowheads show the position of the original surface. Blood vessels are shown by black spots. (A) The original situation. (B) Woven bone, shown by squiggly lines, grows very quickly to form a scaffolding clear of the original surface. (Sometimes this "woven" bone may be much more like parallel-fibered bone.) (C) Lamellar bone, shown by fine lines, starts to fill in the cavities. (D) As more lamellar bone is laid down, so is another scaffolding of woven bone. (E) By the time the first row of cavities is filled in, forming primary osteons, the outer surface of the bone is far away.

chapter 3, woven bone is almost certainly inferior to lamellar bone in its mechanical properties. The undesirable mechanical results of having a bone made from woven bone are partially obviated by the production of fibrolamellar bone. Essentially, an insubstantial scaffolding of woven or parallel-fibered bone is laid down quickly to be filled in more leisurely with lamellar bone (fig. 1.4).

In bovine bone, each lamina is about 200 μm thick. In the middle a two-dimensional network of blood vessels is sandwiched between layers of lamellar bone. Beyond these layers, on each side, is a layer of parallel-fibered or woven bone, which is more heavily mineralized than the lamellar bone (fig. 1.5). In particular, there is a line in the middle, which is exceptionally heavily mineralized. This is the line at which growth stops for a short while before the next later is initiated. The way fibrolamellar bone is laid down means that there are, in effect, alternat-

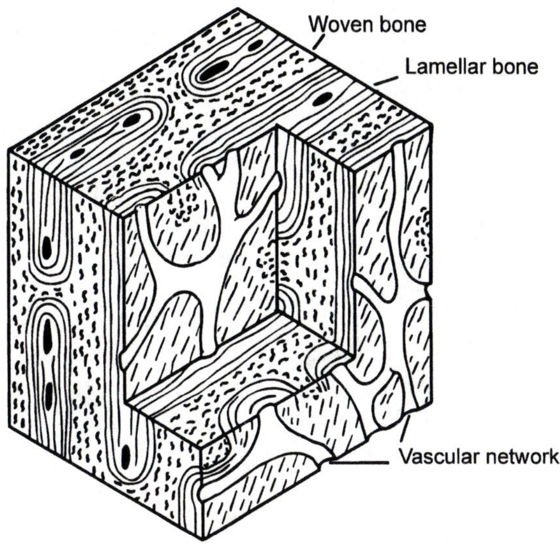

FIG. 1.5 Block diagram of fibrolamellar bone. Conventions concerning histological type as in Figure 1.4 Two-dimensional networks of blood channels, sheathed by lamellar bone, alternate with layers of woven bone.

ing layers of parallel-fibered or woven bone and lamellar bone tissue extending, quite often, for many millimeters, or even centimeters, in the radial direction. Fibrolamellar bone can be laid down vey rapidly. For instance Castanet et al. (2000) report that the femur of a young emu *Dromaius novaehollandiae* can be added up to 80 μm per day on the subperiosteal surface.

This description is of particularly neatly arranged laminar bone. Frequently the blood channels are more irregularly disposed or do not form a network, and the laminar arrangement gives way to one in which the blood vessels anastomose in three dimensions, and each is surrounded by more or less concentric layers of lamellar bone. This produces an appearance somewhat like that of Haversian systems, and the structures around the blood vessels are called *primary* osteons. However, there is a most important difference between primary osteons and secondary osteons, or Haversian systems: Haversian systems are *secondary*, that is, they replace bone that has existed previously. There are differences that enable one to distinguish Haversian systems from primary osteons histologically. In particular, secondary osteons are surrounded by a cement sheath, whereas primary osteons are not. Also, secondary osteons appear to drill through the preexisting bone, without regard to its structure, whereas the lamellae round primary osteons merge smoothly with the surrounding bone. The distinction between the two types of osteon is not mere semantic hairsplitting, because differences between primary osteonal bone and Haversian bone correlate with differences in mechanical behavior, as we shall see. Haversian bone is weaker than primary bone. The distinction is frequently not made, and is a grave source of muddled thinking.

The kind of primary bone laid down depends on the rate of accretion. Castanet et al. (1996) related the changes in the histology of the bone of the mallard duck *Anas platyrhyncos* to the rate of accretion, which they determined by periodic labeling with a fluorescent dye. Different bones grow at different rates, and they were able to see the different kinds of histology being laid down at the same time, so what they found was not an aging or maturation effect. The humerus grew fastest and, initially, at seven weeks of age, had a rate of accretion of about 25 μm a day; the bone was completely fibrolamellar. As the rate of accretion diminished, the fibrolamellar bone gave way to anastomizing primary osteons. Finally, when the accretion rate was only about 1 μm a day, the blood vessels were sparse, all parallel to the long axis of the bone, and the bone consisted of circumferential lamellae. The phalanges, on the other hand, which had a much lower accretion rate right from the start, never showed any sign of fibrolamellar histology.

Stover et al. (1992) show that in bone that grows very fast in the radial direction, such as in the young foal, the outer sheet of woven bone may initially be connected to the rest of the bone by bony struts so sparsely, if at all, as to be effectively lying lie free in the periosteum. The outer sheet becomes connected to the rest of the bone only when mineralization is well underway. This process, which they call "saltatory primary osteonal bone formation," allows growth to be even more rapid than the process I described above. Indeed, Sue Stover tells me that, very occasionally, two layers of free-floating bone can be seen. However, it is obviously a somewhat risky process, because a blow to the periosteum would cause the relative positions of the outer sheet and the main bone to become deranged.

1.5 Primary and Secondary Bone

Primary bone is replaced by secondary bone in two ways: The bone can be eroded away at its surface, and then new bone can be laid down, or else Haversian systems can be formed. Enlow (1963, 1969) gives very clear descriptions of these processes. It is often quite difficult to tell when the former has happened, and the effects, if any, of such replacement on mechanical properties are uncertain. The adaptive reason (if it is adaptive) for the formation of Haversian systems, which have a somewhat deleterious effect on mechanical properties, is obscure. The common explanation half-heartedly held by much of the bone community for a long time was that Haversian systems form when the bone mineral has, from time to time, to be released into the blood system for purposes of mineral homeostasis (Hancox 1972). Nowadays, mechanical explanations are more fashionable. There are various problems associated with such explanations that need not concern us yet, because we can accept Haversian remodeling as a fact and explore its mechanical consequences. I shall say much more, although rather inconclusively, about the function of Haversian remodeling in chapter 11.

The formation of Haversian systems tends to lead to the production of more Haversian systems. Each Haversian system is bounded by its cement sheath, and the passage of canaliculi across these cement sheaths is variable in amount (fig. 1.3), and is often rather sparse (Curtis et al. 1985). When it is poor, blood vessels will be separated from some of their catchment area, and osteocytes outside this area may find it difficult to obtain nutrients and are more likely to die (Currey 1960, 1964b). It is probably for this reason that the formation of a few Haversian systems in a region is often followed by the formation of many more in the immediate vicinity. Haversian systems often occur in clusters more often than would be expected by chance (Bell et al. 2000). Eventually, a region of bone may be completely occupied by Haversian systems and by luckless *interstitial lamellae*, little bits of bone that are separated by cement sheaths from all blood vessels, and so tend to be dead. However, death of bone cells by no means always leads to the formation of new bone to replace the old.

Human bone is like that of many primates and carnivores in that primary fibrolamellar bone is laid down initially, but this bone type is soon replaced by Haversian bone. However, this is not the case in many other mammalian groups. In most bovids (cattle) and cervids (deer), for example, the long bones keep their primary, fibrolamellar structure all through life, with only small regions, usually under the insertion of strong muscles, becoming Haversian. Many smaller mammals show no remodeling at all (Enlow and Brown 1958), the bone being fibrolamellar or, often, mainly composed of circumferential lamellae.

1.6 Compact and Cancellous Bone

At the next higher order of structure there is the mechanically important distinction between compact and cancellous bone. Compact bone is solid, with the only spaces in it being for osteocytes, canaliculi, blood vessels, and erosion cavities. In cancellous bone there are large spaces. The difference between the two types of bone is visible to the naked eye. The material making up cancellous bone of adults is usually primary lamellar bone or fragments of Haversian bone. In young mammals it may be made of woven or parallel-fibered bone.

The structure of cancellous bone varies in three ways: in its fine-scale structure, in its large-scale structure, and in its porosity. At the lowest level, cancellous bone is usually made of lamellar, not woven bone. However, the lamellae usually do not usually run precisely parallel with the external surfaces of the trabecular struts and so they come out to the surface, rather like rocky strata coming to the surface of the earth, at odd angles. Singh (1978) has a convenient description of cancellous bone morphology at the next level. The simplest kind of cancellous bone consists of randomly oriented cylindrical struts, about 0.1 mm in diameter, each extending for about 1 mm before making a connection with one or more other struts (fig. 1.6), usually roughly at right angles. In a variation of this pattern the cylindrical struts are replaced by little plates. The amount of variation ranges from cancellous bone in which there is just the occasional plate among the struts to cancellous bone in which there is just the occasional strut among the plates. In other cancellous bone the plates may be considerably longer, up to several millimeters. When this happens there is a higher level of anisotropy: these longer plates are not randomly oriented but are preferentially aligned in one direction. The final form of such cancellous bone is shown in figure 1.6C, where there are parallel sheets of bone with fine struts joining them. Another type of cancellous bone consists almost wholly of sheets, forming long tubular cavities that interconnect by means of fenestrae in the walls. Gibson (1985) produced a somewhat idealized classification, which, though not grounded so firmly in the messy reality of life, is nevertheless convenient for mechanical modeling. This is discussed in Chapter 5. These different versions of cancellous bone as classified by Singh are found in characteristically different places. The type made of cylindrical struts, with no preferred orientation, is usually found deep in bones, well away from any loaded surface, while the more oriented types, made of many sheets, are found just underneath loaded surfaces, particularly where the pattern of stress is reasonably constant. If the trabeculae are more than about 300 μm thick, they often contain blood vessels, usually within secondary osteons (Lozupone and Favia 1990).

The porosity of cancellous bone is the proportion of the total volume that is not occupied by bone tissue. Usually it is filled with marrow, but in birds there may be gas. The porosity varies from being effectively complete, where there is only the occasional tentative strut sticking into the marrow cavity, down to about 50%. If the porosity is less than about 50%, then cancellous bone becomes difficult to distinguish from

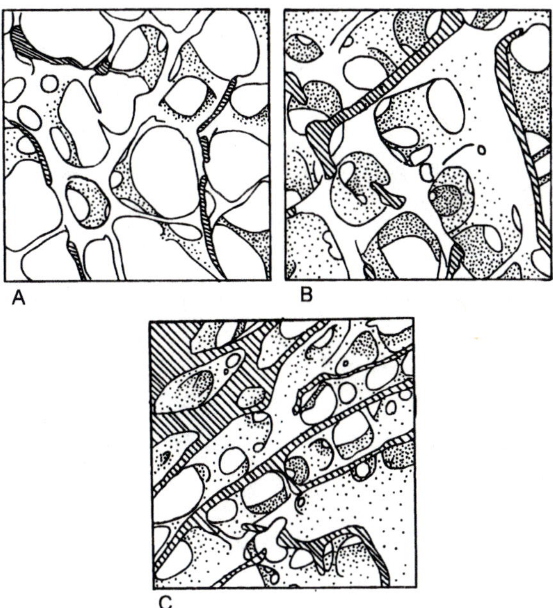

FIG. 1.6 Drawings of cancellous bone, seen by SEM. In each, the hatched parts are at the level of the top of the section. (A) Middle of the human sternum. Rather fine, nearly random network of mainly cylindrical struts. Width of picture 3.4 mm. (B) Human greater trochanter. Many of the elements are plates. Width of picture 3.4 mm. (C) Human femoral neck. The longitudinal plates are very obvious. There are many plates and struts lying orthogonal to them. Width of picture 8 mm (note smaller scale). ([A] Derived from Whitehouse [1975]; [B, C] derived from Whitehouse and Dyson [1974].)

compact bone with many holes in it. However, the change from compact to cancellous bone is usually clear and takes place over a small distance, and bone with a porosity of between 50 and 15% is uncommon. The mechanical reasons for this are discussed in chapter 5.

Bone grows by accretion on preexisting surfaces. Long bones have cancellous bone at their ends for reasons discussed in section 7.6. As long bones grow in length, cross sections that start off near the ends (and as the bone grows, move relatively closer to the middle of the length of the bone) usually undergo a *reduction* in diameter. This is because the ends (the epiphyses) are wider than the middle (the diaphysis) and although the diaphysis is slowly growing in diameter, the metaphysis (the part of bone just underneath the epiphysis) is usually reducing in diameter. The geometry of the situation is such that, quite often, compact bone has to be formed in a region where cancellous bone already exists. Here the old cancellous bone is not replaced; new bone is merely wrapped around the trabeculae, producing an extremely confused structure, with no obvious grain, called *compact coarse-*

cancellous bone. The effect of bone growth on bone histology and many other aspects of bone growth and structure (but not fine structure) are very clearly discussed by Enlow (1963, 1975).

1.7 A Summary of Mammalian Bone Structure

- *Tropocollagen molecules* (wrapped in a triple helix) lined up in files, and bonding side to side to form:
- *Microfibrils*, which aggregate to form:
- *Fibrils*, which are impregnated by and surrounded by the mineral hydroxyapatite or, somewhat more accurately, dahllite.

These fibrils appear in three different forms:

Woven bone	**Parallel-fibered bone**	**Lamellar bone**
Fibrils 0.1–3μm in diameter, arranged fairly randomly	Intermediate	Fibrils 2–3 μm in diameter, arranged in sheets (lamellae) 2–6 µm in thickness

Bone has bone cells, enclosed in *lacunae*:

Roughly isodiametric ~20 μm in diameter		Oblate spheroids, 5:1 ratio of major and minor axes. Major axes 20 μm

Canaliculi, about 0.2–0.3 μm in diameter, are channels containing cell processes that connect the cells with each other and with the nearest blood channel. Each osteocyte has about 60 canaliculi.

The bone is organized, at the next higher level, in four different ways:

Lamellar bone	**Woven bone**	**Fibrolamellar bone**	**Secondary osteons (Haversian systems)**
Often found in large lumps in reptiles. Found as circumferential lamellae in mammals and birds. *Primary and secondary*	Found in large lumps in young animals and in fracture callus. *Primary*	Alternating sheets of lamellar and woven bone/ parallel-fibered bone, with 2-dimensional nets of blood vessels. Ca. 200 μm between blood vessel nets. *Primary*	Cylinders of lamellar bone, solid except for a tube in the middle for blood vessels. Ca. 200 μm in diameter. *Secondary*

(continued on next page)

The bone is further organized into two different types of bone:

Compact bone	**Cancellous bone**
Solid, only porosity is for canaliculi, osteocyte lacunae, blood channels and erosion cavities.	Porosity easily visible to the naked eye. Rods and plates of bone, multiply connected, never forming closed cells.

1.8 Nonmammalian Bone

The bone I have discussed so far is mammalian bone. Rather little is known about the mechanical properties of nonmammalian bone, but I shall say something about its structure because of its interesting similarities and dissimilarities with mammalian bone. Many years ago Enlow and Brown wrote a useful summary of fossil and recent bone of all the vertebrates (1956, 1957, 1958). De Ricqlès produced a massive survey in ten parts of the histology of tetrapod bone, mainly that of reptiles. These papers are all cited in a comprehensive bibliography of 618 references, chiefly on bone histology, in de Ricqlès (1977). A (somewhat) shorter account is in Francillon-Vieillot et al. (1990).

The bone tissue of birds is like that of mammals, although at the naked eye level there are important differences in the proportions of wall thickness to overall diameter, which I shall discuss in section 7.3.2. However, the lamellae are usually less well developed than in mammalian bone, and the canaliculi have a much more wandering course (Rensberger and Watabe 2000). Some reptile bone is like mammalian bone; dinosaurs, in particular, often had well-developed fibrolamellar bone, Haversian systems, and a particularly rich blood supply (Currey 1962a). The bone of pterodactyls (pterosaurs) is more like that of birds (de Ricqlès et al. 2000). However, in many reptiles the bone is poorly vascularized and, indeed, is often avascular, although it does contain living bone cells. This poor vascularization is presumably possible because of the low metabolic rate of many reptiles. A characteristic of reptiles is the *lamellar–zonal* structure. This is bone principally made of parallel-fibered or true lamellar bone. It has poor vascularization and is particularly characterized by zones where growth comes to a halt then starts again. This pattern is characteristic of ectotherms, which often stop growing in the winter, and leads to lively debate among paleontologists trying to determine the physiology of extinct groups from their bone histology. The modern amphibia tend to have a rather simple, often avascular, bone structure. However, the earlier amphibia, such as the Embolomeri, which were quite large, show ill-developed lamellar–zonal or fibrolamellar bone, and also Haversian systems.

In the lower (less derived, in modern biology-speak) teleosts and in lungfish there are bone cells, although there is a tendency for bone to be replaced by cartilage in these groups. However, the bone of most modern bony fish—the advanced teleosts—has no bone cells (Moss 1961b). This is a remarkable fact whose significance, physiological or mechanical, is obscure. The acellularity is brought about in different ways in different groups. In some the bone cells form in the ordinary way from osteoblasts, are incorporated into the bone, and then die, the lacunae they leave being filled up with mineral (Moss 1961a). In other groups the osteoblasts avoid being incorporated in the bone at all (Ekanayake and Hall 1988). Another remarkable feature of the bones of these fish is the way in which they hardly remodel. Fish bones appear to spread from centers of ossification in almost straight lines. (Cod skulls are cheap. Boil one for a while till all the flesh drops off, and you will see the striking difference between "ordinary" bones and the cod's skull bones. The bones themselves are extremely graceful.) In many species osteoclasts have never been observed, though they may be induced to develop by playing physiological tricks on the body chemistry (Glowacki et al. 1986), and sometimes rather peculiar osteoclasts are found naturally (Witten and Villwock 1997). If bones do not remodel, there must be considerable constraints on their functional adaptation.

Fish bone, despite being acellular, and in general not remodeling, can, as always in biology, produce splendid exceptions. The rostrum of the swordfish, which is made of true bone and probably has some hydrodynamic function, is acellular. Parts of it are intensely remodeled, showing dense Haversian tissue and sometimes little groups of secondary osteons surrounding a large blood vessel (Poplin et al. 1976). The lamellar structure characteristic of secondary osteons is not very well developed, and the whole tissue is completely devoid of osteocytes. This strange tissue's adaptations are a mystery.

Very little is known about the mechanical properties of fish bone tissue or of whole bones. One reason for this ignorance is interesting: it is very difficult to obtain specimens of teleost bone that are not pervaded with large, elongated cavities. These cavities, of course, make it difficult to prepare useful test specimens. The anatomy and physiology of teleost bone is a neglected subject. Since most vertebrate species are teleost fish, this is a shame.

Mineralized tissues in the vertebrates have been evolving for a long time; 450-million-year-old fossils of what are probably vertebrate tissues have been discovered in the late Cambrian (Young et al. 1996). The mineralized tissue of "lower" vertebrates, including extinct groups, is discussed by Ørvig (1967) and Francillon-Vieillot et al. (1990). The considerable range of histological structures seen in the nonmammalian vertebrates is a challenge, because so little is known of their mechanical properties. Undoubtedly, when fully investigated they will turn out to have instructive similarities to, and differences from, mammalian bone.

Chapter Two

THE MECHANICAL PROPERTIES OF MATERIALS

2.1 What Is Bone For?

ANY BIOLOGICAL material has an enormous number of mechanical properties that may be measured and may also be tested by natural selection. Not all are likely to be of importance. I shall here discuss a question that will recur, by implication, frequently in this book: What are bones *for*? It is not worthwhile discussing at length the philosophical question of whether bone can be said to be designed. As a convinced Darwinist, I believe that living organisms have, except for some instructive exceptions, been nicely "designed," or fitted, by the blind force of natural selection, for the conditions that their ancestors of the last few million generations have lived in. Organisms have evolved organs that help them to survive and to pass on their genes to the next generation. It is one of the jobs of biologists to find out what these functions are and how the organs perform them. For a famous, entertaining, polemical, and contrary view of the perfection of organisms see Gould and Lewontin (1979).

Often the function seems obvious, and discovering it seems trivial. However, this apparent triviality may be misleading. It would usually be possible, with the materials available, to design an organ that would perform the apparent function better. A heart would pump blood more smoothly, and with more reserves of power, if it were larger. Therefore, the fact that hearts are the size they are, and not larger, must imply that there is some disadvantage in having a larger heart that would outweigh the hydraulic advantage. It may be, for instance, that the reserve of power is, in selective terms, not worth the metabolic cost of keeping a larger heart healthy. Recently, dissecting a laboratory rat and a wild-caught weasel *Mustela nivalis*, my students were amazed how much larger the heart of the weasel, the smaller animal, was than that of the rat. Life in a laboratory cage puts no great demands on the heart; life in the hedgerows where the weasel runs after its supper is a different matter altogether. I shall particularly discuss this problem in chapter 7 when I discuss minimum mass analysis and in chapter 10 when I discuss safety factors. It will be seen that we have to assume certain functions, and then see whether bones perform them efficiently. What are these functions? Baldly, they are to be stiff enough, and not to break under

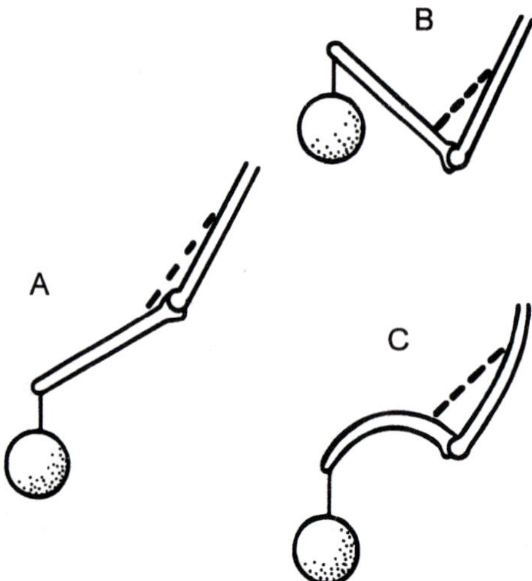

FIG. 2.1 The main function of the bone is to be stiff. (A) The joint is extended. (B) The muscle (interrupted line) shortens; the stiff bone lifts the weight. (C) The flexible bone does not lift the weight.

various kinds of loading. There are some other functions, usually less important, which I shall touch on later.

It is perhaps natural to think that the *strength* of a bone, that is, the load it can bear before breaking, is its most important mechanical attribute. However, it must surely be that bones function mainly by not *deforming* much under load. It is customary to think of bones, especially limb bones, as acting as levers. The muscles, by their contraction and relaxation, determine the distance between fixed points on different bones, and so the bones are constrained to move or to remain fixed. If the bones were floppy, they would not be constrained in the same way and the movements of the muscles would be futile (fig. 2.1). On the other hand, if the bones are stiff, but break, they become useless, and so the strength is of great but secondary importance.

The stiffness of a bone and its strength depend on two factors: the stiffness and strength of the bone material itself and the build of the whole bone. By build I mean the amount of the bone material and how it is distributed in space. The question of the build or architecture of bones will be discussed in chapter 7.

2.2 Mechanical Properties of Stiff Materials

In this chapter I shall be concerned with the basis of the whole book: the mechanical properties of stiff materials in general. Because many readers may not be too clear about the mechanical matters to be discussed, I shall run through various basics now. Later, slightly more recondite mechanical concepts will be introduced. Only the minimum necessary for the understanding of this book will be given. Readers wishing to go further should read such things as Fung (1993), Vincent (1990), Wainwright et al. (1982), Cowin (2001a), and a good book on strength of materials.

2.2.1 *Stress, Strain, and Their Relationship*

First, stress and strain. Consider a bar acted on by a force F tending to stretch it (fig. 2.2). If its original length is L, it will undergo some increase in length, ΔL. Usually it will also get thinner, and its breadth B will decrease by an amount ΔB. The proportional changes in length, $\Delta L/L$ and $\Delta B/B$, are called *normal strains*. They are often given the Greek letter ε. Note, first, that strains refer to changes in length in particular directions, and, second, that the strains tell us nothing, directly, about the forces causing them. By convention, if a dimension increases in length the strain is positive and if it decreases it is negative.

There is technical point that should be mentioned. Strain as defined above is often called *engineering* strain. This term should really be used

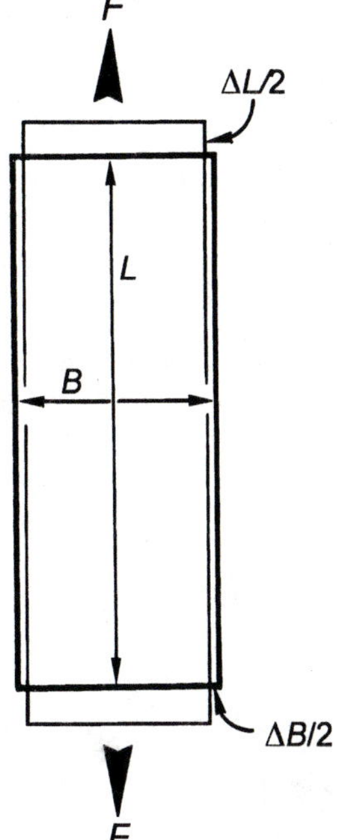

FIG. 2.2 Deformations produced by the application of a load F to a block of material of length L and breadth B. The undistorted shape is shown by the thicker lines. The block gets longer by an amount ΔL and narrower by an amount ΔB.

only for small strains. Suppose we have a specimen of length 100 mm and extend it to 110 mm. The strain, by definition is 10/100 = 10% or 0.1. If we next extend this specimen to 121 mm: 11/110 = 0.1, then we have given the specimen another strain of 10%, or 0.1. So the two strains added together are 20% or 0.2. But another way of looking at this is to say that what we have done is to extend a 100-mm specimen to 121 mm, which is a strain of 21%, or 0.21. These are apparently two different results. Engineering strains are not additive, and this can cause problems in some situations.

There is a way round this problem, which is to define a *natural* strain: $\ln(L_1/L_0)$, where ln is the natural logarithm, and L_1 and L_0 are the lengths before and after deformation is applied. (The fact that it is the natural logarithm rather than the common logarithm is irrelevant; that is just the way engineers like it.) What happens in our example when we use natural strain? $\ln(110/100) = 0.0953$; $\ln(121/110) = 0.0953$; $0.0953 + 0.0953 = 0.1906$. But, also, $\ln(121/100) = 0.1906$; natural strains are additive. Furthermore, they do not run into difficulties when the strains are compressive, which quickly becomes the case with engineering strains if the strains are at all large. (An engineering strain of -0.5 halves the length of a specimen, but a strain of $+0.5$ does not double it.) Fortunately, for most of what this book deals with, the strains are small, so the difference between the two kinds of strains is not large, and I shall use engineering strain. The biomechanical literature often has the term *microstrain*, symbolized by $\mu\varepsilon$. This is strain multiplied by a million. A strain of 0.01 is a microstrain of 10,000

A material can also distort in such a way as to cause changes in *angles* between imaginary lines in the material (fig. 2.3). A block of

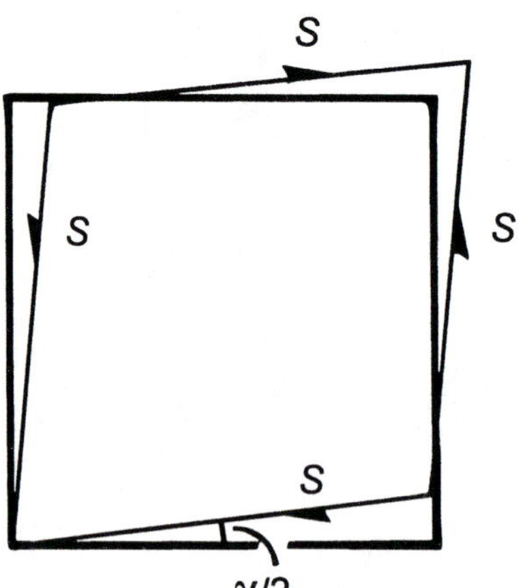

FIG. 2.3 Shear strain γ produced by a shearing force S. The undistorted shape is shown by the thicker lines.

material, which is part of a larger block, is acted on by a shearing force S. (For reasons of equilibrium, if S acts on the top and bottom faces, it must also act, in the way shown, on the right and left faces. This does not mean that forces have to be applied externally; if the block is in a mass of material that is not spinning round like a top, the material takes care of it itself.) The shearing force will tend to distort the block in the manner shown in figure 2.3, and it is undergoing *shear strain*. Shear strain is quantified as the change in angle undergone by two lines originally at right angles. The angle is measured in radians, and is frequently denoted by γ. In bone we are usually dealing with rather small strains, ε less than 0.005, γ less than 0.1. Figure 2.4 shows some loading situations in bones together with the kinds of strains that are produced.

The basic idea of strain is fairly simple, that of stress less so. Stress is best thought of as the *intensity of a force* acting across a particular plane. Imagine an area in some plane in a body that is small enough for the forces acting across it to be essentially uniform. The force can be represented by a vector F (fig. 2.5). This vector can be resolved into three mutually perpendicular vectors, each in the direction of one of the three axes we have set up. The component normal to the plane and parallel to the Z axis is F_{zz}. The other two components are in the X-Y plane and are shear forces F_{zx} and F_{zy}. There is no necessary relation between these forces; their relative magnitudes depend on the angle that F makes with the X, Y, and Z axes. If our little area has a cross-sectional area A, then we define the stresses acting on it as a normal stress F_{zz}/A, and two shear stresses F_{zx}/A and F_{zy}/A.

We cannot talk about the stresses acting in a small volume of material without specifying what plane we are discussing; the stresses depend on the orientation of the plane we consider. To see this, look at a very simple loading system: a bar loaded in tension by a force F. The bar has a square cross section of area A. The vector of the force acting on the cross section has no shear components, so there are no shear stresses across the section. The only stress is the normal stress F/A (fig. 2.6, *left*). However, there is nothing in the situation that we have set up that necessarily makes the cross-sectional plane the important one. Consider instead a plane B lying right along the length of the bar (fig. 2.6, *middle*). There are no normal forces and so no normal stresses. Nor are there any shear stresses, because the forces do not act upward on one side of the plane and downward on the other, which is necessary if shear stresses are to occur. Therefore, there are no stresses acting across the B plane. What about planes at intermediate angles? Suppose the plane C is inclined at an angle θ to the cross section A (fig. 2.6, *right*). The force F can be resolved into a normal force $F \cos \theta$ and a shear force $F \sin \theta$. The area of C is greater than that of A; it is $A/\cos \theta$. The

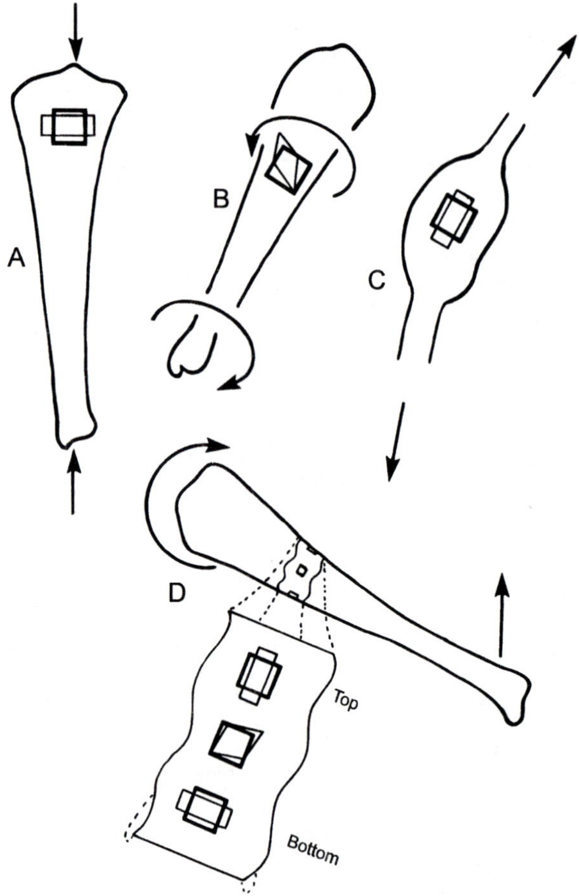

FIG. 2.4 Highly schematic diagrams of loads on bones, shown by arrows, and the resulting deformations. Thick lines are the undistorted shape, thin lines the distorted ones. The deformations, shown much exaggerated, are (A) compression; (B) torsion, producing shear; (C) tension, as in the patella; (D) bending. In (D) the deformations are shown on a piece of bone unwrapped from the whole bone. The bottom part shows tension, the middle shear, the top compression.

normal stress will be $(F\cos\theta)/(A/\cos\theta) = (F\cos^2\theta)/A$. The shear stress will be $(F\sin\theta)/(A/\cos\theta) = (F\cos\theta\sin\theta)/A$. The normal stress is greatest when $\theta = 0°$, as seems obvious, and declines rapidly as θ increases beyond about $\theta = 30°$. The shear stress is, less obviously, greatest when $\theta = 45°$, where it has a value half that of the maximum normal stress.

We have envisaged here a very simple loading system. In fact, a small block of material in a body can be subjected to complex loading situa-

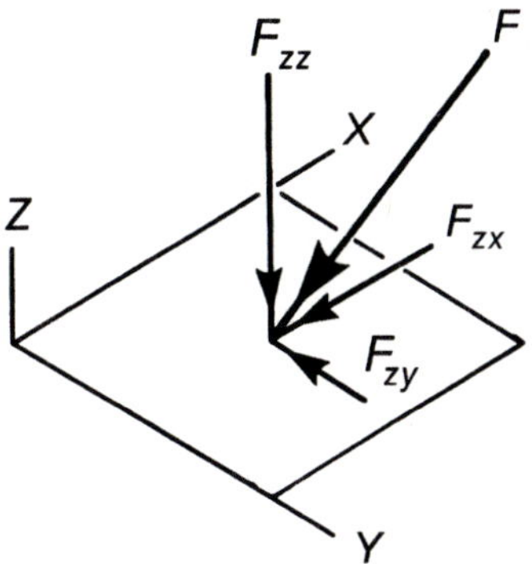

FIG. 2.5 A small area in the *X-Y* plane is acted on by a force *F*, which can be resolved into three forces at right angles, corresponding to one normal and two shear stresses acting on the area. The first and second subscripts of the stresses refer to the axis normal to the plane, and the direction in which the force is acting, respectively.

tions: tension on some faces, compression on others, and shear in various directions on all faces. Nevertheless, the stress across any plane in the block is always resolvable into just three stresses, at right angles, one normal, and two shear. Furthermore, it always possible, mathematically, to rotate this little imaginary block in such a way that in some orientation there are no shear stresses, and there remains just one normal stress on each face. These remaining normal stresses are called the *principal stresses*. Tensile stresses are considered positive, compres-

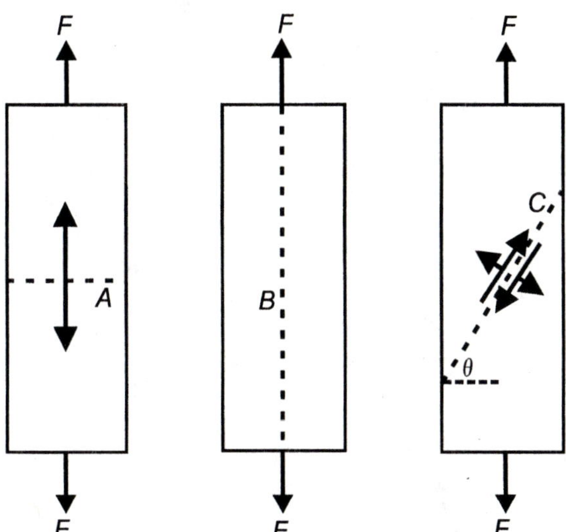

FIG. 2.6 A bar of cross-sectional area *A* is loaded in tension. The stresses across different planes are different. The various planes considered in the text are shown as interrupted lines.

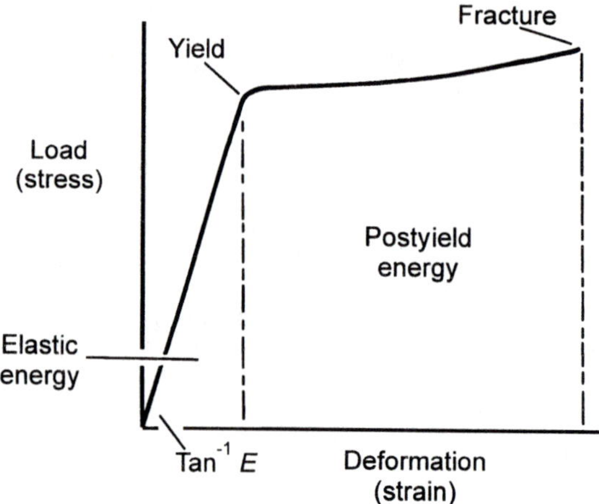

FIG. 2.7 A load–deformation curve of a bone specimen loaded in tension at a middling strain rate (0.01 s^{-1}). For a tensile specimen the load–deformation curve has the same shape as the stress–strain curve. The area under the curve from the origin to the vertical interrupted line dropping from the yield point gives the amount of energy (work) absorbed elastically. Postyield energy, between the two interrupted lines, is usually much greater. If the load and deformation are normalized to stress and strain, the angle made by the elastic part of the curve and the abscissa is an angle whose tangent is Young's modulus of elasticity (E).

sive stresses negative. There are similar conventions for shear strain and stress, but it is not necessary to describe them here.

Now that we have some idea of stress and strain, we can consider the relationship between them. Suppose we load a small specimen of bone in tension until it breaks. We can do this by gripping it at the ends and pulling the grips apart, meanwhile measuring the load on the grips, which will be the same as the load on the specimen. There are devices for measuring the extension of a portion of the length of the test specimen, and the output can be displayed, digitized, and stored in a computer for later analysis. The output will show a curve of load as a function of deformation. A tensile load–deformation curve of bone will look something like figure 2.7. Starting from the origin, there is a part where the load varies linearly and proportionally with the deformation. At the so-called yield point the curve flattens considerably, and now increasing deformation involves little extra load. Eventually the specimen breaks.

A load–deformation curve is all very well for giving a rough idea of what is going on, but does not allow us to put any values on the variables that will tell us anything about the mechanical properties of the material, bone this case, because the load and deformation will depend on the size and shape of the test piece. However, in a simple tensile test this is easily remedied. The force/(cross-sectional area) is the normal stress acting across the cross section, on which there are no shear stresses. Similarly, the increase in length of the part of the test piece being measured (the gauge length), divided by the original length, is the strain. So, the load–deformation curve can, with minimal arithmetic, be turned into a stress–strain curve. (Such simplicity does not hold for such loading systems as bending or torsion.)

The stress–strain curve gives a considerable amount of information. The linear part, which goes through the origin, shows that there is a region where stress and strain are proportional to each other. For many biological materials this is not true. The stress–strain curve for cartilage, for instance, has no linear region; the curve is always curved, right from the origin. The greater the stress associated with a particular strain, the stiffer the material. (People worry that stress appears on the *Y* axis and strain on the *X* axis, because they feel that the stress "causes" the strain. This is a nice, almost philosophical point I shall not pursue. However, most testing machines, in fact, apply a deformation to a specimen and measure the load necessary to do so, so the conventional arrangement of the axes is rational.) The *modulus of elasticity*, or Young's modulus of elasticity, is defined as stress/strain in the linear region of the curve of an ordinary tensile or compressive test. It is expressed in pascals (Pa), which are newtons per square meter, and in bone it has a value of roughly $1–2 \times 10^{10}$ Pa, or 10–20 GPa. It is usually denoted by E. The steeper the initial part of the curve, the stiffer the material and the greater the modulus of elasticity. In fact, the modulus of elasticity is the tangent of the angle made by the curve to the abscissa. Although Young's modulus of elasticity is often referred to as *the* modulus, we shall see in the next section that a number of independent moduli are needed to define completely the elastic behavior of a material. There is no reason why Young's modulus should have preference except that it is the most straightforward to measure and the easiest to understand.

It is found, near enough, that if a specimen of bone has been loaded only into the linear part of the curve, and the strain is then reduced, the force falls to zero when the strain falls to zero. If a material returns to its original size and shape when loads are removed, it is said to behave *elastically*. If the material is also like bone in that the first part of the stress–strain curve is straight, it is said to be *linearly elastic*. In fact, if a material is not linearly elastic, it cannot be said to have a particular values of Young's modulus. Beyond the place where the curve bends over, something happens to the bone because beyond this point it is found that, if the specimen is unloaded, the stress will fall to zero before the strain falls to zero (fig. 3.15). In metals this residual strain is called plastic strain, and the metal is said to be behaving *plastically*. As we shall see in chapter 3, so-called plastic behavior can be caused by a variety of phenomena at the microscopic level, so in bone it is better to use neutral words like *postyield behavior* and *irrecoverable strain*.

Where the deformation changes from being linear is called the *yield point*, or *yield region*, depending on how localized it is. Although a bone reaching this point may still have far to go before it breaks, it is damaged to some extent, and healing will have to take place if it is to regain its pristine condition. It is therefore not surprising that, as we shall see in Chapter 10, bones seem to be designed so that the loads placed on them in life usually load them into the elastic region only. In the postyield region the bone extends with little extra stress. Therefore, the difference between the fracture stress, which is usually considered to be the strength of the bone, and the yield stress is usually rather small. So, if a bone were subjected to an ever-increasing load, it would make little difference to the load at failure whether the bone broke at the end of the linear part of the curve or at the end of the long postyield region. (Although this is true for tension, if the bone is loaded in bending, the postyield region does help to increase the load necessary to cause fracture [Burstein et al. 1972], see section 3.10.)

A material that breaks without showing any irrecoverable, postyield deformation is said to be brittle. Characterizing a material as brittle gives no information about whether it is weak or strong in the sense of how much stress it can bear. However, the presence or absence of postyield deformation is an extremely important feature of the mechanical properties of a material. Materials that show a reasonable amount of postyield deformation are very often "tough." Toughness is a concept that can be described very informally or very formally. Informally, a tough material is one that is resistant to the propagation of a crack (Ashby 1999, p.26). This results in it not being weakened by small scratches or cracks, and usually being able to absorb a great deal of *energy* before it breaks.

Toughness is important in determining how bones break in life. What usually happens is that the bone is subjected to violence. This can be direct, as when a small boy falls off a wall onto his head, or it can be indirect, as when a galloping horse puts its hoof down a rabbit hole. In the latter case the bone is not struck directly, like the boy's skull. Instead, the mass of the horse's body, moving on while the leg is stuck fast, produces a bending load on the leg. Sometimes, as in the spiral fracture of the tibia in a skiing fall, the bone is loaded by the ligaments, which may not themselves break, attaching it to its neighboring bones.

In these cases the bone, and the surrounding tissues, are given a quantity of mechanical energy. This energy has to be dissipated in the bone

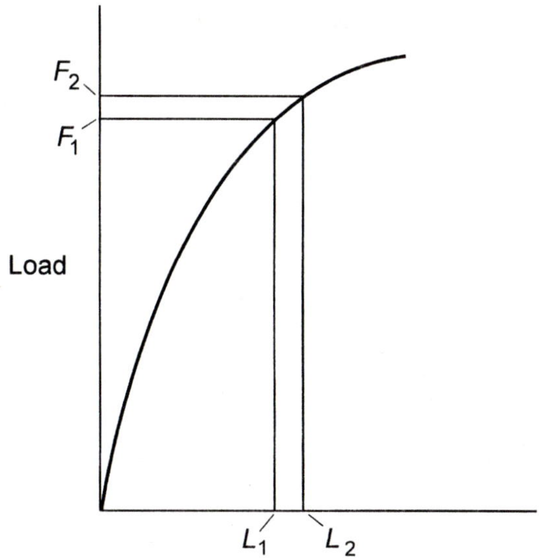

FIG. 2.8 A load–deformation curve. A load F_1 produces a deformation L_1; a load F_2 produces a deformation L_2. If the vertical area is narrow enough, its area $\approx F_1(L_2 - L_1)$, and this represents the work done in deforming the specimen from L_1 to L_2.

and surrounding tissues without anything breaking. Now the energy absorbed by a material is proportional to the area under the load–deformation curve. The reason for this is shown in the curve in figure 2.8. If we take a very narrow vertical strip, its area is nearly $F_1(L_2 - L_1)$. A force F_1 has acted over a distance $(L_2 - L_1)$, and therefore, the amount of work done is $F_1(L_2 - L_1)$. If these strips are summed over the whole curve, therefore, they give the value of the work done, or energy absorbed. It turns out, using similar reasoning, that the area under the stress–strain curve is the work done per unit volume. So, the flat top of the curve of bone, although having little effect on the *load* at which the material breaks, can greatly increase the *work* that has to be done on it to break it.

2.2.2 *Anisotropy*

The information given by a stress–strain curve is considerable, but by no means exhausts what can be found out. Bone, like most real materials, is mechanically anisotropic. A material is anisotropic if its proper-

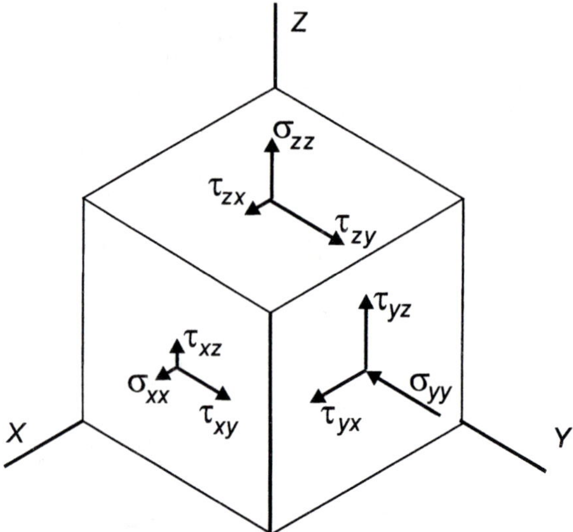

FIG. 2.9 Stresses that can act on the faces of a small cube.

ties are different when measured in different directions. If the properties are the same, then the material is isotropic. In this book we are really interested only in mechanical anisotropy, though anisotropy can occur in many physical properties, such as electrical resistance and optical refractive index. The stress–strain curve of an anisotropic material gives the value of Young's modulus in one direction, but it may be different in a different direction.

Consider a small cube of bone, in equilibrium under the forces acting on it. The forces on each face, divided by the area of the face, give the stresses (fig. 2.9). Because the cube is in equilibrium, we can ignore the equal and opposite normal forces on the opposite faces. As in figure 2.5, for each force the first subscript refers to the face on which the force is acting (≡ the axis normal to the face), and the second subscript refers to the axis parallel to which the force acts. Thus, τ_{zy} refers to the stress caused by the force acting in the *Y* direction on the face normal to the *Z* axis. It is a shear stress. σ_{xx}, σ_{yy}, and σ_{zz} act normally to the faces and are normal tensile or compressive stresses. Conventionally, normal stresses are represented by σ, shear stresses by τ. There are notionally nine stresses that can act on the cube. In fact, the stresses are not all independent because, for reasons of equilibrium, $\tau_{yz} = \tau_{zy}$; $\tau_{zx} = \tau_{xz}$; $\tau_{xy} = \tau_{yx}$. Nevertheless, there remain six independent stresses acting across the faces of the cube. Similarly, we can consider the strains in the cube. There will be three normal strains: ε_{xx}, ε_{yy}, and ε_{zz}. For instance, ε_{xx}

says how movement of a point in the x direction is dependent on its position on the X axis. There are also shear strains such as γ_{xy}. γ_{xy} tells how movement of a point in the x direction depends on its position on the Y axis. Fortunately, as with stress, not all the shear strains are independent; there are only six independent strains.

If the material is isotropic, the relationships between stress and strain are fairly simple. Young's modulus is given by $E = \sigma/\varepsilon$; the shear modulus is given by $G = \tau/\gamma$. If a normal stress σ acts in one direction and is the only stress acting on the specimen, the strain in this direction will be $\varepsilon = \sigma/E$, and there will, usually, be a strain $-\nu\varepsilon$ in directions at right angles to this. If the specimen is, say, a cube that is being stretched in one direction, it will shrink in others. This is called the Poisson effect and ν, Poisson's ratio, shows the ratio of the shrinking to the stretching. The shear modulus and Young's modulus are related by:

$$E = 2G(1 + \nu)$$

For an isotropic material we need only determine two out of E, G, and ν to be able to characterize its elastic behavior completely. Any textbook on strength of materials will list the relationships between the stresses and strains in various directions, and I shall not repeat them all, but I shall give one as an example to make a general point of some importance:

$$\varepsilon_{xx} = \frac{\sigma_{xx} - \nu(\sigma_{yy} + \sigma_{zz})}{E}$$

The normal strain depends on *all three* normal stresses. If a specimen is being loaded in tension in the x direction, it will tend to elongate in that direction, and shrink in the y and z directions. If the specimen is free to shrink in the y and z directions, there will be no stresses σ_{yy} and σ_{xx} and the expression reduces to $\varepsilon_{xx} = \sigma_{xx}/E$. Strain is proportional to stress divided by Young's modulus, as stated previously. However, if the specimen is also being loaded in tension in the y and z directions, these will not shrink so much, and so the specimen will not elongate so much in the x direction as would otherwise happen. The converse of this is that if a specimen is being stressed in one direction and, for whatever reason, does not shrink in the other directions (for instance, because the stressed region is at the tip of a long notch in a wide specimen, which will not shrink much sideways) stresses will develop in the other directions. When this happens, unexpected (until one learns about them) sideways stresses appear in the specimen, preventing it from shrinking in the expected directions. It is for this reason that, in equations relating to stresses in the neighborhood of notches, Poisson's ratio ν keeps popping up, as in section 2.2.6.

In fact, most materials, including bone, are not isotropic. That is, their elastic behavior, and, indeed, other kinds of mechanical behavior, will vary according to the direction of loading. More information is needed before the elastic behavior can be described completely. The general remarks about the importance of the Poisson effect still apply, but the numbers involved will be different. Obtaining the full expression for the elastic constants for bone is technically extremely difficult and has been attempted only a few times. In the most complicated symmetry possible it would be necessary to measure twenty-one independent elastic properties, an almost impossible task with present-day techniques, and simplifying assumptions are always made to bring the measurements required down to a manageable number. For instance, bone is often assumed to have orthotropic symmetry. A material has orthotropic symmetry if it has three mutually perpendicular planes of mirror symmetry. If this is the case it would have three Young's moduli, three shear moduli, and six Poisson's ratios, of which only three are independent (Cowin 2001b). Even this system requires nine measurements, which is difficult. The chapter by Cowin in Cowin (2001a) is an excellent introduction for those who wish to start taking the theory of elasticity seriously.

2.2.3 *Viscoelasticity*

In all this discussion about elasticity, I have assumed that the elastic properties of the material are fixed. In fact, in many materials, if one measured the Young's modulus while loading the material very slowly, one would get a lower value than if one measured it when loading very fast. In other words Young's modulus is to some extent *strain-rate dependent.* (Strain rate is change in strain divided by the time over which the change occurs, and has the units of reciprocal seconds, s^{-1}. It is frequently symbolized as $\dot{\varepsilon}$, the dot being the old Newtonian symbol for differentiating with respect to time.) This is just one manifestation of the phenomenon of *viscoelasticity*, shown by most materials, including bone. Strain rates in highly strained parts of human long bone during vigorous activity are of the order of $0.03\ s^{-1}$ (Burr et al. 1996) and strain rates as high as $0.08\ s^{-1}$ have been reported in galloping horses (Rubin and Lanyon 1982). There are other results of viscoelasticity: if a viscoelastic material is strained and then held at that strain, the stress induced by the strain will decline over time at an ever-decreasing rate, a process called *stress relaxation*; if a viscoelastic material is loaded to some stress and then held at that stress, it will continue to deform, to show increasing strain, at an ever-decreasing rate, a process called *creep*; if a viscoelastic material is loaded and unloaded, the shape of the unloading curve is different from the shape of the loading curve, a process called *hysteresis*.

The mathematics of viscoelasticity can be seriously unpleasant, but fortunately we need little here. A good description of it is in Sasaki (2000). For the moment it is sufficient to know that some solid materials, although elastic, can flow a little bit, but not indefinitely, and that the rate of flow is proportional to the load being imposed, but also inversely proportional to some function of the time that the load has been imposed.

2.2.4 Modes of Loading

A small volume of bone may have all kinds of forces acting on it, and I have considered so far only a rather idealized state of affairs. The stress–strain curve of figure 2.7 is for a bone specimen loaded in tension; a bone specimen loaded in compression would show a different curve. The point at which the curve became nonlinear in compression would be at a greater stress and strain, and final fracture would be a less sudden event; rather it would be a more prolonged disintegration. Fracture in tension involves the pulling apart of two surfaces; once they are apart, that is the end of it. However, fracture in compression involves the shearing of surfaces past each other; the bone cannot collapse into itself, as it were. This shear failure can occur at many places in a specimen at the same time, and is in general a less definite affair. Testing of bone in compression is quite tricky. The loaded ends have to be quite flat and parallel, otherwise the stress will be concentrated at the high spots. This is much less of a problem in tension, because tensile specimens can be easily be prepared to have flared shoulders that lead the stress smoothly from the grips into the central test portion.

The reason that tension tends to get most of the attention is that bone is usually weaker in tension than in compression (table 3.2) and when a bone fails in bending, which is the usual mode of fatal loading, the fatal crack starts on the tensile side of the bone. It is therefore reasonable to consider tensile failure to be the important mode. However, bending is anything but a straightforward mode of loading. Bending strength is an easy property to measure, but a difficult property to understand. Many studies have reported the bending strength of bone (see, for instance, section 3.10) and, of course, bones very often fail in bending in life. When a bone or a bone specimen is bent, one side is loaded in compression, the other in tension, and there is a continuously varying strain between the two. Formulas for the stresses and strains in beams loaded in bending are well known, and so one can calculate the stresses at any point in the beam as a function of the applied load. Unfortunately, such formulae are valid only as long as the bone behaves linearly elastically. Once it yields, usually in tension first, the formulae become dependent on the exact form of the stress-strain curve. This is because the ordinary beam formulae assume that stress is proportional to strain. Once the outermost part of the bone yields, it may undergo a considerable increase in strain with little increase in stress. This matter was discussed, with some experimental verification, by Burstein et al. (1972) and is also taken a bit further in section 3.10. Reported bending strengths depend much more on the geometry of the test specimen, and on its post-yield behavior, than do tension and compression tests.

There is an interesting point about bending that will crop up once or twice in this book: It is possible to bend very *thin* sheets or rods into quite tight curves without their experiencing large, potentially dangerous strains and stresses. Consider a straight rod of length X bent into a circular curve of radius R (fig. 2.10). The neutral axis will remain of length X, and the rod at a distance ΔR away from the neutral axis will have some length $X + \Delta X$. Now $\Delta X = X.\Delta R/R$, and therefore the strain $\Delta X/X = \Delta R/R$. Therefore, if ΔR is very small, then it will be possible to make R small; that is, it will be possible to produce a tight bend, without the strain becoming unduly large.

Most materials loaded in compression actually break up in shear. Bones are also loaded in shear by being twisted. Figure 2.4 shows how torsional loading will induce shear in a bone. Bone is not particularly strong in shear, and twisting of long bones is a fairly frequent cause of failure showing itself, for instance, in the characteristic spiral fracture of the tibia in skiing accidents.

2.2.5 *Fracture and Toughness*

If a material is stiff enough, the next question is whether it is strong enough to stand up to the conditions of service. In the nineteenth century, materials testing had advanced sufficiently for it to be possible to find out, with great repeatability, the tensile and compressive strengths of materials. It nevertheless was apparent that something was wrong. Bridges and, more particularly, ships were breaking up under conditions that, it could be shown, imposed stresses in the material, usually iron or steel, that could easily be borne by the material when it was tested in the laboratory. The reason for this was almost completely obscure until great light was shone on the matter in 1920.

The problem is this. It is possible, though difficult, to calculate for

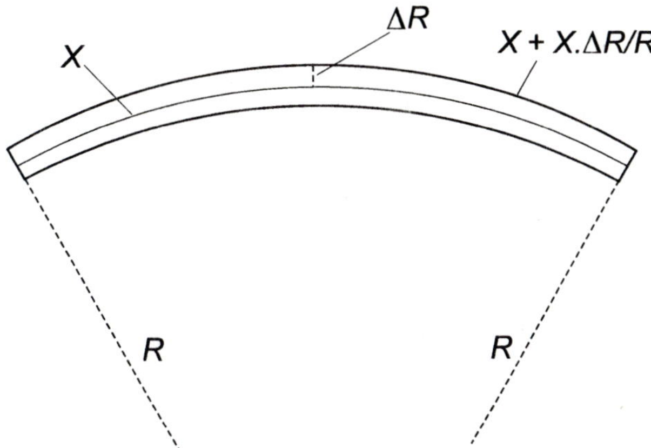

FIG. 2.10 A straight beam of original length X is bent into a circular curve of radius R. The neutral axis will remain of length X. The outermost part of the beam is a distance ΔR from the neutral axis. Its length will be $X + X.\Delta R/R$. The strain in this outermost part is therefore $\Delta R/R$.

simple crystals the force theoretically needed to separate two planes of atoms from each other. However, the answers appear to be too high, by some orders of magnitude, when compared with the results of laboratory tests. Inglis (1913) proposed, as had others, that in engineering structures cracks developed from the corners of hatches and rivet holes because of stress concentrations. Figure 2.11 shows an elliptical hole in a large plate (technically an "infinitely large" plate) loaded in tension. The lines represent, roughly, the way the force is distributed around the hole. The force is not distributed evenly; it is more concentrated near the ends of the hole. There is, therefore, a higher stress and strain near the ends of the hole than farther away from it. Inglis took this simple case and showed that, if the general stress in an infinite plate was σ, then the stress in the material at the ends of the hole is $\sigma(1 + 2C/B)$, where C is the length of the ellipse normal to the stress and B is the length of the ellipse in line with the stress (fig. 2.11). There is a *stress concentration* of $1 + 2C/B$. If the ratio C/B is very large, that is, if the ellipse is like a crack, then the maximum stress is $\sigma(2C/B)$. It is clear that cracks can be very potent producers of stress concentrations. However, these results were not taken seriously for two reasons. The first was that the formula for the stress-concentrating effect of a hole was merely a ratio; there were no sizes in it. Therefore, the size of the ellipse or crack should make no difference. But it was a matter of common though vague experience that long cracks were more dangerous than short ones. The second was that if, on the other hand, one assumed that

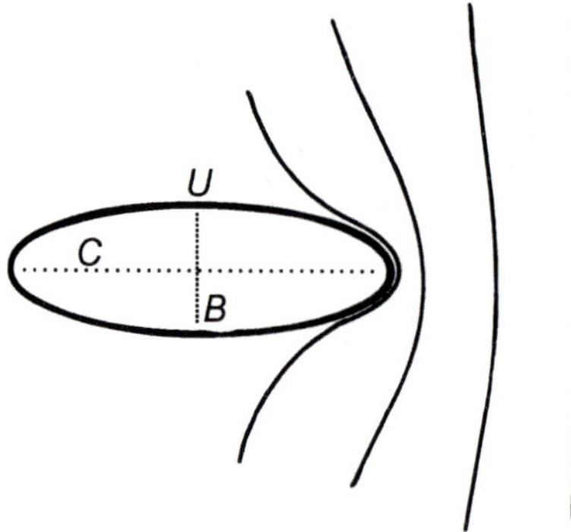

FIG. 2.11 Stresses that can act near the tip of an elliptical cavity. The force is acting up and down the page. The inverse of the distance between the lines of force is a measure of the stress. It is very high near the tip. The region near *U* is nearly unstressed.

a sharp crack had a length *C*, which might be a few millimeters, and *B* was roughly the interatomic spacing, then the values of stress concentration would be so high that everything that was not perfectly smooth should fall apart under its own weight. So the analysis of Inglis, though, in fact, correct, seemed to be of no practical help.

CRACK SPREADING IN ELASTIC MATERIALS

This situation was remedied in 1920 by Griffith. He made the following analysis, in a more rigorous form, of course. Suppose we have a small crack in an infinite plate, which is subjected to a general stress σ. Suppose also that the material is linearly elastic and completely brittle. Consider the energy changes that must occur if the crack is to spread. The surfaces of the crack have a certain amount of surface energy U_s. Surface energy is caused by the surface atoms having fewer near neighbors and, therefore, less negative bond energy than atoms in the bulk of the material. Any increase in the length of the crack must result in an increase in its total surface energy, directly proportional to the increase in crack length. The plate also has strain energy U_ε. Strain energy is the potential energy a material possesses because its interatomic bonds are strained. Strain energy can be released to do work, as in the uncoiling of a watch spring. In an elastic material the amount of strain energy available at any strain is equivalent to the area under the stress–strain curve up to that strain. Finally, in this list of energies, any displacement of the edges of the plate, caused by the loads on them, will make the loads do work W_L, which will be negative if the work is done on the plate. The total energy of the plate U can be expressed as $U = W_L + U_\varepsilon + U_S$.

Under what conditions will the crack spread? Let us suppose that it does, by some tiny amount. The relevant load–deformation curves are shown in figure 2.12. The plate has been loaded along *OA* up to a load

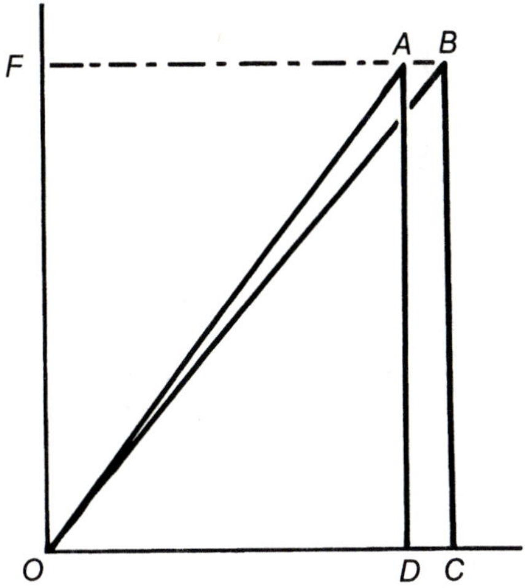

FIG. 2.12 Changes in the strain energy of a specimen with a spreading crack. Explanation in the text.

F, equivalent to a general stress σ, and to a deformation D. Then the crack extends a little. This will make the plate more compliant, because there is less material at the level of the crack to bear the load, and so the plate will extend a little amount DC. The force remains the same, so the work done on the plate as it is extended is force times distance $= F \times DC$, given by the area of the rectangle $ABCD$. The elastic strain energy of the plate was OAD, and is now OBC:

$$OBC - OAD \text{ (the change in strain energy)}$$

$$= ABCD - ABO = \frac{ABCD}{2}$$

Therefore, the work done on the plate $= ABCD$, the extra strain energy is $ABCD/2$, and so there is $ABCD/2$ available to feed into the surface energy of the crack.

Similar geometrical arguments show that if the crack extends under the "fixed-grip" condition, that is, without any increase in the value of OD, then the strain energy of the plate decreases, rather than increasing. However, the same amount of strain energy is made available for the surface energy of the crack.

The critical question is how can the strain energy released be expressed in terms of the surface energy needed? So far we have merely said that the extension of the crack makes the plate more compliant, but not how much more. It turns out that we do not necessarily need to know. Consider the state of stress near the tip of a crack (fig. 2.13). At the tip the stress is very high; far from the tip it is σ; behind the crack

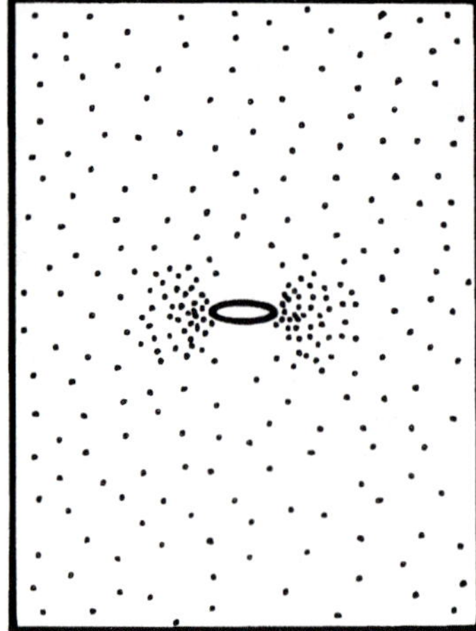

FIG. 2.13 General state of stress around a crack in a large plate. The spacing of the dots indicates (the inverse of) the level of stress.

tip there is a region, very roughly circular, that is effectively relieved from stress. As long as the crack is small relative to the plate, the *shape* of the area relieved from stress, which has consequently given up its strain energy, is not affected by the *length* of the crack. Therefore, because the area of any shape is proportional to the square of the linear dimensions, the area of this stress- and strain-free region, or its volume if we consider the plate to have uniform thickness, is proportional to the *square* of the crack length.

Griffith calculated the loss of strain energy caused by the presence of a crack of length $2c$ to be $\pi c^2\sigma^2/E$, where σ is the general state of stress, and E is the modulus of elasticity of the material. What is the surface energy of the crack? If the surface energy per unit length is γ, then the total surface energy is $4\gamma c$. (The factor of four comes in because the crack length is $2c$, and it forms two new surfaces as it extends.) Surface energy is conventionally given the letter γ; this is not the same as shear strain, which, unfortunately, is also called γ. Therefore, when the crack is just on the point of spreading, the strain energy released by a differentially small increase in crack length must just equal the surface energy needed: $d(4\gamma c)/dc = d(\pi c^2\sigma^2/E)/dc$. Notice that the release of strain energy is proportional to the *square* of the crack length c, whereas the surface energy needed is proportional only to its length. From the equation above, $\sigma = (2\gamma E/\pi c)^{1/2}$. The value of stress given by this analysis is called the Griffith stress. The important point about it is that the stress needed to cause a crack to spread is less if the crack is longer. If we rearrange the equation, this point is brought out: $c = 2\gamma E/\pi\sigma^2$. Thus, for a given general stress σ, there will be some length of crack c_{crit}, which will be sufficiently long to be able to spread.

We have used here a particular loading situation—tension on a thin plate—but the basic theory is applicable to all kinds of tensile loadings. Griffith tested his ideas using glass fibers, for which it was easy for him to get a reasonable idea of γ, the surface energy, and the tests worked out well. However, glass is rather a special material, just because it is so brittle. For glass, and materials like it, it is almost true that the only energy that has to be fed into the material near the crack tip is that required for surface energy. However, many materials, including bone, show nonelastic deformation, and we need to consider this.

CRACK SPREADING IN NONELASTIC MATERIALS

Nonelastic deformation, whose presence is shown by a severely flattened region in the stress–strain curve, can be caused by various kinds of events at the molecular or microscopic level. In metals, plastic deformation is often caused by the movement of dislocations. These are faults in the uniform crystalline structure of the metallic lattice, which will move if the material is stressed sufficiently. In other materials, including bone, nonelastic deformation is mainly produced by the development of tiny cracks in the material, which allow much more deformation than would be possible if the deformation were caused by the increasing of interatomic distances. These cracks may heal spontaneously, the bonds reforming with new partners, though this is not usually the case, and we certainly do not know whether it can happen in bone. The inability of the cracks to close up neatly, because the fracture surface has become deranged, shows itself as a permanent deformation.

All such processes require a force and, because they involve movement, work has to be done to bring them about; they absorb energy. The nonelastic energy-absorbing deformation that occurs when a material is deformed has two components. One is the deformation that may occur generally in the body of the material, quite far from the crack that may eventually break it in two. The other part is specifically associated with the crack and is a necessary concomitant of its travel. Even in glass, which is remarkably brittle and shows no general nonelastic deformation, a few layers of atoms under the fracture surface become irreversibly reordered.

Consider first the nonelastic deformation associated with the crack tip itself. Every differentially small increase in the crack length will require surface energy (γ) and the energy required to bring about nonelastic deformation. These are both linearly related to crack length, and can be combined as a term W. The extent to which W exceeds γ depends on the material and may be very large. γ is roughly 1 J m^{-2} for the great majority of brittle materials. The value for W in bone is in the region of 4×10^3 J m^{-2}, though much of this may be absorbed well away from the crack. It is clear that the surface energy term in bone is trivial compared with other energy-absorbing events.

Another effect of nonelastic deformation may be to blunt the crack. In a material with a low Young's modulus, crack blunting will take place even though the material remains elastic. One can see this easily by cutting a small slit in the cuff of a rubber kitchen glove and then pulling at right angles to the slit. The very sharp crack becomes completely rounded. Even in a fairly stiff material like bone some macroscopic blunting occurs, though it is nonelastic, unlike the kitchen glove. This blunting may cause the local stress to fall significantly. This crack blunting is shown clearly in a poorly mineralized bone such as deer's antler. A sharp cut put into a tensile specimen of antler becomes wide and round-ended before the eventually fatal crack reluctantly starts to travel from its end. For a crack to spread in a brittle material, two conditions must be met. The energy balance must be right; that is to say, the strain energy released by the spread of the crack must be sufficient to produce enough surface and subsurface energy per unit increase in length. Furthermore, the crack must have a sufficiently high strain at the tip to break the bonds right at the crack tip. Nonelastic deformation has an effect on both these conditions, and tends to make it difficult for a preexisting crack in a material to spread. This is of the utmost importance in any real material, which is likely to have a host of tiny cracks on the surface.

If the "worst" crack in the material can be prevented from spreading until the stress at which general yielding takes place is reached, then the material may undergo much more postyield deformation before failure. Figure 2.14 shows this. Suppose the stress–strain curve of the material has the shape *OABC* but the specimen has a crack in it that will cause fracture at a general stress in the specimen of S_1. The energy absorbed by the specimen, per unit volume, before the crack spreads, is proportional to *OAE*. If some mechanism can raise the general stress at fracture to S_2, then the amount of energy absorbed before failure will be proportional to the area *OABF*. Thus a rather small increase in the general stress at failure will cause a great increase in the energy absorbed. This will be of great help during impact loading.

A further complication must be mentioned before these arguments can be applied to bone. We have so far seen that the energy-*absorbing* processes are associated with the growth of the fracture surface. The energy-*releasing* process is the release of strain energy. The Griffith equation assumes that all the energy put into the material in straining it is recovered when it becomes unloaded. But this is, in general, not the case. Some of the strain may be true plastic strain, some may be viscoelastic strain, and some (or, indeed, much, as we shall see) may be

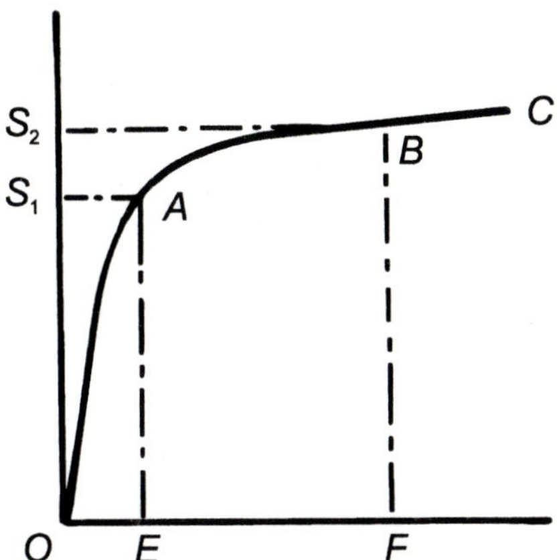

FIG. 2.14 A small increase in the stress at fracture (from S_1 to S_2) may produce a great increase in the energy absorbed (from *OAE* to *OABF*)

associated with microcracking. The energy used up in straining the material plastically is irrecoverably locked up in it, and cannot be released. A little energy is lost as heat. In the case of viscoelasticity, although the strain energy is eventually recoverable, it is so only if the unloading takes place slowly relative to the loading. Often, in life, the unloading tends to take place much more rapidly than the loading. In such cases, the energy put into the material is not all available to help the crack spread. In the case of microcracking the cracks may not "heal" and therefore will have residual strain, and strain energy, associated with them.

2.2.6 Fracture Mechanics

The situation of fracture in the real world looks so complicated, with so many apparent variables, both of the material and of the geometry of the situation, to account for that it might appear difficult to make any predictions about it. However, the area of research called *fracture mechanics* attempts to accomplish just this, by doing two things. One is to characterize the tendency of materials to fracture, in terms independent of the actual geometry of the loading system, that is, to obtain a pure *material property*. The other is to determine how the actual geometries will affect the fracture behavior. (These are two sides of the same coin, of course.) For instance, if we are carrying out a tensile test on a brittle sheet of length L, breadth B, and thickness T, with a transverse flaw in it of length f, the length of the flaw and the ratio of B to T are both important in determining at what load the sheet will rupture. The standard strength of materials approach would be merely to consider the cross-sectional area of the sheet at the level of the crack: $T(B - f)$. The length of the flaw appears only in its effect in reducing the cross-sectional area, which may be trivial compared with its actual effect in reducing the strength of the specimen. The fracture mechanics approach, on the other hand, if successful, allows the engineers to know how dangerous will be flaws of particular length and orientation in any structure they may design.

There are two main ways of characterizing materials and cracks in them: (a) The crack will spread when the energy released from the material as the crack spreads exceeds the energy needed to extend the crack (the approach discussed above), which is called the critical strain energy release rate G_c; and (b) for a given geometry, the crack will spread when the stress at the crack tip reaches a critical value that overcomes the cohesive strength of the atoms just ahead of the crack, which is called the critical stress intensity factor K_c. These two properties might not appear to be related to each other, but for fairly brittle materials they are quite simply related. For instance, for a *thick* plate loaded in tension, $G_c = K_c^2(1 - \nu^2)/E$. Lawn (1993) gives a clear and suitably rigorous introduction to the subject of fracture mechanics.

The ways in which these properties are determined are fairly complex and are not suitable for small specimens. This makes them inappropriate for many bones. The applicability of fracture mechanics to bone is, as yet, disappointingly limited (Melvin 1993 and table 3.5) and it is not, I think, appropriate to say much about the methods used. The difficulty for the fracture mechanics approach is that it necessarily makes some assumptions that are far from being true for bone. The full name of the standard fracture mechanics theory is *linear elastic fracture mechanics*. The assumption is that any part of the material will behave linearly elastically until it ruptures. This is certainly not true for bone. The basic assumption can be relaxed somewhat, as long as the awkward behavior is confined to a small region in the vicinity of the crack tip. Another, related way of characterizing the toughness of a material is through the use of the J integral (Rice 1968). I shall not discuss it in detail, because its implementation is rather difficult and has barely been applied to bone. It attempts to make allowance for the possibility of the zone around the crack undergoing very nonlinear deformation. It is a more general indication of toughness, but, because it is more general, it is more complicated. It seems, unfortunately, that the fracture mechanics approach does not measure material properties that are truly independent of the geometry of the test or the size of the specimen. It is possible that the fracture mechanics approach will become important in time as developments make it more applicable to materials like bone.

Another method of analysis that gives some idea of the toughness of bone is merely to prepare a specimen in such a way that it does not fail catastrophically when it breaks. Such a specimen is a beam loaded in three-point bending, with a deep notch in the middle of the beam (Tattersall and Tappin 1966). The area under the load–deformation curve shows how much work was done in breaking the specimen in two, and the fracture surface area can be measured. In this way the *work of fracture* W can be calculated. The calculated work of fracture of a material does vary somewhat with specimen size and geometry; nevertheless, the value can be quite useful in comparisons between types of bone that vary considerably in their toughness.

2.2.7 Creep Rupture

Many materials (including bone) show the phenomenon of creep rupture. A material is loaded to some stress, and then held at that stress. It does not break immediately, but does so eventually. It is found that there is an inverse relationship between the stress and the time to failure; the nearer the stress is to the failure stress in monotonic loading, the shorter the time before failure. This subject will be dealt with more fully in section 3.5 and later.

2.2.8 Fatigue Fracture

In most materials, repeated loading of a specimen to stresses (or strains) lower than the failure stress (or strain) found in a single loading to failure can result in fracture. Such a fracture is called a fatigue fracture. A comprehensive account of fatigue in nonbiological materials is given by Suresh (1992). Fatigue fractures occur in bone. The common way of testing the fatigue behavior of a material is to construct a so-called S-N diagram. A set of specimens is subjected to various stress amplitudes. The stress amplitude S is then plotted against the number N of cycles required to cause the specimen to fracture. Figure 2.15 is a set of regression lines showing fatigue behavior of different types of bone under various loading conditions. They will be discussed in section 3.6. This diagram shows the log of stress as a function of log cycles to failure. This is an illogical way of showing the relationship, because clearly the cycles to failure are a function of the applied stress, rather than vice

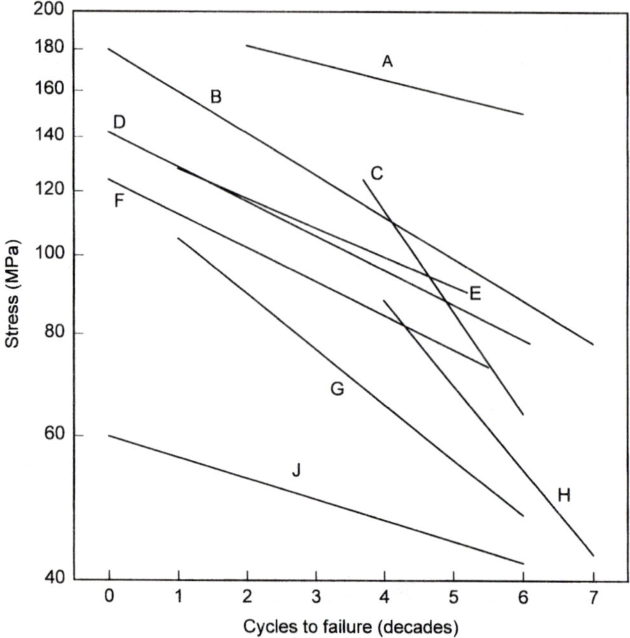

FIG. 2.15. A set of *S-N* curves for compact bone. The abscissa is in powers of ten, for example "4" means 10^4 or 10,000. The separate lines are the lines representing the equations produced by regression analysis. In all the cases the distributions were reasonably linear. Taken from various experiments, with different loading conditions. (A) Choi and Goldstein (1992), human, 4-point bending, room temperature, 2 Hz. (B) Gray and Korbacher (1974), human, compression, room temperature, 30 Hz. (C) Carter and Hayes (1976a), bovine, rotating cantilever, room temperature, 30 Hz. (D) Zioupos et al. (1996a), bovine, tension, room temperature, 2 Hz. (E) Zioupos et al. (1996b), human (27 years) tension, room temperature, 2 Hz. (F) Zioupos et al. (1996a), red deer antler, tension, room temperature, 2 Hz. (G) Carter and Caler (1983), human (69 years) tension, 37°C, 2 Hz. (H) Swanson et al. (1971), human, rotating cantilever, room temperature, 70 Hz. (J) Currey et al. (2001), *Mesoplodon* rostrum, 3-point bending, room temperature, 2 Hz.

versa. Unfortunately, though, this way of doing things is deeply embedded in the fatigue literature. Although the lines are widely separated, they share the characteristic that the lower the stress amplitude, the greater the number of cycles to failure.

The mechanisms of fatigue failure in metals have been the subject of intense study, particularly after beautiful British Comet aircraft started showering themselves like confetti over the Mediterranean. It seems that the origin of fatigue in *metals* is that stress concentrations on the surface or, less commonly, within the body of the specimen cause the material to be loaded locally into the plastic region, even though most of the material is in the elastic range. The alternating size or sign of the stress makes the plastically deformed defect enlarge by the process of slip. Slip is of great interest to metallurgists, but not to us because there is no evidence that it is of importance in bone. Eventually, the slowly traveling defect so weakens the specimen that it fails catastrophically.

After this introduction to the mechanical properties of solids, we turn to the mechanical properties of bone itself.

Chapter Three

THE MECHANICAL PROPERTIES OF BONE

THIS CHAPTER deals with the basic mechanical properties of bone and with attempts that have been made to model this behavior. So much material has been gathered in the last few decades about the mechanical properties of bone that it would be futile to try to list it all. Fortunately, there is good agreement about most of the properties, so I shall take as datum points a few papers, and discuss the main variants in chapter 4. The mechanical properties I shall deal with are elastic properties (mainly Young's modulus), strength (in tension, compression, and shear), fracture mechanics properties, creep rupture, and fatigue. The first sections will be mainly bald statements of results; I shall discuss their implications in later sections. A great deal of the variation found in the literature is real variation and is not, I think, caused by sloppy experimental work. I shall discuss some of the reasons for this variation below, insofar as it relates to the kind of test that is being carried out, and other differences will be discussed in chapter 4. A large and reasonably comprehensive book on methods of testing bone material mechanical properties, and the results of such tests, is edited by An and Draughn (2000). Most surprisingly, and unfortunately, it says very little about measuring the fracture mechanics properties of bone. Turner and Burr (1993) give an authoritative, and more succinct, account of the mechanical testing of bone.

Three characteristics of the loading situation in particular may have a considerable effect on the measured properties: the orientation of the specimen in relation to the bone from which it came, whether the specimen is wet or dry, and the rate at which the strain is applied to the bone (the strain rate). Dry bone has very different characteristics from wet bone, it is stiffer and much more brittle and, since bone in the physiological state is wet, I shall barely discuss the properties of bone tested dry again.

In this book all logs are to the base ten, unless specifically stated otherwise.

3.1 Elastic Properties

There are two main ways of measuring the elastic properties of bone: (1) by applying a load to a specimen and calculating the elastic properties from the resulting deformations (or, frequently, by applying a deformation and calculating the elastic properties from the load necessary to produce the deformation) and (2) by measuring the velocity of sound waves in bone. The velocity of sound in a medium V is obtained from $V = \sqrt{E/\rho}$, where E is Young's modulus and ρ is the density of the medium. So, in theory, Young's modulus can be calculated from a knowledge of the velocity of the sound and the density of the bone. However, in reality there are complications; in particular, the formula above holds only for isotropic materials and it is more complex for anisotropic materials like bone. The sound used is of high frequency, and I shall call one *mechanical* and the other *ultrasonic*, although both are at root mechanical, of course.

Mechanical testing has various advantages: it is relatively straightforward; Young's modulus can be determined in a variety of directions quite simply; cancellous bone and bone full of cavities can be tested without there being too many worries about what precisely is being measured (this is particularly unclear with ultrasonic testing of cancellous bone); and the effect of strain rate on mechanical properties can be investigated.

Ultrasonic testing is less straightforward and there are some methodological problems when it is applied to cancellous bone (Haire and Langton 1999). It cannot be used to determine strength characteristics. However, ultrasonic testing can make possible the derivation of all the stiffness coefficients and, with difficulty, their determination all from the same specimen. Also, it can be applied to complexly shaped specimens, and under some circumstances can be used in vivo.

Table 3.1 gives values for moduli as determined by mechanical testing by Reilly and Burstein (1975). These values are for human and bovine bone. The table also gives values for moduli derived from ultrasonic tests. The ultrasonic values are from Ashman et al. (1984). Kim and Walsh (1992) compared the elastic properties of a bovine femur both ultrasonically and mechanically, using the same specimens. They found variation round the bone and along its length. They also found, when they tested the bones along the main axis of the anatomical bone, that the ultrasonic values were well correlated with the mechanical values, but the ultrasonic values were much higher than the mechanical ones (fig. 3.1). The reason for this discrepancy is not clear.

3.1.1 Orientation Effects

Both the ultrasonically and the mechanically determined values for the properties show considerable variation as a function of orientation,

TABLE 3.1
Elastic Moduli for Bone (in GPa) determined by Reilly and Burstein (1975) and Ashman et al. (1984)

	Ashman et al.		*Reilly and Burstein*				
					Bovine		
	Canine ?	*Human Haversian*	*Human Haversian*		*Haversian*		*Fibrolamellar*
	Ultrasound		*Mechanical*		*Mechanical*		
			Tension	*Compression*	*Tension*	*Compression*	*Tension*
E_1	12.8	12.0	12.8	11.7	10.4	10.1	11.0
			3.0 (25)	1.01 (5)	*1.6 (5)*	*1.8 (8)*	*0.17 (25)*
E_2	15.6	13.4	12.8	11.7	10.4	10.1	11.0
			3.0 (25)	*1.01 (5)*	*1.6 (5)*	*1.8 (8)*	*0.17 (25)*
E_3	20.1	20.0	17.7	18.2	23.1	22.3	26.5
			3.6 (38)	*0.85 (4)*	3.2 (3)	4.6 (5)	*5.4 (6)*
G_{12}	4.7	4.5	—	—	—	—	—
G_{13}	5.7	5.6	3.3		3.6		5.1
			0.42 (10)		*0.25 (22)*		*0.39 (6)*
G_{23}	6.7	6.2	3.3		3.6		5.1
			0.42 (10)		*0.25 (22)*		0.39 (6)
ν_{12}	0.28	0.38	0.53	0.63	0.51	0.51	0.63
			0.25 (24)	*0.20 (5)*	*0.24 (5)*	*0.12 (8)*	*0.23 (6)*
ν_{13}	0.29	0.22	0.41	0.38	0.29	0.40	0.41
			0.15 (26)	*0.15 (4)*	*0.08 (3)*	*0.21 (5)*	*0.23 (10)*
ν_{23}	0.26	0.24	—	—	—	—	—
ν_{21}	0.37	0.42	0.53	0.63	0.51	0.51	0.63
			0.25 (24)	*0.20 (5)*	*0.11 (5)*	*0.12 (8)*	*0.23 (6)*
ν_{31}	0.45	0.37	0.41	0.38	0.29	0.40	0.41
			0.15 (26)	*0.15 (4)*	*0.08 (3)*	*0.21 (5)*	*0.23 (10)*
ν_{32}	0.34	0.35	—	—	—	—	—

Source. Material in tables 3.1, 3.2, 3.3, and 3.5 has been reproduced (with modifications) from the *Proceedings of the Institution of Mechanical Engineers, Journal of engineering in medicine*, Part H, **212**, 1998, pp. 399–411 by J. D. Currey by permission of the Council of the Institition of Mechanical Engineers.

Note. Reilly and Burstein did not test in all directions, and assumed that the bones were transversely isotropic. Subscripts 1, 2, and 3 refer to the radial, circumferential, and longitudinal directions relative to the long axis of the bone. Shear values (G) are placed midway between Tension and Compression merely for convenience. The values for G_{13} and G_{23} are assumed equal, and also pairs of Poisson's ratio (ν), ν_{12}, ν_{21} and ν_{31}, ν_{13}, are assumed equal. (Assumptions of symmetry underpin the assumption of equality.) In fact, fibrolamellar (= plexiform) bone is almost certainly *not* transversely isotropic.

General remarks about the tables 3.1–3.3. and 3.5. The tables come from publications in the general literature. The number of papers on mechanical properties of bone is very large and growing rapidly, and the tables do not purport to be comprehensive. However, they do give a reasonable idea of the state of affairs at the time of writing. The values are not always as given in the original publication. For instance, I may take the mean of values of specimens from several bones, which the original authors have not done. In these tables, though not elsewhere in this book, I give estimates of statistical error so that the reader can have some idea of the variability of the data. It was not always possible to derive sample sizes and standard deviations from the published papers, and in some cases it was, for various reasons, inappropriate. The three values usually given—X, Y, and *(Z)*—are mean, standard deviation, and sample size, respectively. Occasionally (still!) authors do not state whether '±Y' refers to the standard deviation, the standard error, or the 95% confidence interval. This sloppy practice, of which I have been guilty, should cease.

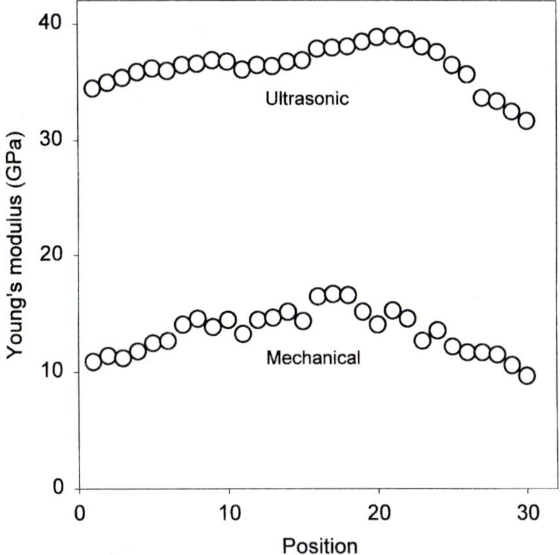

FIG. 3.1. Values of Young's modulus of elasticity determined in compression along the long axis of the bone. Position: position along the length of the bone, from proximal to distal (arbitrary units). Values derived ultrasonically and mechanically are shown. (Derived from data given in Kim and Walsh [1992].)

with the stiffness measured along the length of the bone being about 1.6 to 2.4 times as great as that measured at right angles to this. Burstein and Reilly measured at some intermediate angles and found, not surprisingly, intermediate values. Bonfield and Grynpas (1977), Pope and Outwater (1974), and Liu et al. (2000) found similar variations. Currey et al. (1994), investigating narwhal dentin, found a twofold difference in the modulus measured along and across the grain.

3.1.2 Strain Rate Effects

Bone is slightly viscoelastic, and its measured Young's modulus is to some extent strain-rate dependent. Carter and Caler (1983), basing their results mainly on the work of Wright and Hayes (1976), suggested that Young's modulus is proportional to (strain rate)$^{0.05}$. Currey (1988) produced results suggesting that the effect is somewhat smaller. Assume that Carter and Caler are right; how important would the effect be? A thousandfold increase in strain rate will result in an apparent increase of Young's modulus of 40%; the effect is marked, though not huge. However, it does show that mechanical tests should, if possible, be conducted as close to real-life strain rates as possible. Ultrasonic tests, of course, test at very high strain rates. Nevertheless, it is satisfactory that, except for the results of Kim and Walsh (1992), the results are not much different from those of mechanical tests. Although I said above that mechanical tests should be carried out at physiological strain rates if possible, this is not always appropriate. It is sometimes possible to get information about *mechanisms* by testing at more than one strain rate. This is true, for instance, of fatigue resistance, as we shall see in section 3.14.

3.2 Strength

The strength of a material is usually easy to measure. One measures the load at which a specimen breaks and calculates the strength directly; there is no absolute necessity to measure deformation as well as load, though normally this is done to give Young's modulus and information about postyield deformation. Nowadays there is a view that the "tensile strength" of bone is not a useful concept, because such a measure gives no idea of the toughness of the material. Such doubters say that fracture mechanics parameters give one a much better idea of what bone is really like. A proponent of this view is Bonfield (1987). Although the fracture mechanics approach is important, and some results will be discussed below, there are some difficult problems in applying fracture mechanics concepts to a material like bone. Relatively simple determinations of strength do have several advantages. First is simplicity. Fracture mechanics techniques usually require rather large specimens with precut notches, or various other awkward geometries. These requirements are difficult enough, but added to them is the requirement, often, to measure the length of cracks. Second, although the usual fracture mechanics tests will measure the behavior of the bone when the crack surfaces are pulling apart, they have little to say about how bone behaves in compression. Third, *pace* Bonfield (1987), the coefficients of variation (standard deviation/mean) achieved in well-conducted ordinary fracture tests are not large. Indeed, they are, if anything, smaller than those found in fracture mechanics experiments and allow one, for example, to find fairly easily differences between specimens with different histology or between bones from individuals of different age.

Tables 3.2 and 3.3 give values for various fracture-related properties of human and bovine bone. They are taken from three classic papers (Reilly et al. 1974; Reilly and Burstein 1975; Cezayirlioglu et al. 1985). I chose these papers because they report carefully carried out work in which the compression tests and tensile tests are readily comparable, being made on the same-shaped specimens loaded at the same strain

TABLE 3.2
Strength Properties of Bone (strength values in MPa)

	Human Haversian		Bovine					
			Haversian			Fibrolamellar		
	Longitudinal	*Circumferential*	*Longitudinal*	*Circumferential*	*Radial*	*Longitudinal*	*Circumferential*	*Radial*
TENSION								
Strength	133	53	150	54	39	167	55	30
	15.6 (21)	10.7 *(20)*	*11 (10)*	*5.8 (4)*	*4.7 (6)*	*8.8 (6)*	*9 (31)*	*3.2 (6)*
Yield stress	114	—	141	—	—	156	—	—
	7.1 (21)		*12 (10)*			*7.9 (6)*		
Ultimate strain	0.031	0.007	0.020	0.007	0.007	0.033	0.007	0.002
	0.006 (21)	*0.0014 (20)*	*0.005 (10)*	*0.004 (4)*	*0.002 (6)*	*0.0049 (6)*	*0.0013 (31)*	*0.001 (6)*
COMPRESSION								
Strength	205	131	272	171	190	—	—	—
	17.3 (20)	*20.7 (8)*	*3.3 (3)*	*25 (8)*	*18.0 (5)*			
Ultimate strain	0.019	0.050	0.016	0.042	0.072	—	—	—
	0.003 (20)	*0.011 (8)*	*0.0015 (3)*	*0.01 (8)*	*0.014 (5)*			
SHEAR								
Strength		67 *3.5 (12)*		70 *9 (7)*			64 *7 (12)*	

Source. Values were taken directly from, or from information in, Reilly et al. (1974) and Reilly and Burstein (1975).

Note. Specimens cut longitudinally, circumferentially, or radially with respect to the long axis of the bone. Shear strengths are placed arbitrarily in the columns. The shear specimens were twisted about their long axes. See also general remarks with table 3.1.

TABLE 3.3
Strength Properties of Bone (strength values in MPa)

Property	*Human Haversian*	*Bovine Histology Undetermined*
Tensile strength	158	162
	8.5 (13)	*14.2 (27)*
Tensile yield stress	128	132
	11.2 (13)	*10.6 (27)*
Ultimate tensile strain	0.042	0.049
	0.0085 (13)	*0.0042 (27)*
Compressive strength	213	217
	10.1 (9)	*26.8 (25)*
Compressive yield stress	180	196
	12.5 (9)	*18.5 (25)*
Ultimate compressive strain	0.026	0.033
	0.0056 (9)	*0.0056 (25)*
Shear strength	71	76
	7.8 (10)	*6.3 (21)*
Shear yield stress	53	57
	7.7 (10)	8.4 (21)

Source. Taken from Cezayirlioglu et al. (1985).
Note. See general remarks with table 3.1.

rate. Most other values in the literature, in which one has confidence, are somewhere near these values. Nevertheless, readers are warned against comparing different tables and figures in this book while expecting complete consistency. For instance, ultimate tensile strain for bovine fibrolamellar bone in table 3.2 agrees with the stress–strain of figure 3.8, because they were derived from the same data. However, bovine femur in table 4.3 has a much lower value. The values are, I hope, correct; it is the specimens that are different. This apparent inconsistency is especially the case for values of ultimate strain, which, as will be explained later, is particularly strongly affected by mineral content.

3.2.1 *Orientation Effects*

Table 3.2 shows the great differences in the strength properties when the bone is loaded in different directions. When bone is loaded in tension, it is stronger and, in particular, it has a much higher strain to failure when it is loaded longitudinally rather than circumferentially. Similarly, Liu et al. (2000) found that osteonal bone of baboons was four times stronger when loaded along the long axis of the bone than when loaded normal to this direction (289 MPa vs. 71 MPa). They also found a corresponding 3¾ difference in the strain at failure. These two differences imply a roughly 16-fold difference in the work necessary to fracture a specimen. Currey et al. (1994), investigating narwhal dentin, found a threefold difference in the tensile strength measured along and across the grain, the specimens measured across the grain being weaker. The anisotropy of strength of bone in compression is not as marked as it is in tension. The ultimate strain is greater in the circumferential than it is in the longitudinal direction, the opposite of what happens in tension.

3.2.2 Strain Rate Effects

The tensile strength of bone increases with strain rate, at least up to rates producing fracture in a time not much less than that occurring in life, say 20 ms. Torsional tests, in which the bone is loaded in shear, produced a ⅓ increase in ultimate strength over a 10,000 increase in strain rate (Sammarco et al. 1971). I found an increase of tensile strength and tensile yield stress of about ½ with a thousandfold increase in strain rate, from 0.0001 to 0.1 per second (Currey 1975). The strain rate of 0.1 per second is rather low; violent fractures probably occur at strain rates of between 0.3 and 1 (Currey 1988). More satisfactory, in that it incorporated strain rates including physiological rates, is the study of Wright and Hayes (1976). They tested femoral cortical bone over a series of strain rates from 0.001 to 100 per second. The stress at failure continued to increase at all the strain rates they investigated. The effect was not very strong, tensile strength being proportional to strain rate to the 0.07 power.

There have been tests at very high strain rates, which are certainly unphysiological, but might occasionally by produced by missiles. Saha and Hayes (1976) have loaded bone in tension in impact at very high rates (about 1000 per second, which took about 1/10000 second to fracture). Although they found a weakening effect of Haversian systems, they were unable to obtain the very uniform load–deformation curves that, for instance, Burstein and his co-workers consistently obtained at lower strain rates. This may well be because of the shock waves that would be set up by such extremely rapid loading. These waves would bounce off interfaces, cavities, and the edges of the specimen, reinforcing and canceling each other out in unpredictable ways. Despite the fact that these tests were performed at unphysiologically high rates of loading, the disparity of the shapes of the curves they found and of those produced at lower strain rates is a warning: static tests are useful, but they may not give a clear idea of what goes on at very high loading rates.

TABLE 3.4
Some Values for Young's Modulus and Strength Determined in Bending

Authors	*Species* *Bone*	*Histology*	*Young's modulus (GPa)*	*Bending strength (MPa)*
Martin and Boardman (1993)	Bovine Tibia	Fibrolamellar	21	230
		Osteonal	19	217
Currey (1999)	Many species Many bones	Various, well mineralized	18–30	200–300
McAlister and Moyle (1983)	Goose Femur	—	17–21	230–280
Bigot et al. (1996)	Horse Metapodials	—	14–16	200–220

Note. The values for Currey (1999) are taken, rather arbitrarily, from a set in table 4.3.

McElhaney (1966) tested small blocks of bovine and human bone in compression and found that, between strain rates of 0.001 to 1 per second, there was an increase in compressive fracture stress. Bovine bone increased from 180 to 250 MPa, human bone from 150 to 220 MPa. McElhaney's specimens showed reasonable postyield deformation, the ultimate strains at strain rates of one per second being 1.2 and 1.8% for bovine and human bone, respectively.

3.2.3 *Modes of Loading*

Table 3.2 shows that bone is stronger when loaded in compression than when loaded in tension. Table 3.4 gives some values for bending. In general, bending strength values are greater than those in tension and compression. I discuss the reasons for this in section 3.10. Here it must suffice to say that if the stress at failure is calculated from beam formulas, which do not allow for the effect of postyield deformation, bending strengths for ordinary bone are of the order of 50% greater than the corresponding tensile strengths (Burstein et al. 1972).

3.3 Inferring Bone Material Properties from Whole Bone Behavior

All the tables in this chapter relate to the properties of bone tissue. There have been a number of attempts to back-calculate the properties of the *tissue* from the mechanical properties of the *whole bone*. In studies of small animals, such as rats and below in size, this may be the only method available. Sometimes, conversely, people have attempted to calculate the properties of whole bones from the behavior of isolated samples (I have done it myself! [Brear et al. 1990a]). There are considerable problems with these procedures, which it is well to bear in mind when reading papers that report results from using them. Consider first the process of back-calculating tissue properties from whole bone properties. The usual procedure is to load the bone in bending, and calculate stiffness (load/deformation) and strength (load at failure). Then histomorphometric measures are made. These are usually the second moment of area (I) of one or a number of cross sections, and also the distance of the furthest point from the neutral axis (c) at the section that broke. Then, considering the bone as a beam with varying cross-sectional properties, one can calculate the Young's modulus of the material that would produce the observed stiffness. One can also, similarly, calculate the notional stress at failure by making use of cross-sectional properties at the point of break. Calculating whole bone properties from specimen properties is the converse of this, except that one has more mechanical information; instead of having just one load–deformation curve, one has as many curves as one has prepared specimens.

Estimating strength is more reliable than estimating Young's modulus. One can obtain a reasonably precise value for the bending moment at failure, and also obtain morphometrically a good value for c/I at the point of break. The maximum stress σ in a beam at a particular cross section is given by $\sigma = Mc/I$, where M is the bending moment at that section, I is the second moment of area, and c is half the depth of the section. This formula breaks down if the section is very asymmetrical about the neutral axis. So far so good. However, as I explain in section 3.12, this formula is valid only insofar as the material remains linearly elastic up to the point of failure. This is never true for ordinary bone. If, however, one calls the stress *bending strength* and not *tensile strength* and the bones being compared are similar in their cross-sectional shape, then one is roughly comparing like with like, but one should be aware of the limitations.

The problems are much more severe for calculations of Young's modulus. Unlike the case for strength, the deformation of the beam is a function of *every* section between the supports. It is a complicated function, because the sections near the outer supports have less effect than the sections near the middle. However, if one takes a fair number of sections it is possible to estimate the behavior of the beam. Many studies have assumed that the second moment of area is that of the center section, or is the mean of the sections that have been examined. These simplifications will not really do unless the bone is unusually uniform in its cross-sectional properties. A greater problem with estimating Young's modulus is that beam theory assumes that Young's modulus is a constant or, if it not, it must be taken into account. But it cannot be taken into account, because the investigator does not know the Young's modulus—this is why he or she is doing the back-calculation. Young's modulus is certainly not constant around sections (Batson et al. 2000; Kim and Walsh 1992) and even less along the length of the bone. The metaphyseal bone, for instance, nearly always has a lower Young's modulus than bone near the middle of the shaft (Evans 1973). Kim and Walsh's study on bovine femora is particularly thorough, relating mechanical and ultrasonic testing; see figure 3.1, for example.

Another difficulty one sees in the literature is that, when investigating small, perhaps rather delicate bones, workers do not take into account that the bone may deform *locally* under the loading points. Any such local loading will increase the apparent overall deformation of the specimen, and hence reduce the apparent Young's modulus, if the deformation is measured from cross-head travel of the testing machine. This particular problem is obviated, however, if an LVDT is used to measure the deformation.

Despite these difficulties, sometimes one has no choice if the bone is small. Increasing use of finite element analysis, and of micro-computer tomography (μ-CT), will increase the morphometric accuracy of these investigations, but the differences in Young's modulus in different parts of the specimen will prove less tractable.

The problems of back-calculating the behavior of whole bone from knowledge of test specimens relate less to estimates of stiffness than to strength. Because one can measure Young's modulus all over the bone, one can, with tedious morphometry and calculation, obtain a good estimate of the stiffness of the whole bone. Estimating load at failure is more problematical, because the discrepancy between the behavior of whole bones and the behavior of specimens is not at all well known. However, again, as long as one is carrying out work on roughly similar-shaped bones, then estimated differences between bones may well be reasonably accurate, though the absolute values may be wrong.

3.4 Fracture Mechanics Properties

As I mentioned above, fracture mechanics properties are derived by rather complex testing methods. The values G_c and K_c in table 3.5 are, respectively, the critical strain energy release rate and the critical stress intensity. These values are mainly given in Behiri and Bonfield (1984) and Melvin (1993).

TABLE 3.5
Fracture Mechanics Properties of Some Bony Tissues

Bone	*Direction*	K_c *(MN* $m^{-3/2}$*)*	G_c *(J* m^{-2}*)*	*Test*	*Rate*	*Authors*
Bovine femur	Long	3.21 *0.43 (12)*	1380–2560	SENT	Slow	Melvin and Evans (1973)
Bovine femur	Long	5.05 *0.51 (12)*		SENT	Fast	Melvin and Evans (1973)
Bovine femur	Trans	5.6 *0.52 (12)*	3140–5530	SENT	Slow	Melvin and Evans (1973)
Bovine femur	Trans	7.7 *0.77 (12)*		SENT	Fast	Melvin and Evans (1973)
Bovine femur	Long	3.62 *0.73 (40)*		CT	Slow	Wright and Hayes (1977)
Bovine femur	Long	2.4–5.2	920–2780	CT	Slow	Bonfield et al. (1978)
Bovine tibia	Long	2.8	630	CT	Slow	Behiri and Bonfield (1984)
Bovine tibia	Long	6.3	2880	CT	Fast	Behiri and Bonfield (1984)
Bovine tibia	Long	3.2		CT(g)	V. slow	Behiri and Bonfield (1989)
Bovine tibia	Trans	6.5 *1.2 (7)*		CT(g)	V. slow	Behiri and Bonfield (1989)
Human tibia	Long	2.4–5.3		CT	Slow	Bonfield et al. (1984)
Human tibia	Long	3.7	360	CT	Slow	Norman et al. (1991)
Bovine tibia	Long	7.2		CT	Slow	Norman et al. (1991)

TABLE 3.5 (*cont.*)

Bone	*Direction*	*K_c (MN $m^{-3/2}$)*	*G_c (J m^{-2})*	*Test*	*Rate*	*Authors*
Bovine tibia	Long	8.0 *(36)*	940 *(36)*	CT	V. slow	Norman et al. (1992)
Bovine femur	Trans	2.2–4.6 *(8)*	780–1120	SENT	Slow	Bonfield and Datta (1976)
Bovine femur	Trans	*5.7 1.4 (21)*		3–PT	Slow	Robertson et al. (1978)
Bovine tibia	Trans	11.2 2.6 *(75)*	7960 WoF	SENB	Slow	Moyle and Gavens (1986)
Human tibia	Long	4.0–4.3 *0.7 (10)*	590–830 *0.2 (10)*	CT	Slow	Norman et al. (1995)
Bovine femur	Long	4.9		CNT	Slow	De Santis et al. (2000)
Bovine tibia	Long	6.2–6.7 *1.5 (30)*	900–990 *0.13 (30)*	CT	Slow	Norman et al. (1995)
Human femur	Trans	6.4 *0.34 (30)*	3400 *0.43 (3) WoF*	SENB °	Slow	Zioupos et al. (1997)
Antler	Trans	5.4		SENT	Slow	Sedman (1993)
densirostris rostrum	Trans	1.3	91 WoF		Slow	Zioupos et al. (1997)
Narwhal tusk	Trans		1800–3100 WoF		Slow	Zioupos et al. (1997)

Note. "Direction" refers to the direction of crack travel. Many workers have used a longitudinal direction of crack travel, even though this is not the usual mode of failure in life, because it is easier to get specimens large enough to load in this direction. The ranges given in the results columns refer to various experimental variables that were altered, and do not refer simply to the range of values found for one type of experiment. For instance, the ranges in Norman et al. (1995) refer to the values obtained with different lengths of notch. "Test" refers to the type of test configuration used. CT, compact tension; CNT, chevron- notched tension; SENT, single edge notch tension; SENB, single edge notch bending; (g), specimen grooved to force the crack to travel in a particular direction; 3-PT, 3-point bending with a single notch, WoF, work of fracture as determined by three-point bending in a Tattersall-Tappin type test; ° Regression on age, normalized to 35 years. See general remarks with table 3.1.

The wide variation seen between values found for the same type of bone loaded in the same direction seems to be rather wider than the variation found in the literature in ordinary strength tests. Melvin (1993) shows that problems produced by size effects are by no means solved. The table also shows that specimen orientation and crack velocity have strong effects on the measured properties. Bonfield and Behiri (1983) find that fracture toughness of bone increases with crack velocity at rather low velocities, up to 10^{-3} m s^{-1}, but is lower at the catastrophic rates seen when a bone breaks in life.

Although we have a fairly good idea of the fracture mechanics properties of "standard" bone, we are much less advanced in our understanding of what features of bone affect these properties than we are, for instance, in the case of Young's modulus (see next chapter). The interaction of traveling cracks with the structure of bone is a complex process, and is certainly not well understood at the moment. It is such interactions that determine the fracture mechanics properties.

3.5 Creep Rupture

The strains undergone by a specimen of bone (and many other materials), deformed by a constant load that is not large enough to break it at once, fall into four periods. First, the specimen deforms essentially instantaneously. (There is an effect here that should be taken into account if the load is applied by a dead weight. A load applied truly instantaneously produces a strain, and therefore a stress, in a specimen that is up to twice the strain that would be produced if the load were applied sufficiently slowly for the specimen to be in static equilibrium with the load all the time. This happens because the suddenly applied load causes a wave of deformation to travel through the specimen [Young 1989]. Usually, with tests carried out with standard testing machines the loading is quasi-static and the effect can be ignored.) Second, there is a period of increasing strain, but the rate of increase of strain decreases with time. This is called *primary creep*. Then one of two things happens. If the load is less than some value the bone will hardly deform any more, and the bone and the load are in equilibrium. If the load is greater than this value, then the bone continues to show increasing strain, at a constant rate. This is called *secondary creep*. Finally, there is a region, called *tertiary creep*, in which the strain rate increases, and fairly soon the specimen breaks.

Figure 3.2 (derived from Mauch et al. [1992], including data from Carter and Caler [1983]) shows the relationship between the "normalized" tensile stress (the stress divided by the Young's modulus) and

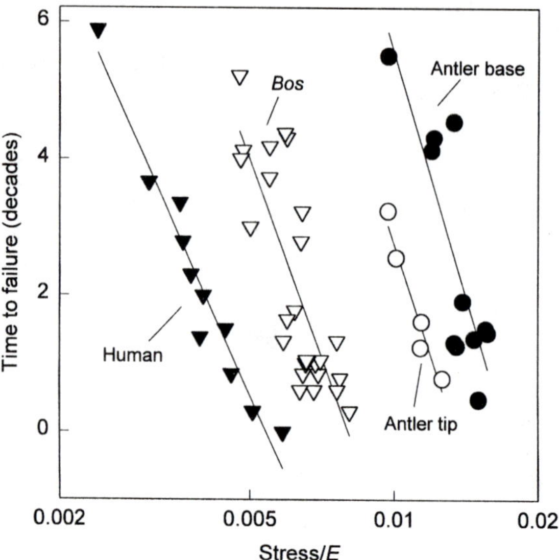

FIG. 3.2 The relationship between the time to failure in tension of bones loaded to different values of initial strain (stress/Young's modulus). Human data from Caler and Carter (1989); the rest from Mauch et al. (1992). The sloping lines are the regression lines for each slope. Note that both axes are in log scales. For any initial stress/E the antlers took much longer to fracture than do the bovine and human bones. However, the antler bone has a Young's modulus about 50 times larger than its tensile strength, whereas in ordinary bone the ratio is 150–200. Therefore, the stress vs. time to failure (in seconds) results suggest that antler is weaker than ordinary bone in creep. The relationships for each bone, however, are much less clear with stress, rather than stress/*E*, as the explanatory variable.

the time to failure of four bony tissues: human bone, bovine ulna, and the base and the tip of red deer antler. Both the axes are on a logarithmic scale. There is a great deal of spread in the data, but each individual bone behaves in a rather uniform way. The time to failure decreases very rapidly as the normalized tensile stress increases. The equations for the relationships are given in table 3.6. For completeness I give a cancellous bone equation as well. Figure 3.2 gives only the time to failure at a particular stress. Fondrk et al. (1988) showed the way in which, once the bone passes a certain threshold, the rate of secondary creep increases steadily. The threshold stress, below which secondary creep occurs at a negligibly low rate was different for human and bovine bone. Human bone started to creep at about 65 MPa, whereas bovine bone did not creep until a stress of about 100 MPa was imposed.

The significance of the data reported in this section may seem a little obscure. However, I hope that when I discuss the damage process in

TABLE 3.6
Creep Behaviors of Various Bony Tissues

Human femur	Log (t) =	-36.4	-17.8 Log (σ^*)	Compression	1
Human femur	Log (t) =	-35.8	-15.8 Log (σ^*)	Tension	1
Bos ulna	Log (t) =	-34.6	-16.7 Log (σ^*)	Tension	2
Antler base	Log (t) =	-47.8	-26.9 Log (σ^*)	Tension	2
Antler tip	Log (t) =	-39.3	-21.0 Log (σ^*)	Tension	2
Bos cancellous	Log (t) =	-32.0	-16.2 Log (σ^*)	Compression	3

Note. t, time to failure in seconds; σ^*, stress (in MPa) divided by initial Young's modulus. Note that σ^* is, in effect, equivalent to strain. However, in these creep tests the stress remains constant, but as the bones underwent damage their effective Young's modulus decreased and so the strains became greater with time. 1, Carter and Caler (1983); 2, Mauch et al. (1992); 3, Bowman et al. (1994).

bone, in section 3.15, I shall be able to show that a whole set of experiments will fall into place.

3.6 Fatigue Fracture

Fatigue in bone is best documented in humans, but it is not well understood. Orthopedic surgeons rather confusingly call fatigue fractures "stress fractures." This usage is unfortunately well dug into the literature. It comes from the idea, which is correct, that such fractures can result from stresses imposed by the muscular system during locomotion. The clinical symptoms of fatigue fracture are usually that of pain developing in a bone when activity has been increased over the normal. This may happen, for instance, when a young man joins the army and is set marching, when golfers greatly increase their training schedule (Lord et al. 1996), when an athlete restarts training, or when an elderly widow has to walk to the stores, having previously been driven there by her husband. Often the radiograph shows nothing, though radiodensity develops later, as repair starts. No obviously excessive loading has taken place, yet the repetition of somewhat higher than usual loads has resulted in imminent or actual failure. (An interesting case is reported by Singer et al. (1990), who found three identically positioned stress fractures in two Israeli soldiers, the same time after they had been recruited to the army. The interesting thing is that the soldiers were identical [monozygotic] twins!) Fatigue fracture may also develop in different circumstances, such as when the bone is loaded strongly by muscular action, and fracture takes place without premonitory pain. The difference between these two modes will be discussed below. A comprehensive discussion of human fatigue fractures, including epidemiology and treatment, is in Burr and Milgrom (2001).

Evans and his co-workers were the first to study the fatigue properties of bone in the laboratory (Evans and Lebow 1957; Lease and Evans 1959; King and Evans 1967). More recently Carter and his co-workers (Caler and Carter 1989; Carter and Hayes 1976a, 1977b,c; Carter et al. 1976, 1981a,b,c), Schaffler et al. (1989, 1990); Michel et al. (1993). Zioupos et al. (1996a,b; 2001), and Gibson et al. (1995), among others, have carried things further.

The laboratory tests showed bone to have an *S-N* curve (section 2.2.8) like that of other materials. Figure 2.15 shows the results of nine experiments using various tissues. There is considerable variation. Some of these differences may be attributed to different loading systems. Most of the tests were at room temperature, and it is likely that such tests would have rather larger numbers of cycles to failure than tests at body temperature. Only Carter and Caler (1983) tested at body temperature. Their curve is rather low and steeply declining compared with most of the rest. It is interesting that the most steeply declining curves (C and D) are for specimens that were rotating cantilevers, which are loaded alternately in both tension and compression; that is, the stresses are *reversed* in each cycle. (The extremely low values for *Mesoplodon* rostrum are put in figure 2.15 merely for interest, not enlightenment. The rostrum is very highly mineralized, and its properties are quite unlike those of most other bone. See section 4.3.)

To see if the fatigue tests have any relationship to the situation in life let us take a value of a million cycles. A migrating animal might load its leg once a second for 12 hours a day. It would take about a month to load its leg about a million times. According to the results summarized in figure 2.15, the various estimates of the stress that would induce failure after a million cycles are about 150, 85, 83, 75, 70, 63, 53, and 50 MPa (and 42 MPa for *Mesoplodon*). Alexander et al. (1979a) have calculated the maximum stresses in various long bones of animals traveling fast (see chapter 10 for details). The calculated peak stresses vary from about 60 to 150 MPa. The very rough match between these two sets of figures is misleading, for various reasons. First, Alexander's animals were traveling fast; our hypothetical migrator would not be galloping, and the stresses in the bone would be correspondingly less. Second, the value of 60 MPa is a stress *amplitude*; that is, it refers to movements away from zero stress equally in both directions. Although the maximum stresses calculated by Alexander's group will diminish and even change sign at other periods in the gait cycle, it is certain that, because of the geometry of the limb, the stress will not be so great on the other side of zero. Third, and very important, is the fact that here we are comparing situations in vivo with laboratory tests. Living bone has the ability to repair fractures, and so may be capable of repairing the tiny cracks caused by the fatigue process. Devas (1975) shows great bone reorganization taking place when a bone has not completely fractured. It is unknown, however, whether this repair can take place before gross bone damage occurs. When gross damage does occur, the repair process will be obvious, but we do not at the moment have any real idea of how the race between microdamage and microrepair goes. Martin and Burr (1982) have suggested, indeed, that Haversian systems have a function in preventing the spread of fatigue cracks. Their proposed mechanism is dealt with in chapter 11.

Most studies on the mechanical properties of bone are carried out at room temperature. For tests lasting a few seconds or less, the fact that the temperature of mammal bones may be 20° above room temperature (in English rooms, anyhow) is probably unimportant (Evans 1973). However, for long-lasting tests temperature does become important. This is seen in tests on viscoelasticity (Currey 1965) but also, strikingly, in fatigue. Carter and Hayes (1976a) showed that between 21 and 37°C there was a fall of roughly one-half in the fatigue life (the number of cycles required to fracture a specimen at a particular stress amplitude). Carter et al. (1976) went on to show that two features of the specimen itself had an effect on the fatigue life of the bovine bone they tested: its density and its histological structure. The greater the density, the greater the fatigue life; the more of the specimen that was occupied by Haversian systems, the shorter the fatigue life. These two features are difficult to disentangle because Haversian remodeling itself reduces mineralization. Nevertheless, by use of a large number of specimens the effects of these two variables were distinguished. The authors produced a formula relating experimental and specimen variables to the number of reversals before failure of the form: log (no. reversals) = A log (stress) − B temp + C density + D histology (histology having arbitrary values 1 to 4). This equation explains an extremely high proportion of the variance in fatigue life. The effect of histological structure is strong; specimens with 100% primary bone have a fatigue life about five times greater than those that are completely Haversian, even when the effect of mineralization has been removed. This last is a remarkable result, indicating as it does a severe disadvantage of Haversian remodeling.

Taylor et al. (1999) show that the *volume* of bone stressed can be an important feature, as it may indeed be in any test. Because specimens vary in strength through their volume, large specimens will have more weak points (strong ones are irrelevant) than small ones. This matter is discussed in section 3.10. Taylor et al. use the volume effect to normalize the differences seen between different tests and to predict the extent

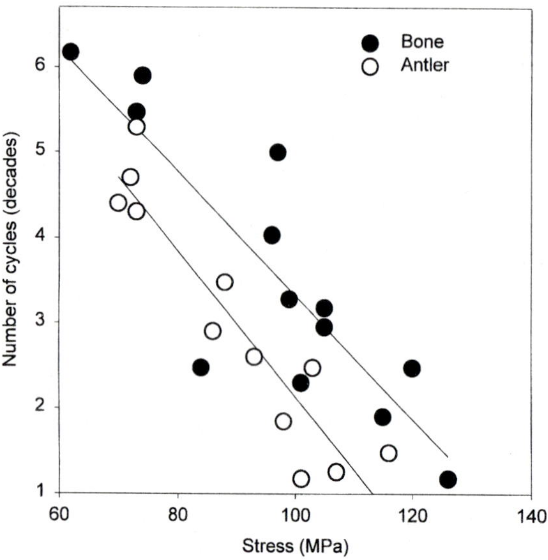

FIG. 3.3 The number of cycles of tensile loading needed to produce a 10% reduction of modulus in bovine femur and red deer antler. The sloping lines are the regression lines.

to which whole bones might be more prone to fatigue than isolated specimens.

Carter and Hayes (1977b) showed that when bone had been loaded sufficiently hard for it to be on the way to failing by fatigue, it had a lower modulus of elasticity and lower strength than unloaded bone. Figure 3.3 derived from data in Zioupos et al. (1996a) shows how as the applied stress increases, the number of cycles before a particular loss of stiffness is reached reduces. It also shows how two different types of bone (bovine long bone and deer antler) behave differently. At any given stress the bovine bone takes about ten times as many cycles to reach the same reduction in stiffness. At low stresses there is very little decrease in stiffness until nearly the end. At high stresses the modulus is reduced almost from the beginning (Gibson et al. 1995; Pattin et al. 1996; Zioupos et al. 1996a).

There is a problem about fatigue loading that has not yet been resolved. Is it the number of *cycles* of loading that are important, or the *time* over which loading takes place? Section 3.5 on creep rupture shows that bone may eventually break at a load that does not initially cause it problems. Could it be that what is causing the bone to break is the additive effect of all the time it has been loaded to fairly high stress, and that the reversals are irrelevant? The finding of Lafferty (1978) that increasing the frequency of loading from 30 to 125 Hz reduced the number of cycles required to fracture the bone to a half or a third could almost as well be explained by the fact that the time to failure is not much altered. Papers by Carter and Caler (1983, 1985) go some way to deciding this, and it is dealt with section 3.15 below, where I discuss more generally the question of damage in bone.

Drs. Burstein and Frankel suggested to me that the cause of fatigue fracture in bone may be of two types. One is "standard," the loads being fairly high, but a large number of cycles occurring before symptoms appear. The other type is when, through excitement, fear, or other intense emotion (and their equivalent in racehorses, greyhounds, and, presumably, antelopes), the normal neuromuscular inhibitions to the overloading of bone are overridden. In such circumstances the bone may be loaded into the postyield region, and rather few reversals will cause failure. Yoshikawa et al. (1994) found that muscle fatigue in dogs markedly increased the strain the femur suffered. They ran the dogs on a treadmill until the quadriceps showed electromyographic signs of fatigue. The gait of the dogs was essentially unchanged. However, the distal tibia showed an increase in maximum shear strain of about 35%, and also large changes in the principal stresses. Most interestingly, the amount of increase in strain in different dogs was well correlated with the amount of fatigue of the quadriceps as measured by change in the electromyogram (EMG) signal (including one superdog whose EMGs showed "negative" fatigue, and a concomitant *reduction* in maximum strain.)

I mentioned above the results of tests performed on human cortical bone in tension by Carter and his co-workers (Carter et al. 1981b, c). They were concerned to show whether it was the range of strain that is important, or the maximum strain. They loaded at a low strain rate, 0.01 s^{-1}, which is characteristic of strain rates produced by locomotion during life. The fatigue life they found was about 3½ decades less than other workers have found. They attribute this primarily "to the use of uniaxial rather than rotating bending and flexural tests." They also loaded at lower strain rates and at higher temperatures than others had used before. Even so, it is a little difficult to credit that these variables can have had such an extremely large effect on the fatigue life of the bones.

A suggestion has been put forward by Otter et al. (1999) that fatigue fractures as seen in vivo are caused by the results of damage to the osteocytes by transient ischemia and reperfusion. If this is the case, then the actual stresses induced in the bone may be almost irrelevant to the causation of fatigue fractures. I shall return to this diverting suggestion in section 11.10.6.

In summary, fatigue fractures occur in bone in life. Laboratory specimens show the characteristic *S-N* curve, and fatigue with loads near those experienced in life. There is, however, the evidence of Carter et al. that the fatigue life of bone may be much less than previously thought. If this evidence is corroborated, we shall have to reconsider drastically our ideas of the rate at which fatigue damage can be repaired in life.

3.7 Modeling and Explaining Elastic Behavior

One of the never-failing pleasures of being a biologist is finding out how well organisms are designed. It will become clear later that whole bones and skeletons are, indeed, designed well for their functions. However, before being able to say that natural selection is wonderful and has done it again, one has to find out what exactly has been done. Doing this for bone material, as opposed to whole bones, is difficult, because materials scientists have not yet developed a comprehensive framework for explaining the mechanical properties of a material like bone. In this section I shall try to show how we are beginning to be able to model some of the mechanical properties of bone. We still have far to go.

As I said at the beginning of the book, I think the most important feature of bone material is its stiffness. The operations of the skeleton cannot be described without at least the implication that these functions could not be performed by flexible structures. On the other hand, there are no particular implications that these structures should be strong, or tough, or good stores of calcium. I shall therefore start consideration of the properties of bone with stiffness.

Bone is essentially a collagen framework packed with calcium phosphate mineral, with a fair deal of water around. Collagen is widespread in the animal kingdom, and it is surprising that no other phylum has adopted this method of producing a stiff skeletal material. In fact, no other animals, except some crustaceans, seem to have such an intimate association of organic material and mineral. Lowenstam and Weiner (1989) list the minerals used by living organisms, and it is apparent from this that although the number is considerable (about 60 at that writing), easily the most widely used ones are calcium carbonate, silica, and, a poor third, calcium phosphate and its relatives. Calcium carbonate is nearly always found in a composite of organic and mineral parts, but the volume fraction of organic material is very low, nearly always less than 5%, usually much less. Calcium carbonate may also occur as isolated spicules. These spicules are usually large, visible to the naked eye. Muzik and Wainwright (1977) provide an account of the spiculation found in a number of soft corals. That paper gives a good idea of the kind of variation one finds in animals, and how different mechanical properties are produced by different volume fractions and arrangements of spicules. Jeyasuria and Lewis (1987) have quantified the effect of calcareous spicules on the Young's modulus and torsional modulus of soft coral skeletons.

Silica usually occurs as isolated spicules, though it may form the entire skeleton of some protozoans (Simkiss and Wilbur 1989). It never behaves as a composite in which there is tight bonding between the organic and mineral parts. Most of the animals with calcium carbonate skeletons have collagen, but do not deposit mineral in it. Stiffening in the invertebrates is produced by having very high volume fractions of fairly large mineral crystals, or spicules, or by tanning (cross-linking) organic materials. This last method is adopted by the arthropods, in particular, whose cuticle consists of fibrils of chitin bound together by a more or less highly cross-linked protein. The resulting composite is probably mechanically superior to bone in every respect on a per-weight basis, an embarrassing fact for people who think that bone is the ultimate in biomaterials, which I shall not pursue further, except to say that this mechanical superiority is bought at the cost of considerable developmental complexity. During growth the cross-linked cuticle cannot be resorbed, and so must be shed periodically.

The Young's modulus of tendon, which we can take to represent collagen in a fairly uniformly aligned arrangement, is about 1.5–2 GPa (Bennett et al. 1986). The Young's modulus of hydroxyapatite is about 110 GPa. The problem, which has been attacked by several workers, is to obtain a model that will explain the elastic behavior of bone using this and other information, such as the volume fraction of mineral, anisotropy, and the detailed arrangement of the microscopic constituents of bone.

Katz and his co-workers were the first people to make a serious attempt to resolve this problem (Katz 1971). Katz started by modeling bone as if the mineral and the collagen formed a multilayer sandwich with indefinitely large layers (fig. 3.4). If such a sandwich is loaded parallel to the layers, the *stiffnesses* of the two components are additive; this is the so-called Voigt model. If the sandwich is loaded normal to the layers, the *compliances* are additive. (Compliances can be considered to be simply the inverse of the stiffnesses.) This is the so-called Reuss model. If the bone is considered to be a multilayered composite material of modulus E_b, made of layers of collagen and layers of mineral of modulus E_c and E_m, and of relative volumes V_c and V_m, respectively, then in the case of loading parallel to the layers $E_b = E_c V_c + E_m V_m$. This is the Voigt model. For the Reuss model; $1/E_b = V_c/E_c + V_m/E_m$.

Let us look at the predictions of the two models. Assume that the

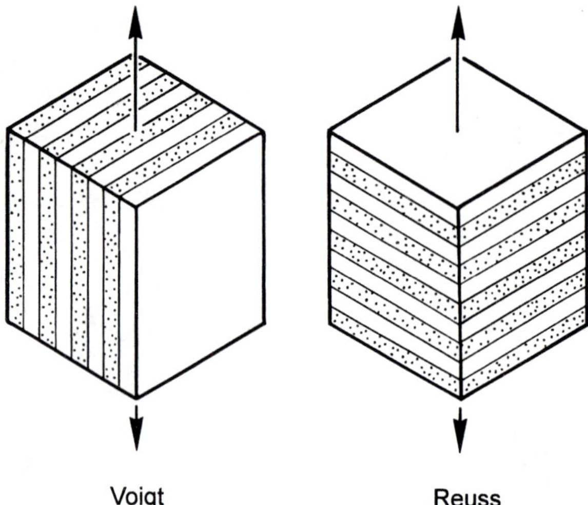

FIG. 3.4 The Voigt and Reuss models of composites. The plain and stippled layers are the two components of the composite and have different Young's moduli. In reality, there is no requirement that the volume fractions of the two components be equal, as drawn here.

collagen and mineral have equal volume fractions, and Young's moduli of 2 and 100 GPa, respectively. For the case of loading parallel to the layers;

$$E_b = 2 \times 0.5 + 100 \times 0.5 = 51 \text{ GPa}$$

For the case of loading normal to the layers;

$$\frac{1}{E_b} = \frac{0.5}{2} + \frac{0.5}{100} = 0.255, \text{ so } E_b = 3.9 \text{ GPa}$$

In the Voigt model the Young's modulus is roughly the stiffness of the apatite multiplied by its volume fraction, while in the Reuss model the compliance is roughly the compliance of the collagen multiplied by its volume fraction. In each case the stiffness of the whole is completely dominated by the stiffness or compliance of one or other of the components.

Although bone, and enamel, do fit within the upper and lower bounds, the bounds themselves are so far apart that this model of Katz can hardly be considered to be satisfactory (fig. 3.5). Piekarski (1973) came to a similar conclusion. Of course, by using some arbitrary mixture of the Voigt and the Reuss models, it would be possible to get quite a good fit. However, it is then found that the modulus for loading at angles to the long axis is not well predicted. A satisfactory model must

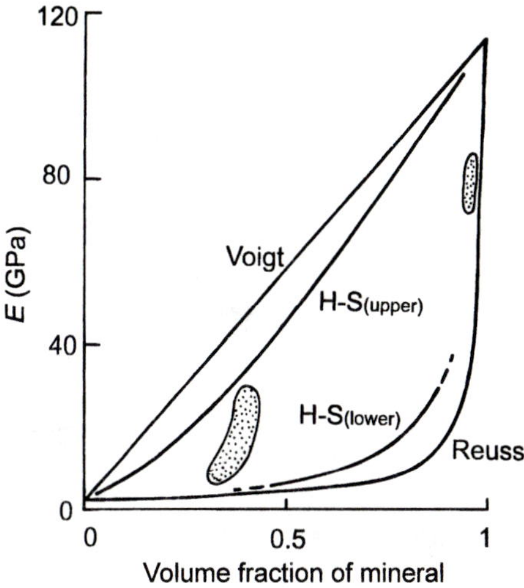

FIG. 3.5 The early modeling of Katz. Ordinate: Young's modulus; abscissa: volume fraction of mineral. The upper and lower curves are the predictions of the Voigt and Reuss models. The inner lines (H-S_{upper} and H-S_{lower}) are the upper and lower bounds of another model (Hashin-Shtrikman) studied by Katz. The lower stippled area represents the actual value for dentin and bone; the upper stippled area represents enamel. (From Katz [1971].)

incorporate detailed consideration of the disposition of the two components.

Later, Katz refined his 1971 model by assuming that the apatite crystals are oriented along the collagen fibrils and tightly bonded to them. The collagen fibrils, however, will be in various orientations in various lamellae. Following Krenchel (1964), Katz (1981) developed the equation

$$E_b = \frac{E_c V_c (1 - \nu_c \nu_b)}{1 - \nu_c^2} + \sum E_{ha} V_{ha} \alpha_n (\cos^4 \varphi_n - \nu_b \cos^2 \varphi_n \sin^2 \varphi_n) \quad (3.1)$$

Subscript c refers to collagen, ha to apatite, ν is Poisson's ratio, α_n is the fraction of the apatite crystallites that lie at an angle φ_n from the direction of stressing. The first part of the equation refers to collagen, which is assumed to have the same modulus in all directions. This, though not true, does not matter, because the modulus of collagen is so low. The second part gives the effects of apatite. The important point to note here is the $\cos^4 \varphi_n$ term, which implies that the effectiveness of the mineral as a stiffener will fall off rapidly as it becomes misaligned. ($\cos^4$ 10° is 0.94, $\cos^4$ 20° is 0.78, $\cos^4$ 45° is 0.25.) Making some reasonable assumptions about the orientations of the crystals, Katz was able to compare theory with experiment, using his and other people's data. The fit was not very good, particularly when the grain of the bone was at a large angle to the loading direction. At large angles the observed values for Young's modulus were considerably greater than the model predicted.

There have been a number of more recent attempts to model the elasticity of bone material, using more detailed information about the size, shape, and orientation of the mineral in the tissue. Pidaparti et al. (1996) did not attempt to model individual lamellae. They measured the elastic properties of the bone at different angles, using acoustic microscopy and found that their modulus results fitted best with the assumption that most of the mineral crystals lay outside the collagen fibrils and were aligned along the length of the bone, whereas the collagen fibrils were oriented at about 30° to the long axis of the bone.

Sasaki et al. (1991) and Wagner and Weiner (1992) modeled bone's elastic behavior starting at the lamellar level. To determine the angular dependence of the Young's modulus of a single lamella, which can be considered uniform, they first had to make various reasonable assumptions for the Young's and shear moduli of the two components (collagen and mineral), the volume fraction of the mineral, and the geometry of the mineral phase. Sasaki et al. and Wagner and Weiner then used versions of the Halpin and Tsai (HT) equations to predict the contributions of these components to the moduli of elasticity in tension of the lamella along its principal directions, and its modulus in shear in the plane of the lamella. (The Halpin and Tsai equations predict the longitudinal and transverse moduli of lamellae reinforced with platelike, as opposed to infinitely long, inclusions; these are collagen fibers and apatite plates in our case.) Wagner and Weiner assume platelike crystals with a breadth-to-thickness ratio of 10 and a length-to-thickness ratio of 25 in the two orthogonal directions, whereas Sasaki et al. (1991) assume rodlike crystals, with various ratios of length to diameter, varying from 2.5:1 to 10:1. These models have been discussed by Currey et al. (1994).

Currey et al. (1994) determined the elastic behavior of narwhal tusk, which is, in fact, a tooth. The choice of this seemingly bizarre experimental tissue was guided, as well as by a certain urchin pleasure in using something so unusual, by the fact that the tusk is made of highly oriented layers of dentin, so that it is possible to determine the angular dependence of Young's modulus on a tissue that is more uniform than the bony specimens used previously. There are some other differences between bone and dentin, particularly in relation to the placing of cells, but any model that aims to predict the elastic behavior of bone should be able to cope with dentin and, indeed, because the dentin is more highly oriented and has a very uniform mineral content, to predict its properties better. We found that Young's modulus measured in tension declined smoothly from a value of 12 GPa when loaded along the fibers to 5.5 GPa when loaded at 85°. We tried to fit the models of Sasaki et al. and Wagner and Weiner to these data. The rod model favored by Sasaki et al. cannot be made to get close to the narwhal results along the compliant direction (about 5–6 GPa) except by choosing unreasonably high mineral volume fractions (>0.5). The plate model of Wagner and Weiner produces values for Young's modulus that are too high for a reasonable volume fraction, but the ratio of the transverse and longitudinal Young's moduli is about right if a suitable aspect ratio of the apatite plates is chosen.

I think that the main reason this sophisticated modeling does not give very good answers for the Young's modulus of mineralized tissues is twofold. One is that the mineral crystals are so small, yet have such a high volume fraction, that they distort the stress fields around themselves. This distortion will have an important effect on Young's modulus of the collagen, but is as yet too complicated to be modeled. The other is that we still do not have a very good idea of the actual arrangement of the mineral crystals in the lamellae, crucial though this is to successful modeling.

Jäger and Fratzl (2000) developed a model that took account of the fact that the crystals could overlap in the mineral fiber. The models previously discussed assumed that the crystals lay end to end, thus requiring a rather close packing in the longitudinal direction if the reasonably high modulus of bone were to be modeled. Jäger and Fratzl show how the stiffness of their model will change as the spacing between the mineral crystals is varied both side-to-side and in its overlap. The authors claim that their model has the great advantage over the previous ones that those models have a maximum *strength* equal to that of collagen, because if the crystals lie end to end, then at the longitudinal gaps between the crystals, the collagen will be bearing all the load, in tension. Although their model may be better at modeling stiffness in a single fibril, I think the matter of strength is not relevant, because the strength of collagen molecules when attached very firmly at each end to mineral crystals will be quite different from, and probably much higher than, the strength of bulk specimens of collagen. I assume that Wagner and Weiner, and Sasaki et al. did not model nonoverlapping crystals for computational simplicity rather than through a belief that the crystals do not overlap. It will be interesting to see the Jäger and Fratzl model tested by being loaded off-axis, which is a major test of any model of bone elastic behavior.

Katz and his co-workers developed a model to account for the behavior of human Haversian bone. This is moving up one level in the hierarchy of structures. They suggest that the Haversian systems are arranged roughly in hexagonal packing, and that the moduli of the systems themselves and of the interstitial lamellae are different. They include in the effective modulus of the interstitial lamellae the stiffness of the cement lines surrounding the Haversian systems, which they claim is rather low. Frasca et al. (1981) have tested isolated groups of secondary osteons from human bone in shear and find a marked size effect: the greater the number of osteons in the sample (from one to eleven), the more viscous the sample becomes and the lower its shear modulus of elasticity. Although the scatter of the data is rather large and the specimen was immature (from a 12-year old), these results do show the possibility that the cement lines are allowing relative movement of the osteons, and that they are more viscous than the rest of the bone.

Even if this model does seem to account accurately for the behavior of Haversian bone, it has two shortcomings. One is that it seems, if anything, to lead us even further away from an understanding of fibrolamellar bone. For in fibrolamellar bone there are no cement lines, yet this type of bone is more anisotropic in its elasticity than Haversian bone. The other shortcoming is that the basic model does not account for the observation that the Young's modulus of bone increases very markedly with mineral content over a small range of mineral values. Equation 3.1 shows that, if the modulus of the mineral is much greater than that of the organic material, then the modulus should be roughly proportional to the volume fraction of mineral. Yet, a number of studies have shown that the modulus increases much more rapidly than the volume fraction of mineral (Currey 1969, 1990; Ramaekers 1977), although this effect may be hidden if there are large changes in porosity (Schaffler and Burr 1988; Hernandez et al. 2001). Katz assumes that the apatite needles are bound rigidly to the collagen, but makes no allowance for their length; the equations he uses are for very high aspect ratio fibers. Making allowances for the length of fibers is difficult, and the problem has not been completely solved theoretically yet, except for very simple cases.

The analyses of these various workers have all assumed values for the Young's modulus of collagen tested on its own. However, McCutchen (1975) proposed that the mineral phase of bone "straightjackets" the collagen and prevents it from straightening under stress. Lees and Davidson (1977) proposed a similar scheme. Hukins (1978) proposed a modification of McCutchen's model, in which the collagen fibrils are considered to be arranged as liquid crystals whose habitual reorientation under stress is prevented by the mineral. All these suggestions are ingenious, not to say amusing, but until they can be quantified it is probably best to think of them as suggestions that should be continually borne in mind when considering models such as those of Katz et al., Frasca et al., Sasaki et al., Wagner and Weiner, and Jäger and Fratzl. Certainly, when there is such an intimate relationship between the collagen and the mineral at the nanometer level, it is dubious to what extent one should consider the collagen to be behaving just like collagen tested on its own.

I mentioned in section 1.1 that we are rather ignorant of the actual bonding between the mineral and organic phases. There has been a huge amount of work recently on characterizing the proteins and other organic molecules that may have an effect on initiating, preventing, and controlling mineralization, but almost no work has been done on the characteristics of the final product. A fairly recent, supposedly comprehensive book on bone biology 1398 pages long (Bilezikian et al. 1996) says effectively nothing about the mineral component of bone, the 2200-item index not mentioning *apatite* or *dahllite*, and the seven references to *mineral* are related to the process not the product. This neglect is perhaps unfortunate, but since apatite is not specified by genes, the neglect is not, in today's funding climate, surprising.

Spatially, the mineral within the collagen fibrils and the fibrils themselves are extremely close, and nonspecific van der Waals interactions probably play some part in binding them together. There is also the probability that ionic interactions and even possibly some covalent bonds may bind collagen to the mineral. There are, also, a large number of noncollagenous proteins that can act as intermediates between collagen and the mineral itself (Robey and Boskey 1996; Boskey 2001). Walsh and Guzelsu and co-workers have shown that many mechanical properties of bone can be degraded by soaking the bone in solutions containing fluoride or phosphate ions (e.g., Walsh and Guzelsu 1993, 1994). They suggest, reasonably, that these ions compete with the mineral for sites on the collagen or other intermediate proteins, and thereby reduce the strength of the mineral-organic binding. This has a degrading effect on both the modulus and the strength. For instance, the compressive modulus of bovine femora was reduced by about 45% and compressive strength by about 40% by this treatment.

We have seen that bone shows a certain amount of anisotropy of Young's modulus: it is about twice as stiff in the stiffest direction as in the less stiff direction. This is because bone is like wood in having a marked grain, the structural elements tending to be oriented preferentially in one direction, and it is in this direction that they will be stiff. However, if it were adaptive to have an isotropic material, isotropy could be achieved by the bone having the collagen fibrils and all other elements arranged randomly in space. This is effectively what happens in woven bone. Unfortunately, there is a price to be paid for this isotropy; the greatest stiffness will be reduced. There seem to be no studies on the mechanical properties of woven bone. (They would be difficult to carry out because woven bone usually comes in very small lumps.) We do not know, therefore, how stiff three-dimensionally isotropic bone is. Because the mineralization and microporosity of woven bone is different from that of lamellar bone, such a comparison would not separate the effect of fiber direction from that of mineralization.

In general, one expects materials that are fairly stiff to have the same Young's modulus in tension and in compression. Originally, Reilly et al. (1974) showed this to be the case for bovine bone in a set of very careful experiments. Later work from Burstein's laboratory (Torzilli et al. 1982) showed that, at least in dog's bone, the compressive Young's modulus is greater than the tensile modulus. This result is disturbing and, if not caused by some subtle artifact, will have to be explained by people modeling the Young's modulus of bone.

3.8 Modeling Fracture in Tension

To understand the mechanisms in bone that tend to stop it from breaking, we must try to understand what goes on when it does break. Bone often breaks in tension, and because tensile fracture is easier to understand than compressive fracture I consider tensile fracture first.

In a truly brittle material, if high strength is to be achieved, there must be no flaws on the surface or within the body of the material. The only material I know of, produced by a living organism, that comes in reasonably large blocks and that is also truly brittle is the skeleton of echinoderms (starfish, sand dollars, sea urchins, etc.). This skeleton consists of blocks made of single crystals of calcite. The external surfaces are kept remarkably smooth by a layer of living tissue. Nevertheless, echinoderm skeletons are probably rather low-stress structures (Nichols and Currey 1968) and Gibson and Ashby (1997, page 117) produce good theoretical arguments to show that porous materials, like echinoderm skeleton, made of very brittle materials, like calcite, can never be tough. All other skeletal materials seem to have at least some degree of toughness.

3.8.1 *The Effects of Stress Concentrations*

Bone, in fact, has a mass of potential stress-concentrating discontinuities. Most obvious are the cavities for blood vessels, osteocyte la-

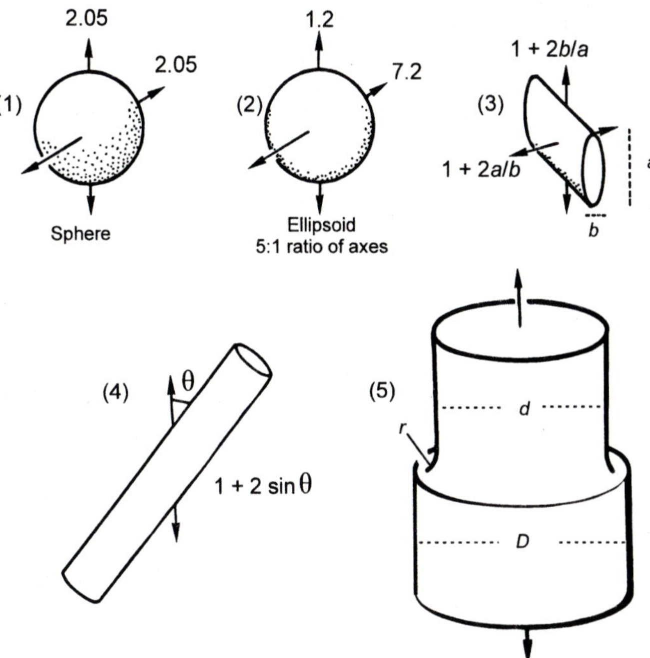

FIG. 3.6 Stress concentration factors for some simple discontinuities. (1) A sphere. The stress concentration is 2.05 for all normal stresses. (2) An ellipsoid of revolution with a ratio of 5:1 in the major and minor axes. If the load is normal to the major axis, the stress concentration is 7.2; if it is normal to the minor axis it is only 1.2. (3) An elongate cavity of elliptical cross section. (4) A cylindrical cavity at an angle θ to the stress direction. (5) A cylinder with a step, loaded in tension. If $r = (D - d)/2$ and $r = d/10$, then the stress concentration is about 1.78. Note that these particular values for stress concentrations apply only to homogeneous, isotropic solids, which, as explained at the end of the section, is not the case for bone.

cunae, and canaliculi. It is easy to calculate their theoretical stress-concentrating effect because they are effectively voids, though filled by fluid (fig. 3.6). Currey (1962b) showed that the stress-concentrating effects of the flattened osteocytes could be as great as 7*S*, where *S* is the general level of stress, in the region of the stress concentrator. Even cylindrical blood channels could have a stress-concentrating effect as great as 3*S*. In fact, these values refer to stress concentrations in isotropic solids. In a material like bone, in which the moduli in different directions are different, the stress-concentrating effects will usually be less, but if the load is in the "wrong" direction, it could be much more. Whether these stress-concentrating effects are important depends on the size of the cavities and their orientation with respect to dangerous stresses.

Fracture mechanics theory shows that a sharp crack will not spread if

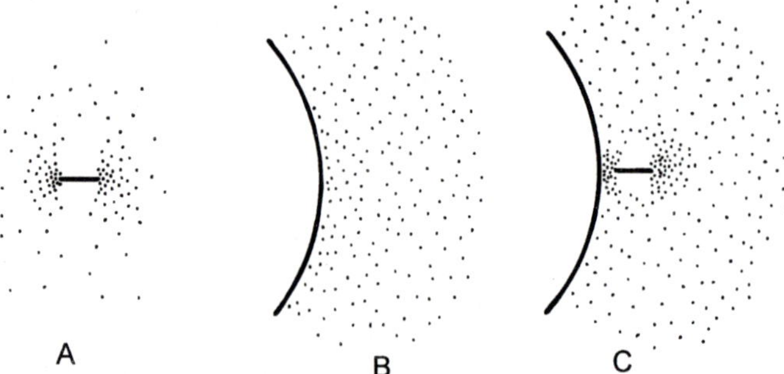

FIG. 3.7 (A) A short sharp crack does not strain a large enough volume of its surroundings for there to be enough strain energy to allow it to spread. (B) A large stress concentrator may not have a sharp enough end to allow it to spread. (C) Together they may be able to initiate crack travel, which in this case would probably run from the small defect into the large one.

it is below a certain size, the size depending on the intensity of the local stress field. A simple-minded way of thinking about interactions is shown in figure 3.7. A large stress concentrator such as an osteocyte lacuna in the region of a crack may allow it to run, even though the average stress in the material is low. The crack on its own could not spread because the amount of strain energy available locally would not be great enough. (The thermodynamic criterion for crack spread would not be met [section 2.2.5].) The large stress concentrator on its own would not be dangerous because, although the stress in its neighborhood is higher than generally, it is still far below that required to initiate a spreading crack. (The stress criterion would not be met.) However, the large stress concentration will produce a stress field around the crack sufficiently intense and extensive for strain energy to be available to drive the crack forward (in this case the crack would probably run from the small defect to the large one [Kachanov 1994]). In fact, the question of the interaction of stress concentrations is one of alarming mathematical difficulty, and some of the results are counterintuitive. A flavor of the matter can be found in Kachanov (1994) and Tsukrov and Kachanov (1997). Such studies at the moment can consider only truly brittle solids, which bone is not, of course, and this adds a further complication to what might happen in bone.

The other potential stress concentrators in bone are morphological features, such as sharp notches or discontinuities on surfaces, and concentrations caused by large differences in modulus, such as may occur at cement sheaths around secondary osteons. Figure 3.6 shows the stress-concentrating effects of some simply shaped discontinuities.

Stress concentrations formed by cavities and discontinuities are, therefore, a potential hazard in bone. Does the overall or histological structure of bone give any indication that it is adapted to this state of affairs? It certainly does. Perhaps the most obvious feature, in long bones anyhow, is the way in which the blood channels are arranged. Blood channels are the largest potential stress concentrators. They run predominantly in a very gently spiralling course through long bones, at quite a small angle to the bone's long axis. In this arrangement their stress-concentrating effect on most ordinary stresses will be very small. The obliquely arranged blood channels have a roughly circular cross section. However, in fibrolamellar bone there are two-dimensional networks of blood channels between each pair of laminae (fig. 1.5). In these networks the blood channels run in all directions in the circumferential direction. Nevertheless, there is a predominance of channels running nearly longitudinally. Also, the channels in the networks are elliptical in cross section, and this will reduce the stress concentrating effect of those that run nearly horizontally (fig. 3.7).

Given the necessity for blood channels in bone, their arrangement seems to be mechanically sound. But this might be an example of mechanical serendipity—the arrangement of blood channels being determined mainly by the requirements of an efficient blood supply. However, the blood supply of long bones would seem a little bizarre if the mechanical effects of blood channels were not important. The efficiency of nutrient and waste transfer would be greatest if the very narrow vessels in the cortex of the bone were short, running directly from endosteum to periosteum. Yet few do.

Long bones usually have one or more quite large nutrient arteries that pierce the cortex and supply the marrow. These nearly always run obliquely through the cortex. There are many arguments about the mechanism producing this obliquity (Brookes and Revell 1998), yet the mechanical advantages of it seem never to be mentioned by anatomists.

The lacunae of osteocytes show a variety of shapes, from nearly spherical in woven bone to flattened oblate spheroids, with a ratio of major to minor axes of 3.3:1 or more, in lamellar bone (Currey 1962b). A tensile stress in the direction of the short axis of such an oblate spheroid would produce a stress concentration of 5.2, while in the direction of the long axis it would be only about 1.3 (Peterson 1974). Lamellar bone is very, very rarely arranged so that the long axis of the bone cuts across the lamellae at a large angle. The osteocyte lacunae lie in the plane of the lamellae, so they also lie with a long axis along the long axis of the bone, mechanically the most advantageous direction.

Canaliculi spread right through bone tissue at all angles and, therefore, theoretically must produce stress concentrations of three throughout the tissue. However, they are very small, about 0.3 μm in diameter (Cooper et al. 1966), and so may not be effective in helping cracks to spread.

Bones very rarely have sharp steps on their surfaces. All necessary changes in diameter are smooth. This is particularly evident in cancellous bone, where the holes in the plates and the junctions between longitudinal and transverse struts are well "radiused" (fig. 1.17). In woven bone the osteocyte lacunae are fairly spherical, and the blood channels are much more variously arranged. However, woven bone is adapted for speed of construction, not mechanical excellence.

So, it would seem, that except at the lowest level, that of the canaliculi, potential stress concentrators in bone are arranged in such a way that minimizes their stress-concentrating effect. This apparently appropriate arrangement of potential stress concentrators is real only if the dangerous stresses run reasonably along the length of the bone, which they usually do. If bone is loaded in other directions, its fracture behavior alters drastically. However, as Steve Cowin, of the City University of New York, pointed out to me, two features need to be taken into account. First, bone is elastically anisotropic, and its modulus in the longitudinal axis is greater than in the directions normal to it. This will have the effect of increasing the stress-concentrating effect above the theoretical value if the bone is loaded, as usual, along its stiffest axis (Green and Taylor 1945). Second, stress concentration is calculated assuming the material is homogeneous. However, examination of the lamellar histology in the neighborhood of stress concentrators shows that the lamellae "flow" round the concentrators like streamlines round a fish. The forces will tend to follow the lamellae, and therefore the effect of the concentrators will be less, possibly much less, than calculated from theory, at least when the forces lie in the direction to which the bone is adapted. Such amelioration does not, however, apply to secondary osteons, which punch brutally through preexisting lamellae and are not adapted to their orientation at all.

3.8.2 The Effects of Remodeling

In bovine bone, which is the only type of bone that has been thoroughly studied in this respect, it is now accepted that primary fibrolamellar (laminar) bone has a higher tensile strength than remodeled, Haversian bone (Currey 1959; Evans and Bang 1967; Reilly and Burstein 1975; Carter and Hayes 1976a; Saha and Hayes 1976; Vincentelli and Grigorov 1985). This is true in static, dynamic, and fatigue loading. The effect is quite marked: A completely laminar specimen can be more than one and one-half times as strong as one made entirely of Haversian systems or their remnants (Currey 1975) (fig. 11.9). Rimnac et al. (1993) showed that creep, leading to creep rupture, is markedly affected by remodeling, such that pure Haversian bone was estimated to creep about 100 times faster than pure primary lamellar bone, a very surprising result.

I carried out some experiments on bovine femoral unnotched bending specimens in impact, and also calculated work of fracture from sharply notched bovine femoral specimens loaded slowly. Impact strength ranged from 3 to 15 kJ m^{-2}. Work of fracture ranged from 1.3 to 2.8 kJ m^{-2}. The highest impact strengths and highest works of fracture were from specimens with clear laminar structure, with no, or virtually no, Haversian remodeling. However, the weakest specimens tended to consist of compact coarse-cancellous bone, with a greater or lesser amount of Haversian remodeling. This compact coarse-cancellous bone is organized in a rather chaotic way; in particular, it is usually oriented in directions at an angle to the long axis of the bone from which the specimens came. This orientation is caused by the grain of the cancellous bone, which is the scaffolding on which the compact bone is deposited. The cancellous bone may be oriented in all sorts of ways relative to the long axis of the definitive bone. When this chaotically oriented bone is replaced by Haversian systems, these are oriented more nearly along the axis of the bone than the bone they replace, bringing order out of chaos to some extent. So, the strongest bone is fibrolamellar bone, but the weakest is compact coarse-cancellous bone, which is also primary, but very badly oriented.

Haversian systems, being necessarily younger than the bone in which they lie, tend to be less well mineralized, so reconstruction and low mineralization tend to go together. There is perhaps a general tendency for tensile strength to increase with mineralization. This is not true, as we shall see in the next chapter, when the amount of mineralization becomes very great, and some types of antler bone are spectacularly strong in tension despite having rather poor mineralization. They achieve this by bearing a large increase in stress *after* they have yielded. No simple statistical model can be used, therefore, to relate tensile strength to mineral, so it is difficult to distinguish the effects of mineralization per se and of reconstruction. However, Carter et al. (1976) have shown, in investigating fatigue strength, that it is possible to distinguish the effects with a large sample size and careful experimentation.

The reasons for the weakening effect of reconstruction are not clear. There is a trivial effect caused by the area of bone tissue in Haversian bone being less than that in a similar-sized piece of well-developed fibrolamellar bone because of the presence of erosion cavities. Another possible factor is the difference in modulus caused by differences in mineralization. The less mineralized bone in relatively young Haversian systems has a lower modulus than the older surrounding bone (Rho et al. 1999). This modulus mismatch can have a stress-concentrating effect, but not an important one, because even if the Haversian system bone had a modulus only half that of the surrounding tissue, the stress-concentrating effect when the bone was loaded in the worst possible orientation would be only a factor of 1½ (Peterson 1974; fig. 3.7).

Nevertheless, for whatever reason, Haversian bone is definitely weaker in tension than primary fibrolamellar bone. Where Haversian systems replace compact coarse-cancellous bone, they probably have a beneficial effect on the strength of bone by improving the orientation of its grain. However, they do not have a good effect when they replace well-organized fibrolamellar bone. Haversian bone might seem to be well designed for toughness in that it has many potentially crack-stopping cement sheaths throughout the tissue. Presumably the deleterious effects of modulus mismatch and the boring of holes more or less at random through previously well-ordered tissue outweigh this possible advantage.

The remarks above refer to bone loaded in *tension*. The work of Riggs et al. discussed later in this chapter does suggest that Haversian systems may under some circumstances replace primary bone and make it stronger in compression. This seems to happen because the fibril orientation is changed considerably, becoming much more transverse, which is presumably beneficial in withstanding compressive loading. I shall consider the question of the function of remodeling again in chapter 11.

3.8.3 Anisotropy in Fracture

Reilly and Burstein (1975) studied the mechanical anisotropy of bone in some detail. Their findings about Young's modulus have already been given (table 3.1). They consider Haversian bone to be approximately transversely isotropic. That is to say, it has an axis of symmetry along the length of the bone and its behavior in all directions at right angles to this axis is the same. Their data do support this supposition. Human and bovine Haversian bone are rather similar, except that the ultimate strain in bovine bone loaded along the axis is rather low (table 3.2). The contrast in the properties in the two directions is striking. The ultimate tensile strength is only about 30–40% as great in the circumferential direction as it is in the longitudinal direction. The ultimate strain is

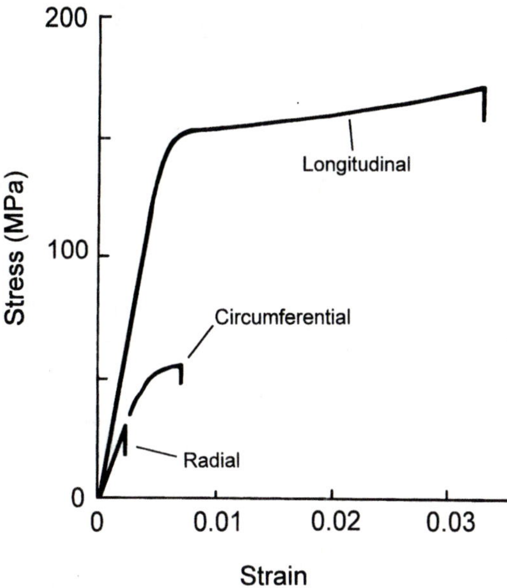

FIG. 3.8 Idealizations of stress–strain curves of bovine fibrolamellar bone loaded in tension. (Derived from data in Reilly and Burstein [1975].)

similarly reduced. This reduction in strain is brought about mainly by a reduction in length of the postyield region.

The fracture behavior of fibrolamellar bone is not transversely isotropic but orthotropic, the behavior in the radial direction being different from that in the circumferential direction (fig. 3.8). (There is a clear description of the various kinds of symmetry, with special reference to bone, in Cowin [2001b].) The tensile strength in the radial direction is pathetic. The amount of energy absorbed by radial specimens before fracture is only about 1/100 the energy absorbed by longitudinal specimens. This is brought about by the combination of low tensile strength and, more importantly, by the almost completely brittle behavior of the radial specimens. The poor performance of radial specimens usually would not matter in life, because it is difficult to arrange things so that bone is loaded radially. When it is loaded radially, as may happen under muscle insertions (the tuberosity on the radius under the insertion of the biceps, for example), the histology of the bone is quite altered, its grain leading smoothly into the line of action of the tendon.

This anisotropic behavior is what we should expect. The collagen fibrils, the cement sheaths, the blood vessel cavities, and, indeed, nearly all the structures in long bones are for the most part oriented roughly along the long axis. They do not cause stress concentrations when

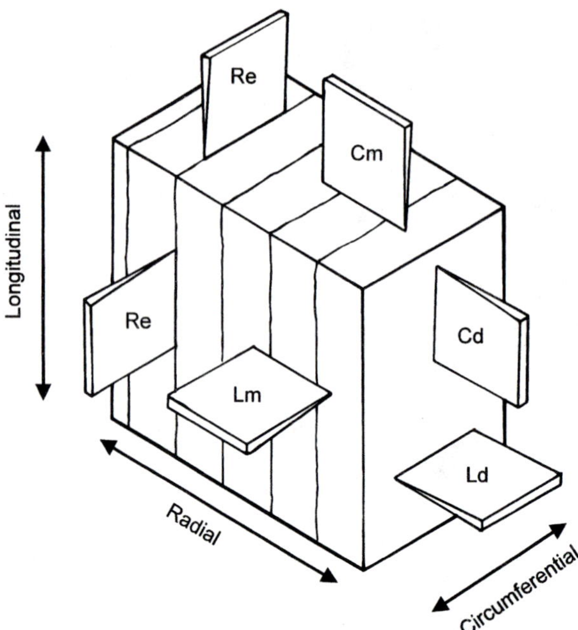

FIG. 3.9 The directions of crack travel in tensile specimens of fibrolamellar bone. R (radial), C (circumferential), and L (longitudinal) refer to the direction of loading that would cause cracks (shown by wedges) to travel in the direction of the wedge. e, m, and d refer to easy, middling, and difficult directions for the crack to travel. The difference between Cd and Cm is probably not great, but in the more difficult direction the periodic interlaminar vascular networks would act as plane of weakness, tending to stop the crack.

loaded longitudinally and, if a crack does start to spread, they may well be positioned to prevent its further spread, depending on the direction from which the crack comes. They will do this by forming a mass of relatively weak interfaces at a large angle to the direction of crack travel. The structures forming these interfaces will delaminate or pull out, increasing the energy needed to continue the crack spread. The crack will also be blunted. In fact, the situations shown in figure 3.9 are for an isolated test specimen. The longitudinal specimen will tend to break, with the crack going in the circumferential, somewhat easier direction Lm, because the crack-interrupting effects of the laminae will not be met. In life, however, the crack will have to travel from the outside of the bone (where the stresses are greatest) inward, in the difficult direction Ld, and the laminae will be more effective.

When the crack is traveling along the line of the structures in the bone, as happens when the test specimen is radially oriented, the situation is trebly different. First, the structures in the bone are now oriented with respect to the stresses so as to produce large stress concentrations. Second, the side-to-side connections between neighboring fibrils, between neighboring lamellae, between laminae, and between Haversian systems and their neighboring interstitial lamellae are relatively weak. As I show below, this is advantageous when the crack is spreading *across* these interfaces but is disastrous if the crack is traveling *along* them. The cement sheaths around Haversian systems are certainly rather weak—cracks have frequently been seen going around them (McElhaney 1966; Carter and Hayes 1977c; Behiri and Bonfield 1980; Martin and Burr 1989; Pidaparti and Burr 1992; and many observations in my laboratory). Laminae are very weak at the level of the blood channel networks. There is considerable reduction of effective area caused by the presence of blood channels all in one plane, and fibrolamellar bone, loaded radially, nearly always breaks at the level of the blood channels. Third, when the crack is traveling along the length of the bony structures, there is little tendency for the crack to become blunted, since any delamination occurs in line with the direction of crack travel.

When fibrolamellar bone is pulled longitudinally in life, all the factors cooperate to maximize the strength and crack-stopping ability of the material. When loaded in the circumferential direction, the general grain of the fibrils is at right angles to the stress. As a result, strength and toughness are much reduced. In the radial direction the lamellae, the blood vessel networks, and the laminae are in the worst possible orientation, and strength and toughness are minuscule.

3.9 Fracture of Bone in Compression

So far, I have talked only about modeling (understanding) fracture in tension, with a brief comment on shear. Compressive fracture is probably less important in life, but compressive failure certainly can occur and, in particular, is likely to result from fatigue. Excessive bending, which will result in tensile fracture, is likely to occur in an accident. In normal loading, as in locomotion, bones are probably loaded more in compression. The mode of compressive fracture is rather different from tensile failure. The load–deformation curve of a compression specimen shows a linear portion followed by a short yield region, the load soon dropping slightly and then continuing at roughly the same level until the specimen finally breaks up (fig. 3.10). This final breakup may occur only after considerable deformation. Yield stress is higher in compression than in tension. Yield damage in compression is microscopically quite different from yield in tension. What we can call "shear lines"

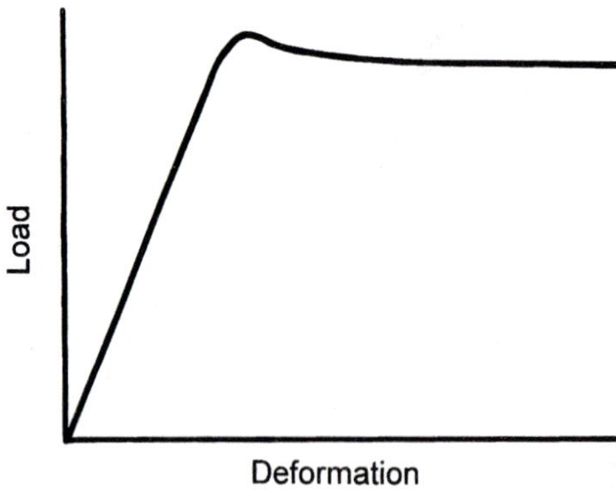

FIG. 3.10 Load–deformation curve of compact bone in compression. After yield the load drops somewhat, and then remains constant for very variable amounts of deformation; ultimate strains may be as great as 5%.

appear. They were clearly described by Tschantz and Rutishauser (1967) and Chamay (1970). These shear lines show a strong tendency to initiate at large stress concentrators such as blood channels, which is not always the case with the diffuse microcracks seen in tensile loading. When the specimen is loaded along the length of the bone, the lines form an angle of 30 to 40° with the long axis. These lines probably represent very small-scale buckling of lamellae or fibrils. The discontinuities initially formed soon coalesce and make small cracks. These spread through the specimen, which eventually disintegrates.

I should explain here that most of the load–deformation curves described in this book are derived from so-called *strain-controlled* experiments. In these the specimen is loaded to greater and greater deformations, and the load required to reach the strains is measured. It is possible, though not so simple, to use *load-controlled* experiments, in which the load is increased, and the resulting deformation examined. Of course, in a load-controlled experiment it would not be possible for the load to drop, as happens in compressive tests. Most single-pull-to-failure tests are under strain control, but fatigue tests are usually under load control, for a reason I shall explain in section 3.15.

There have been a number of studies on the effect of strain rate and other variables on the compressive failure of bone; in general, the same story emerges as for tensile failure. The anisotropy of bone in compression is not as marked as it is in tension (table 3.2). This is what we might expect. When a specimen is loaded laterally or radially, there are more stress concentrations in dangerous orientations than when it is loaded longitudinally. On the other hand, in longitudinal compressive loading the rather weak side-to-side bonding of fibrils, which is an advantage in tension, now becomes a disadvantage, allowing these structures to separate from each other by buckling. This effect is seen strikingly in wood in which, because of the presence of cavities into which the cell walls can buckle, the compressive strength is only about one-half to one-third the tensile strength (Jeronimides 1980). Haversian reconstruction usually makes bone weaker in compression as well as in tension. This was first determined by Heřt et al. (1965). Keller (1994) has tested a large number of human cortical cubical specimens in compression, and concluded that the strength was proportional to the "apparent dry density" raised to the power of about 2. This effect is probably produced by a small reduction in the porosity of the specimens, and a small increase in the amount of mineral in the tissue. In tension the situation is more complex, because some cortical specimens with low mineral have a very large increase in postyield stress (see section 4.2). No such behavior has been reported for specimens loaded in compression.

3.10 Fracture of Bone in Bending

When a bone breaks it will do so usually in tension, although this may be on an unexpected shear plane if it is loaded in torsion. However, if it is loaded in bending the bone may well start to show little fractures on the compression side before the fatal crack starts on the tensile side. For instance, in my laboratory we loaded long slender pieces of bone in compression so that they started to buckle in Euler buckling. (Euler buckling is discussed in chapter 7.) The compressive loads on the specimens were rather low when they started to buckle, and so almost all damage seen in the bone must be the result of the bending of the specimen. Examination of the specimens by scanning confocal microscopy showed that tensile microcracking, which is readily distinguishable from compressive microcracking (and which I consider at length in section 3.14) extended much further along the tensile side of the bone, and deeper into it, than did the compressive microcracking. However, in the middle of the length of the bone, where it fractured, the compressive microcracking almost met the tensile microcracking in the middle of the thickness of the specimen. What effect this compressive damage will have on the loads that the specimen can bear is very difficult to say, but it must be the case that the bone is weaker than it would be if it underwent no compressive damage at all.

As I mentioned in chapter 2, bending strengths are usually calculated as if the bone material remained linearly elastic until it fractured, which is rarely the case. If we have two bone specimens that have the same size and shape, we can compare their bending strengths by finding out what bending moments caused the specimens to break. This leaves open the question of what the actual values of their bending strengths were,

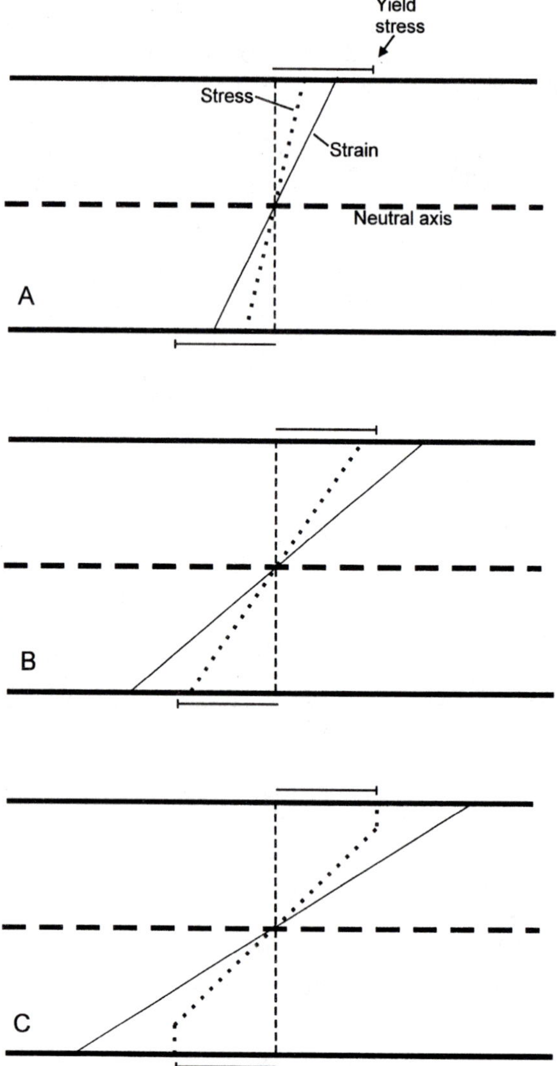

FIG. 3.11 A beam of bone loaded in bending. The neutral axis is shown by the long dashes. The short vertical dashes indicate the section that we are considering. The dotted angled line represents the stress, and the solid angled line the strain. (For simplicity I am assuming that the bone behaves the same in tension and compression.) (A) As a bending moment is imposed, the stress and the strain are proportional to each other, and increase from zero at the level of the neutral axis to a maximum at the outer tensile and compressive surfaces. (B) The bending moment is increased. The stress and strain are both increased, and are still proportional to each other. The stress has not reached the yield stress yet. (C) The outermost region of the specimen has reached a strain that exceeds the yield strain, and stress, of bone. The amount of stress it undergoes cannot increase but it does

but clearly the specimen bearing the greater bending moment would have the greater bending strength. Suppose that in a tension test the materials of both specimens had the same linearly elastic behavior, but that one failed as soon as it reached the end of its linearly elastic region, whereas the other underwent some postyield damage, but without any increase, or decrease, of stress. This behavior is quite like that of real bone (fig. 2.7). Obviously, the tensile strength of these two materials would be the same, although the postyield damaged one would show a greater strain at failure. In a bending test the specimen that was linearly elastic to failure would have a bending strength that could be predicted by beam theory and knowledge of its tensile strength. However, let us consider what happens with the specimen that can undergo some postyield strain.

When the outermost part of the specimen loaded in tension reaches its tensile yield stress it cannot bear a higher stress, but it can maintain that stress. The region underneath will be at a lower stress, and so will not have yielded. As the bending moment increases, the region underneath will begin to reach the yield stress. And so on (fig. 3.11). The postyield behavior of the specimen does not allow any part of it to bear a greater *stress*, but does allow the specimen as a whole to bear a greater *bending moment*, in other words to be stronger in bending. While all this yielding is going on, the neutral plane is moving toward the compressive side and the situation becomes decidedly complex. Sooner or later the compression side will start to undergo damage, and the situation will become even more complex. This matter is analyzed in some detail by Burstein et al. (1972).

Another quite separate effect is important here, which will increase the apparent bending strength compared with the tensile stress. Bone, like nearly all materials, is not completely uniform, and will have weaker and stronger parts. A specimen, or a whole bone, will break when some sizeable volume of the specimen or bone has failed. In a tensile specimen, in which the stress is essentially uniform throughout, this volume will be the weakest volume. However, in a bending specimen the situation is different. Consider a horizontal beam loaded in three-point bending and ignore shear stress. The stress is not uniform, being, at any particular cross section, zero in the middle of the depth and becoming greater toward the top and bottom surfaces. Also, it will be less toward the ends of the specimen, compared with the middle of

---

not break yet and the maximum strain can be increased. Therefore, the bending moment can still be increased, and a greater and greater thickness of the bone will reach the yield stress.

the length, because the bending moment decreases linearly to zero at the ends. Now the probability of any particular volume failing is a product of its weakness and of the stress it feels. Weak volumes may be "hidden" in low stress parts of the specimen. However, the calculation for bending strength assumes that the bone will break when the most highly stressed region reaches the material's strength, which is supposedly uniform. Therefore, failure in bending is a more stochastic process than it is in tension, depending as it does on *where* the weak bits happen to be; in tension it does not matter where they are. If we have two bones, one of which has more fine-grained variability than the other, and prepare many test specimens, then the more variable bone will show on average a greater difference between the results of tensile and bending loading.

Even in tensile loading variability can be important, because it will produce a size effect. A large specimen is more likely to include weak volumes than is a small specimen, even though it is also more likely to include strong volumes. Unfortunately, strong volumes do not help! Therefore, other things being equal, small specimens are likely to appear stronger, on a stress at failure basis, than large specimens. The question of weakest links and their effects on calculated strengths was first analyzed by Weibull (1951). He produced a fairly simple model, showing how the variability of stress at which a sample failed would be a function of the variability of the strength of little bits of the specimen and the size of the specimen. An introduction to the subject is given in Hull and Clyne (1996). Until recently it has hardly been considered in bone, though King and Evans (1967) made an early attempt. Harner and Wilson (1985) and Sadananda (1991) applied Weibull theory to chicken bones (though these bones seemed to be remarkably variable in their strengths). This has been analyzed by Taylor and colleagues (Taylor 1998; Taylor et al. 1999) in relation to compressive fatigue (section 3.6).

In a further analysis Taylor (2000) pointed out that there was a problem because if this size effect were important, larger animals should have bones less able to bear particular stresses than small animals. Yet all bones seem to undergo roughly the same maximum strains and stresses (section 10.1). Taylor shows that, in general, larger animals have bones that are stronger in fatigue, thereby overcoming this particular difficulty. Although Taylor's analysis appears valid for the rather limited data set that he had available, I feel uneasy about this result because it suggests that smaller animals have bone that is less good mechanically than it could be, whereas, in general, as we shall see in the next chapter, bones are usually nicely adapted to their functional requirements. One possibility is that, in life, large animals are able to remodel their bones and take out fatigue cracks (see section 11.10.5),

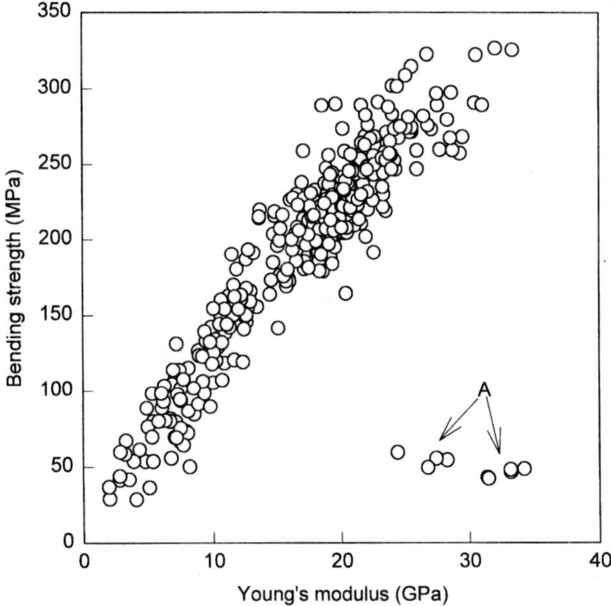

FIG. 3.12 Relationship between the bending strength and the Young's modulus of a set of bone specimens. The provenance and treatment of these specimens is described in section 4.2. The specimens labeled A come from bones with a very high mineral content. (From Currey, J.D. 1999. What determines the bending strength of compact bone? *Journal of experimental biology* 202:2495–503. By Permission of the Company of Biologists Ltd.)

whereas smaller animals cannot do this. Unfortunately, this is not really a way out of the present difficulty, because the specimens included in Taylor's analysis were loaded in vitro, and no remodeling was possible.

Figure 3.12 shows, for a wide variety of bones, bending strength as a function of Young's modulus. There are clearly two distributions. In the main one the trend is extremely clear. As Young's modulus increases, bending strength increases in proportion. (This relationship does not hold well for tensile tests. For instance, in the specimens loaded in tension there were a number of antler specimens that, while having a rather low Young's modulus, had a high tensile strength. This antler bone does not have a correspondingly high bending strength.) The lower, and quite distinct distribution, labeled A, is from two bones with a *very* high mineral content. This shows that there is a level of mineral content at which there must be almost a step change in the bending strength.

Consider only the main distribution for a moment. Bending strength is proportional to Young's modulus. Young's modulus is affected mainly by mineral content and porosity. If all bones behaved as totally brittle materials, then the fact that bending strength is proportional to Young's modulus would imply that bone failed in bending at a particular *strain*. (Because $\varepsilon = \sigma/E$, if σ/E is a constant, as here, then ε is a constant.) As I showed above, bending strength is rather a complex property, depending both on the stress at which yield occurs and how much postyield deformation there is before fracture occurs. One cannot determine yield strength in bending; yield is too gentle a process, affecting only a tiny part of the bone initially. However, I was able to estimate it, by taking the yield strength in tension as a function of Young's modulus of tensile specimens taken from alongside the bending specimens (Currey 1999). Most bony materials yield in tension at roughly the same strain (about 0.002), though it declines very slightly with increasing modulus. Therefore, other things being equal, a high Young's modulus will be associated with a high yield stress, and therefore a high bending moment when the specimen yields. If the material shows a reasonable amount of postyield strain then the bending moment will increase for a while, and the apparent bending strength will be quite high. Burstein et al. (1972) showed that, for a square cross section, if the postyield strain is equal to the elastic strain, then the bending moment will be raised by a factor of about 1.5 compared with a completely brittle material. If the postyield strain is five times that of the elastic strain, then the factor is 1.7. It does not, therefore, take a great deal of postyield strain to reap most of the benefits of not being completely brittle. So, if a bone has a low modulus, it will yield at a rather low stress, and even if it undergoes a large amount of postyield deformation, its apparent bending strength will not be increased much. This combined effect of yield strain and postyield deformation, therefore, may explain the linear relationship between Young's modulus and bending strength. However, the general level of bending strength as a function of Young's modulus is higher than might be explained simply by extrapolation from the results of tensile tests and the effects of postyield deformation. This may be explained by the Weibull effect: all the weak parts of a tensile specimen are found out in a tensile test, but in bending, the high stresses are confined to smaller parts of the specimen, which by statistical chance may not have a weak part.

If, however, the bone is very highly mineralized, the few cracks that appear in the preyield region find it so easy to travel that they become fatal, traveling easily through the material, and so the usual strain at yield, about 0.002, is not achieved, and the material fails in a brittle manner at a low bending moment. This is what is happening, presumably, in the specimens making up the lower distribution in figure 3.12.

3.11 Mechanical Properties of Haversian Systems

Over many years Ascenzi and his co-workers have attempted to discover the mechanisms of the elastic and fracture behavior of bone by investigating single secondary osteons, mainly from human bone. This work has given much useful information, but it does suffer, as these workers acknowledge, from the fact that only one element in a composite is being studied. In particular, the machining out of Haversian systems removes the cement sheath, by which the Haversian system is connected to (or perhaps separated from) the rest of the bone. Also, Haversian systems are not found at all in many bones. The methods used by this group, though not results, are described fully in Ascenzi et al. (2000).

To test Haversian systems in tension, they grind longitudinal sections of bone about 20 to 50 μm thick. This is much less than the diameter of a Haversian system, which is characteristically about 200 μm, and so only longitudinal segments of Haversian systems remain. Where a Haversian system is sectioned so that its central canal is in the plane of the section, one side is isolated as a test specimen. For compressive tests they mill out a cylindrical test specimen about ½ mm long. They tested specimens wet and dry. I shall describe tests on wet specimens only. They consider, usually, two types of Haversian system, apparently distinguishable from each other under the polarizing microscope. I call these *longitudinal* and *alternating*. In the longitudinal specimens the fibrils in the lamellae have a very low angle to the long axis of the system, and there is a small difference in the direction between neighboring lamellae. In alternating systems the fibrils in one lamella will be nearly longitudinal, and almost at right angles to this in the neighboring lamellae. (Many workers think that Haversian systems are not so neatly classifiable as this [Evans 1973; Giraud-Guille 1988; Katz 1981; Marotti 1993].)

The Ascenzi school also distinguishes between fully calcified systems and those that have just started to calcify. Initially, Haversian systems calcify very rapidly, and then the process slows, so the difference in the amount of mineralization between partially and fully calcified systems is not great. The effect of completing calcification is always to increase the strength and modulus. The results of their studies on single fully mineralized systems are given in table 3.7, derived from Ascenzi and Bonucci (1967, 1968) and Ascenzi et al. (1990, 1994). There is an indication that longitudinal systems are stronger in tension and weaker in compression than alternating systems. This indication receives support from the works of Riggs et al. (1993a,b) discussed below.

TABLE 3.7
Mechanical Properties of Isolated Secondary Osteons

Property	*Longitudinal*	*Alternate*	*Authors*
Tension modulus	11.7 GPa	5.5 GPa	Ascenzi and Bonucci (1967)
Compression modulus	6.3 GPa	9.3 GPa	Ascenzi and Bonucci (1968)
Bending modulus	2.3 GPa	2.6 GPa	Ascenzi et al. (1990)
Torsional modulus	22.7 GPa	16.8 GPa	Ascenzi et al. (1994)
Tensile strength	120 MPa	102 MPa	Ascenzi and Bonucci (1967)
Compressive strength	110 MPa	164 MPa	Ascenzi and Bonucci (1968)
Bending strength	390 MPa	348 MPa	Ascenzi et al. (1990)
Torsional strength	202 MPa	167 MPa	Ascenzi et al. (1994)

Three things are striking about these results. First, there are very large differences between the values for Young's modulus in tension and compression, even though they are both low compared with the values found in larger specimens (table 3.1). Admittedly, the values for tension and compression were not measured on the same samples, but it is surely unlikely that the tensile modulus of the longitudinal system should be almost double the compressive modulus, these differences, incidentally, being highly statistically significant. Second, the bending modulus is extraordinarily low compared with the torsional modulus, which is indeed very high. Third, the bending strengths and, particularly, the torsional strengths are very high compared with the tension and compressive strengths.

These varied results are difficult to reconcile, but Lakes (1995) suggests that the torsional modulus of *single* osteons may indeed be as high as 20 GPa or so. He proposes that the lower measured modulus of larger specimens is caused by the individual osteons rotating relative to each other, a process permitted by the supposed viscous nature of the cement lines round each osteon. This produces an apparently low value for the block of bone, although the individual elements may be very stiff. An analogy would be the twisting of a bundle of wires. The bundle would be quite compliant in torsion, although the individual wires might have a very high torsional modulus. Such behavior is described by *Cosserat* elasticity. Frasca et al. (1981) found that specimens became stiffer in shear when they contained fewer secondary osteons; this is at least consistent with Lake's suggestion. I must admit to finding the Cosserat explanation not completely convincing, particularly because Reilly and Burstein found that fibrolamellar bone, which has (*pace* Lakes and Saha 1979) no cement lines, has a shear modulus of about 5 GPa. Nevertheless, the argument is diverting.

The mode of failure in single Haversian systems described by Ascenzi and his co-workers is interesting. The compressive failures involved fissures traveling at 30 to 35° to the long axis of the system. They are very similar to the kind of thing seen in larger test specimens. The fissures show little tendency to deviate or interrupt when traveling from one lamella to the next. In tension the behavior of alternate systems is more complicated (Ascenzi and Bonucci 1976). Above a load of about 30% of the failure load the transversely oriented lamellae break periodically along their length, but the structural integrity of the specimen is maintained by the longitudinally oriented lamellae. This is accompanied by a change in the slope of the load–deformation curve (fig. 3.13). The transversely oriented lamellae crack frequently along their length and eventually the longitudinal lamellae must break. The situation is much as in some engineering laminates, in which alternate sheets are oriented at about 90° to each other. Such a laminate should notionally have a strength, when loaded in the direction of one of the sets of fibers, not much greater than half that of a unidirectional laminate if the matrix is weak. In fact, according to Ascenzi and Bonucci's 1967 study (table 3.6) there is only a 15–20% difference (120 vs. 102 MPa).

This discrepancy results, I think, from the histology being more complex than Ascenzi and Bonucci allow. Even so, these microscopic studies do show why, for unidirectional loading, the alternate arrangement may not be very efficient. The Young's modulus of the transverse lamellae is presumably less than that of the longitudinal lamellae. Also, the sets of lamellae are in parallel. Therefore, for any given strain the stress in the transverse lamellae, and the load borne by them, is less than in the longitudinal lamellae. However, if, as seems to be the case, the difference in strength is greater than the difference in modulus, then the transverse lamellae will break first. As the transverse lamellae break, more and more of the load is thrown onto the longitudinal lamellae. The system is weakened, therefore, by the presence of the transverse lamellae.

In fact, the situation is not nearly so clear as has been made out so far. The spatial arrangement of the lamellae into longitudinal and transverse, neatly alternating, has been questioned. In particular, Frasca et al. (1977) isolated and examined individual lamellae from Haversian systems. Systems the Ascenzi school would classify as longitudinal do indeed have a predominance of longitudinal fibers. However, alternate osteons also have considerable number of longitudinal fibers, according

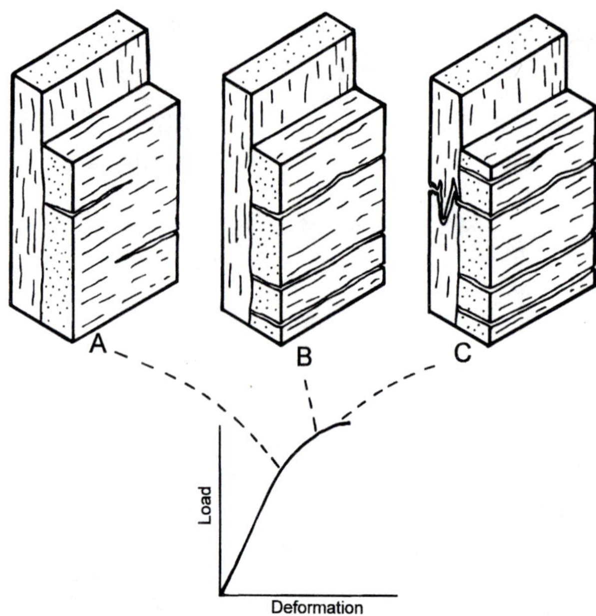

FIG. 3.13 Progressive failure of "alternate" Haversian lamellae, loaded in tension up and down the page. (A) Cracks appear in the transversely oriented lamellae. The load–deformation curve stops being linear. (B) More cracks; the curve flattens. (C) A longitudinal lamella fails; failure of the specimen is imminent. (Based on ideas in Ascenzi and Bonucci [1976].)

to Frasca et al., but now there are a number of "oblique fibers, circumferential fibers and coexisting fibers of various orientations." In general, the studies of Katz's group reveal a great predominance of fibers oriented more or less in the direction of the bone axis and, in trabecular bone, along the axis of the trabeculae (Katz 1981). The group could find no sets of lamellae in which longitudinal and transverse fibers alternated neatly and regularly. The work of Marotti (1993) and Giraud-Guille (1988) also suggests that the arrangement of the collagen and its associated mineral is much less orthogonally arranged than suggested by Ascenzi and his colleagues.

There is good evidence from studies of ordinary testing specimens, rather than single Haversian systems, that the arrangement of fibers in the different types system can have mechanical consequences. Riggs et al. (1993a,b) examined the histology and mechanical properties of bone from the anterior (cranial) and posterior (caudal) walls of the radius of horses. The differences were quite marked (table 3.8).

TABLE 3.8
Mechanical Properties of Specimens Taken from the Anterior (cranial) and Posterior (caudal) Cortices of the Horse's Radius

Loading direction	*Test mode*	*Cortex*	*Young's modulus (GPa)*	*Ultimate stress (MPa)*	*Ultimate strain*	*Energy absorbed ($J\ m^{-3} \times 10^{-5}$)*
Longitudinal	Tension	Cranial (T)	22.1	161	0.028	38.1
		Caudal (C)	15.0	105	0.016	13.5
	Compression	Cranial (T)	18.6	185	—	38.1
		Caudal (C)	15.3	217	—	58.8
Transverse	Compression	Cranial (T)	8.2	—	—	—
		Caudal (C)	10.9	—	—	—

Source. From Riggs et al. (1993b).
Note. T, Loaded in tension in life; C, loaded in compression in life.

Quantification of circularly polarized light images was used by Riggs et al. to determine the predominant orientation of the collagen in the sections with respect to the long axis of the bone. (It is not possible actually to *quantify* the distribution of directions of the collagen using polarized light, but the different sections can be placed in order according to the amount that the collagen fibrils are arranged off-axis.) The caudal cortex was much more remodeled than the cranial cortex, and the predominant direction of the collagen fibrils was farther from the longitudinal axis. The orientation of the fibers was a more important determinant of the mechanical behavior than the difference in mineralization (remodeled bone is usually less highly mineralized than primary bone). These results suggest that converting bone in which the fibers are predominantly longitudinal into bone in which the fibers are on average more transversely arranged reduces Young's modulus when tested in both tension and compression. It also reduces the tensile strength but it increases the compressive strength. In tension the ultimate strain is also reduced. Also, the anisotropy of Young's modulus was much greater in the bone with more longitudinally arranged Haversian systems.

An important point about this work by Riggs et al. is that it accords with the known loading on the horse's radius. During ordinary locomotion the major loading of the radius is bending, with the cranial cortex being loaded in tension and the caudal cortex in compression (Biewener et al. 1983b). Remodeling that has taken place in the caudal cortex probably makes it more able to bear the compressive stresses to which it is loaded, though at the same time making it weaker in tension. This was confirmed by Reilly and Currey (1999). When a beam is loaded in pure bending, theoretically the greatest compressive stress on one side is the same in magnitude as the greatest tensile stress on the other. Reilly and Currey showed that when bone specimens from the horse radius were loaded in bending, bone from the cranial cortex (adapted for tension) showed microcracking first on the compressive side, whereas bone from the caudal cortex (adapted for compression) showed microcracking first on the tensile side.

In a study with the same general objectives Takano et al. (1999) have shown that in the radius of dogs the bone was both mechanically and structurally more anisotropic in regions predominantly subjected to tension in locomotion, compared to regions subjected predominantly to compression. The bone was mainly Haversian, and the authors did not measure the orientation of the Haversian systems; instead, they measured the predominant orientation of the collagen. The obvious explanation for this difference is that the grain of the bone in the tensile side Haversian systems was more longitudinal than in the compressive side, just as in the radius of the horse.

3.12 Cancellous Bone

The mechanical properties of cancellous bone are much less well understood than those of compact bone, although the amount of empirical information is greater. Because some of the ideas necessary to understand cancellous bone, such as buckling, are hardly introduced yet, and because cancellous bone does not occur on its own, but instead virtually always forms structures with compact bone, I shall discuss it later, in chapter 5. Cancellous bone is always much less stiff and much weaker than compact bone.

3.13 Bone as a Composite

Materials can break in various ways. A truly brittle material in tension will simply cleave catastrophically from some preexisting flaw when the energy balance is right. A ductile material such as a mild steel will deform plastically by sending off dislocations from the highly stressed regions. These dislocations will distribute the strains over quite a large volume, far from the highly stressed region. These are shear dislocations, and the result of their movement will be that in a conventional cylindrical test piece in tension the highly stressed region will become thinner, tending to increase the stress on it. This leads to the phenomenon of *necking*, in which a short region of a tensile test piece becomes quite narrow and bears an increasingly higher stress, even though the load on the specimen may be increasing little, if at all. Eventually, small cavities appear in the neck region; these coalesce and the material fails. The fracture surface is pitted, and the broken structure, if fitted back together, would obviously be highly deformed. However, bone is not a ductile material and breaks in a quite different way.

For some reason it appears to be bad form to refer to differences of opinion between the first edition of a book and the second edition, except in the introduction. I shall show bad form here by saying that at this point in the first edition I spent almost six pages trying to explicate the fracture behavior of fibrous composites; this time I shall say rather little. The reason for the change is, I think, instructive. Everyone who knows about the mechanical properties of bone thinks of it as a fibrous composite. The elastic and the strength behaviors of composites are by now fairly well worked out. Unfortunately, all this knowledge seems to have little relevance to explaining the fracture properties of bone. When I wrote the first edition (in 1983) I thought that materials scientists would take the problem on board and come up with some convincing analysis. They have not. Therefore, although the mechanics of the fracture of composites is a rather satisfying intellectual exercise, it does not help much to explain bone. Nevertheless, the general idea of fibrous composites is clearly relevant in some way to bone, so an idea of the basics is necessary.

Composite materials, that is, materials made of at least two different phases, show elastic properties that in some way are a mixture of the elastic properties of the phases. I discussed this in section 3.7. The strength properties, however, may be quite different from the strengths of the individual phases. Often the composite is stronger than either phase on its own. It is also quite possible to have a composite made of two completely brittle materials that itself shows a considerable amount of postyield deformation. How this increase in strength and toughness is brought about depends in detail on the particular components, but, in general, the composites are effective in crack management. As we saw in section 2.2.5, cracks in a brittle material spread more and more easily as they get longer. However, in a well-designed composite material any crack, as soon as it starts to travel, is forced to change its direction, or to pass into a void that makes it blunt-ended, or to give up the strain energy that could be driving it forward to the work required to pull strong fibers out of matrices, or to overcome a whole host of other features that all require an input of energy before the crack can travel further (fig. 3.14). All these process are brought about by the composite nature of the material. The phases of a well-designed composite are held together sufficiently strongly for them to behave as a monolith in the elastic region, and to show a fairly high Young's modulus and shear

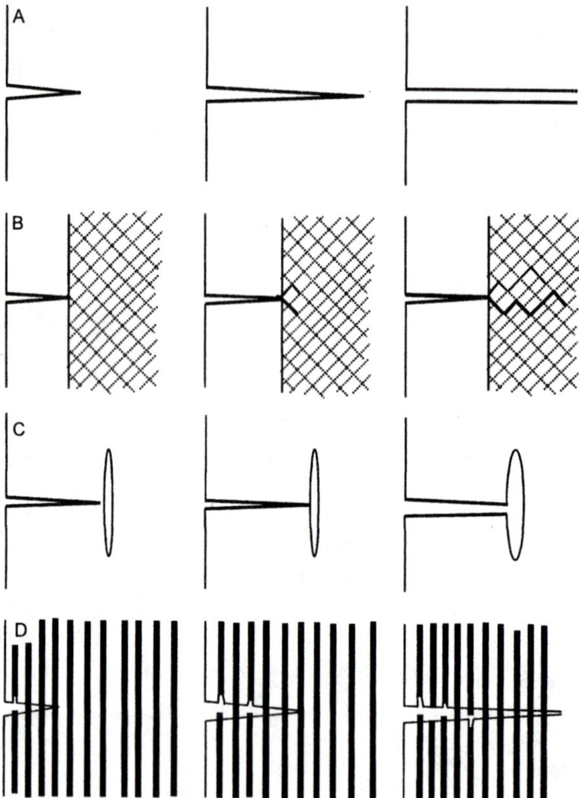

FIG. 3.14 Diagram of crack travel in various materials. (A) The material is completely brittle, and once a crack has started to travel it is unimpeded. (B) The crack passes into a set of layers in which the direction of travel is easy in two different directions, at different depths. As a result, different parts of the crack front go off in different directions in different layers, and then have to break back across in the difficult direction. The two solid lines from the end of the crack represent cracks at two different depths. This mechanism of crack stopping is found in enamel (see fig. 6.2). (C) The material has voids (or, often, weak interfaces) arranged so that a crack running into them has difficulty in traveling further, as the crack tip is blunted. (D) There are strong fibers embedded in the material. As the crack advances it has to break through the strong fibers (which are protected by a relatively weak interface with the matrix) and it also has to expend energy overcoming frictional forces in pulling the fibers out of their matrix.

modulus. However, when cracks start to appear the phases will, where appropriate, separate from each other and frustrate the advance of cracks.

A material like bone, which has a strongly hierarchical structure, has phase changes and material direction changes over many length scales, and these heterogeneities will no doubt account for the ability of bones to contain stable cracks that themselves range over many length scales. Working out the interactions between heterogeneities and microcracks is a fiendishly difficult theoretical problem, and there is an enormous way to go before it is understood properly (for an example of the difficulties involved, see Leguillon et al. 2000).

The difficulty in trying to analyze bone as a composite when considering its failure behavior is that it is difficult to know at what level the processes important to failure occur. Bone has a hierarchical structure, and failure processes may take place at any level in the hierarchy. In general, this is not true of technological fibrous composites, in which it is fairly clear where the important processes are occurring. At one level we could consider the collagen to be the matrix, in which the mineral "fibers" are embedded. At the next level we could consider the collagen fibrils, with their associated mineral, to be the fibers, each more or less loosely connected to its neighboring fibers, all of them making lamellae. At the next level the lamellae are separated from their neighbors by a change in the predominant orientation of the collagen fibers and their associated plates. This discontinuity will be a line of weakness. At the highest level we could consider the osteons or laminae to be the fibers, again more or less loosely connected to each other by cement lines and other structures.

The early analogies between composite materials and bone (Currey 1964a; Mack 1964) concentrated on the lowest level, that of collagen as the matrix and the mineral as the fiber. However, this turns out not to have been very fruitful. The reason is that at such a low level, with such an intimate relationship between the mineral and the protein, it is impossible to have much idea of how the collagen is behaving. We have seen in the discussion on elasticity in section 3.7 that it has been suggested, and indeed there is some corroborating evidence, that the properties of collagen in bone may be radically different from those it has in, say, tendon (McCutchen 1975; Hukins 1978; Lees et al. 1990). At the moment we cannot talk meaningfully about the sizes of the mineral fibers and the collagen matrix because we are not sure of the mineral morphology and know even less about the mode of attachment of the collagen to the mineral. Furthermore, at this level we cannot see the fracture surface clearly enough to make out, for instance, whether the minute mineral blocks pull out from the matrix. It is likely that the composite nature of bone is important, at this level, for fracture as well as for elasticity, but we shall have to wait a good while before we know much about that.

Larry Katz once remarked to me that he supposed the reason that the vertebrates had adopted apatite, rather than calcium carbonate, as the mineral for bone was that apatite was very reluctant to form large crystals, and therefore the vertebrates did not have to adopt elaborate mechanisms as, for instance, the mollusks do, to keep the crystals small. Large crystals are, of course, brittle and, in general, to be avoided in materials that need to be tough. Although this idea may not be correct, it struck me at the time as being such a classic case of the "blinding glimpse of the obvious" (Haldane 1953) that I feel I should introduce it anecdotally. Another suggestion (Ruben and Bennett 1987) is that apatite would be less prone than calcium carbonate to dissolve in the very acid internal environment of an intensely exercising vertebrate.

A difficulty in determining how strength is affected by the mineral and organic components is that, almost inevitably, if they are altered, by disease or experiment, they are both altered, and one cannot distinguish which, if either, is the important effect. For instance, Landis (1995) shows how the crystal form in cases of osteogenesis imperfecta is different from normal. This difference is almost certainly caused by the deranged packing of the collagen fibrils in this disease. However, since both the collagen and the mineral are deranged, it is not possible to say what causes the undoubted degraded mechanical properties of osteogenesis imperfecta bone. At the next higher level things are better because the individual fibers and larger structures can be made out with the scanning electron microscope, which is particularly useful for examining fracture surfaces.

Bone does not undergo plastic deformation like metals, and we need now to consider in more detail the tensile load–deformation curve in figure 2.7. This showed that at the end of the straight elastic region (at a strain of about 0.002–0.005) the stress–strain curve flattens, and the strain increases with little increase in stress. The final amount of strain that the bone undergoes before fracture varies considerably with the amount of mineralization, being less the greater the mineralization (I return to this important fact in chapter 4). In reasonably well mineralized bone the ultimate or fracture strain is about 2–4%, but in some less highly mineralized bony tissues, such as antler, the final strain may be as much as 10%. Superficially, the postyield behavior looks like plastic deformation, but it is caused by a totally different mechanism.

The bone is behaving elastically in the first region, as is shown by the fact that if the bone is loaded and unloaded cyclically, the stress–strain curves are virtually superimposable. However, if the bone is unloaded and reloaded after the stress–strain curve has flattened out, the curves are no longer superimposable (fig. 3.15). The ordinary stress–strain curve to failure and the set of loading and unloading curves have important things to tell us:

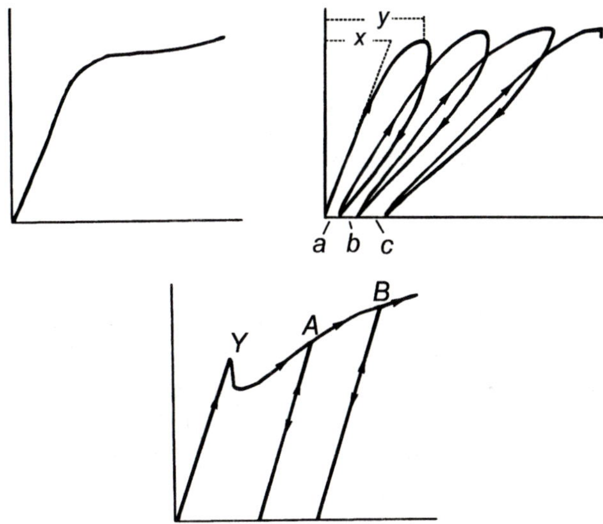

FIG. 3.15 (*Top left*) Stress–strain curve of bone loaded in a single pull to fracture. (*Top right*) A bone specimen loaded and unloaded three times and then loaded to failure. (*Bottom*) Mild steel loaded beyond the yield point *Y* and then unloaded twice, at *A* and *B*.

- Bone is to some extent viscoelastic, and it might be that the residual strains *a*, *b*, and *c* in figure 3.15 would recover with time. Some of this strain is recovered after a few minutes, but most of it is not. This indicates that the strains are to a large extent irrecoverable. Bending a specimen shows that this is correct. A specimen that is bent so that the load–deformation curve flattens out, and is then unloaded, is visibly bent and the bend remains indefinitely.
- The strain at *y* has two components, an elastic component *x* and a component *y–x* that in some way we can attribute to the postyield behavior of the specimen. The fact that *a* is less than *y–x* shows that some of the strain occurring in the yield region is recoverable almost instantaneously on unloading, but some is not.
- The slopes of the loading curves are different, the specimen becoming more compliant each time it is loaded into the yield region.
- The stress achieved does not increase much with strain, the envelope of the curves being rather flat-topped. This is also true of the loading curve that does not involve unloading before fracture.

Before discussing what these observations imply, it is helpful to look at a stress–strain curve for unloading and reloading in a typical metal such as mild steel (fig. 3.15). There is a drop at *Y*, at the end of the elastic region, which is produced by a mechanism that need not concern us. On unloading at *A* and reloading there is no loop, showing that steel is barely viscoelastic. The unloading and reloading curves have the same slope as the initial part of the curve, before yield occurred. Steel behaves as if its yield behavior is in some way a function of strain (for instance, *B* is at a considerably higher stress and strain than *A*, but its elastic behavior is unaffected by yielding). When steel yields, shear dislocations travel through the material, allowing it to distort rather easily. For various reasons these dislocations become trapped and entangled with each other and in so doing become much more difficult to move. That is why the stress in the yield region rises with strain. When the material is unloaded and reloaded again, the elasticity is unchanged, because any dislocations whose easy motion might make the steel more compliant have become trapped and cannot move until the unloading stress is again reached. In particular, there are no cracks in the steel; instead, it has a population of dislocations, which will be able to move if stressed to various levels.

3.14 Microdamage

The critical difference between the behavior of a metal and bone is that in the postyield region, bone is undergoing damage, in the form of microcracking. As the bone extends, a host of little cracks appear, but they *do not spread or coalesce*. In the discussion of toughness in chapter 2, I showed how, in a brittle material, it is the length of the crack that mainly determines how dangerous it is. The tiny cracks that form in bone are too small to be dangerous, at least initially.

3.14.1 *Microcracking Phenomena*

Bone, when loaded, produces many little cracking sounds, each corresponding to the release of some energy. Thomas et al. (1977) Yoon et al. (1979), and Netz et al. (1980) loaded bones in bending and examined and attributed this to the generation of microcracks. We have examined acoustic emission in both ordinary bovine bone and deer antler (Zioupos et al. 1994). We found that in the preyield region the specimen is almost silent, but that as the curve bends over there is a burst of noise, which quiets down somewhat in the postyield region, before appearing again as the final crack spreads through the specimen (fig. 3.16). These final noises are particularly energetic, so agreeing with the findings of Akkus et al. Akkus et al. (2000) have been able to relate

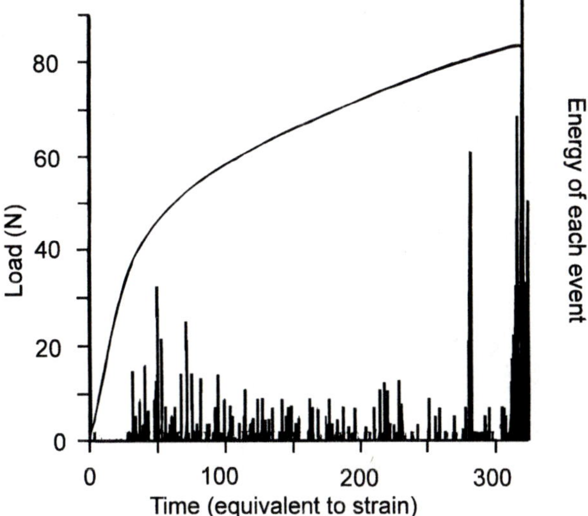

FIG. 3.16 Acoustic emission in antler bone. The top curve is load (in newtons) versus time (in effect strain). Initially the curve is much steeper than it becomes later. Where the curve bends over, microcracking starts. The spikes from the bottom are acoustic emissions from the specimens. Initially the specimen is almost silent. When the curve begins to bend over, acoustic emission becomes steadily more intense, but reduces in intensity in the long part of the load–time curve, which is almost straight. Finally, there is a burst of noise just before the specimen breaks in two. (Derived from *Medical engineering and physics*, Vol. 16, Zioupos, P., Currey, J.D., Sedman, A.J. An examination of the micromechanics of failure of bone and antler by acoustic emission tests and laser scanning confocal microscopy. 1994, pp. 203–212, with permission from Elsevier Science.)

types of acoustic emission events, particularly their energy, to particular types of microcracking damage, the highest energy being associated with the travel of a major crack.

When bone is loaded into the yield region it shows whitening. This is seen particularly clearly when the specimen is illuminated strongly from the front. The whitening appears only when the bone has yielded, appears most strongly around large stress concentrators, such as holes and notches, and to some extent disappears when the bone is unloaded. It is a much more obvious effect in poorly mineralized tissues such as antler and dentine. The change in the bone presumably means that little interfaces have opened up, which scatter the light.

Interestingly, an early observation of this effect comes from one of the Icelandic sagas: Egil's Saga. The hero, Egil, was a ferocious man with a large skull. Long after he died his skeleton was found by a priest, Skapti. Acting in a way less priestly than curious he tried to break the skull. "Þá vildi Skapti forvitnask um þykkleik haussins; tók hann þá handøxi vel mikla ok reiddi annarri hendi sem harðast ok laust hamrinum á hausinn ok vildi brjóta, en þar sem á kom, hvítnaði hann, en ekki dalaði ne sprakk." ("Skapti wanted to find out the thickness of the skull. He picked up a big handax and swung with one hand as hard as possible, and struck the skull with the back of the ax, intending to break it. Where it struck, the skull whitened but was neither dented nor broken.") It has been argued that this was no ordinary skull, and that Egil suffered from Paget's disease, in which the process of remodeling is deranged and the skull and other bones increase in thickness greatly (Byock 1995).

Whitening in itself, however is not conclusive evidence for microcracking; stronger evidence comes from observing the cracks themselves. Cracks can be seen, for instance, by loading a specimen in bending and then immersing it in stain (Currey and Brear 1974). The stain settles on surfaces, particularly rough ones, and the yielded region on the tensile side, which had become opaque, is now picked out in color. The coloring is not uniform but extends in thin wavy lines part way across the breadth of the specimen. Similar diffuse lines can sometimes be made out at the ends of tensile cracks that have failed to break a specimen in two.

There have been an increasing number of reports in the literature of the appearance of microcracks in bone that has yielded but not failed (Carter and Hayes 1977c; Frost 1960; Schaffler et al. 1994; Vashishth et al. 1994; Villanueva et al. 1994) The earlier workers used conventional microscopy, bulk staining the specimens, and were able by various clever tricks to separate artifactual from in vivo damage (Burr and Stafford 1990). It later became possible to visualize these microcracks more clearly, using laser scanning confocal microscopy (LSCM) (Zioupos and Currey 1994; Fazzalari et al. 1998; Huja et al. 1999; Jepsen et al. 1999).

Dye, which settles preferentially on surfaces within the specimen, is excited by the laser, allowing optical slices of the specimen to be made, so a three-dimensional image can be stored on a computer. Using LSCM we (Zioupos and Currey 1994) have studied these cracks in a number of mammalian mineralized tissues, including bovine bone, antler, and dentine. We have imaged the cracks in three dimensions, and find some to be quite large (50 μm or so across) down to many of the order of 5 μm.

3.14.2 The Mechanical Effects of Microcracking

A number of interesting things have emerged from these studies on microdamage. We find the small cracks in bovine fibrolamellar bone tend

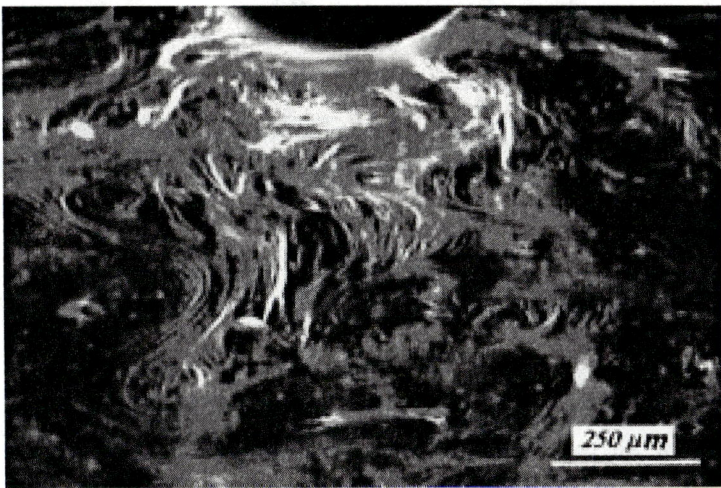

FIG. 3.17 Scanning confocal microscope photograph of cracks induced in a specimen of antler. The specimen was notched at the top and loaded in tension from side to side. Cracks swirl from the root of the notch. Although the brightest arrays look continuous, they are not; higher power magnification would show that they are composed of thousands of tiny microcracks. Width of field: 1¼ mm.

to initiate in the *hyper*mineralized, "woven" bone part of the laminae in the bone and to come to a halt when they reach the lamellar bone. Sometimes the microcrack arrays snake round the lamellar bone, keeping to the woven bone. The general form of the microcracks is consistently different between bovine bone and the less well-mineralized tissue of antler; the cracks in antler are much more convoluted, following the grain of the tissue more closely than they do in bone. The cracks very often end up parallel to the loading direction, in which orientation there will be little tendency for them to spread further (fig. 3.17) (Zioupos and Currey 1994).

In tension, loading and unloading a specimen shows that significant microcracking does not appear until the yield point is reached. As the specimen is loaded further and further into the postyield region the microcracking becomes increasingly pervasive. However, by using the highest powers of the optical microscope it is possible to see that individual microcracks do not get longer; instead, more and more of them appear (Reilly and Currey 2000). There is no really consistent pattern to where the cracks form (fig. 3.18). Sometimes they form near blood channels, and frequently little arrays of cracks are seen associated with osteocyte lacunae (Reilly 2001). However, in most places they appear in

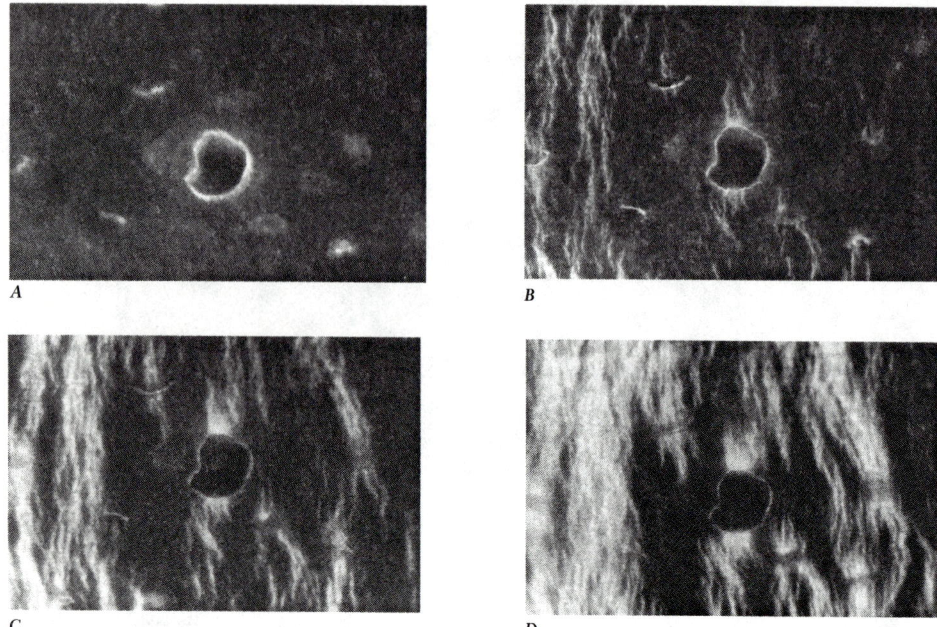

FIG. 3.18 Development of microcracks in a specimen loaded repetitively in bending to higher and higher maximum stresses. The object in the middle is a blood channel. Several osteocyte lacunae can be made out. The loading direction was horizontal, so the plane of the greatest stress is vertical. (A) Stress at the field of view 50 MPa. Nothing seems to be happening. (B) Stress 130 MPa. Some microcracks have appeared, originating from the blood channel and from several of the osteocyte lacunae. There is also a considerable amount of microcracking on the left side of the field of view, which is not particularly related to any obvious stress concentrators. (C) Stress 160 MPa. The development of the microcracking is much more obvious. Microcracking has developed from the osteocyte lacunae, and the microcracking on the left hand side is much more intense. However, some areas (midleft, midright) that were unaffected at 130 MPa still seem to be undamaged. (D) Stress 210 MPa. The specimen broke at this stress, though not in the field of view. Microcracking is very well developed, but still regions to the right and left of the blood channel, and far right, are unaffected. (Photographs by Gwendolen Reilly.)

nondescript parts of the bone. In compression and shear the microcracks seem to form preferentially in the direction of lamellae, and also to follow round the cement lines of secondary osteons (Jepsen et al. 1999).

The formation of this cloud of microcracks may lead to the phenomenon of *microcrack toughening* (Kachanov 1994; Shum and Hutchinson 1990). The precise mechanism of this toughening totally defies simple description. However, it can be thought of as consisting of two parts. One part is the effective blunting of the crack tip, because the strain gradient in its vicinity is diffused over a larger distance. The other part of the toughening comes from the work done in opening up the innumerable microcracks, which are a concomitant of the advance of the crack.

Since there is good evidence that the tensile yield behavior of bone is caused by tiny cracks forming but not growing very much, how does that information allow us to account for the behavior shown in figures 2.7 and 3.15? Each crack in the specimen would increase the specimen's compliance; that is, the specimen would become more flexible. So the increased compliance as the postyield strain increases can be accounted for, as can the fact that the residual strain on unloading is not as great as the strain that has occurred in the yield region. There is some residual strain, which perfectly neat cracks would not produce. However, in such an irregular material as bone many cracks, once formed, would not be able to close up completely when the specimen was unloaded.

The fact that the yield curve is fairly flat-topped is not explained ipso facto by the formation of cracks. However, suppose that microcracks form rather easily and spread a little, but then require a considerably greater stress to spread further. There is little direct evidence for this supposition at the moment, although Akkus and Rimnac (2001) have shown such behavior of cracks in fatigue specimens: cracks start, spread quickly, slow down, and stop. The reason for this reluctance to spread further is not clear. It is a well-known phenomenon in layered composite materials, in which the crack is brought to a halt at the interlayer junction. Whether this is happening with microcracking in bone is unknown. (A feature of cracking in layered materials is that the spacing of the microcracks in a layer is rather uniform. This is caused by the counterintuitive fact that midway between two cracks that are close, relative to the thickness of the layer, compressive stresses are set up that inhibit the formation of a new crack between the two previously existing cracks [Bai et al. 2000]. Although this almost certainly accounts for the behavior of cracks in some mollusk shell materials, I doubt whether it is of significance in bone.)

In any specimen we can imagine many elementary volumes (without specifying what level of organization we are imagining). Some will be weaker and some stronger. When a weak element cracks, the crack will spread a little way, but will then be stopped by some crack-arresting mechanism. The stress in the cross section in which the element broke will be increased a tiny amount because its effective area has been reduced by the crack. This effect will be trivial at first. Usually, the distribution of strength through the volumes in the specimen will be of a few weak ones, many more somewhat stronger ones, and many, many more volumes somewhat stronger. These will be distributed randomly throughout the specimen. The first volume to break is unlikely to have any nearly equally weak volumes around it, and so the formation of one crack is unlikely to cause another. Now the load is increased slightly. Soon another volume will crack, but this could occur anywhere in the specimen. The tendency of any volume to crack will be related to (1) its strength, (2) the amount of the cross section that has already cracked (because this will raise the stress *generally* across the section), and (3) the presence near it of any cracks (because these will modify the stress *locally*, though whether that will increase or decrease the local stress depends on the exact position and orientation of the other cracks). However, when a substantial number of microcracks have formed, a small increase in load will induce, in some section that by chance had a large number of weak volumes, more and more volumes of the section to crack through phenomenon 2, and sooner or later a cascade will occur and the section will break.

Such a set of events would explain why the *stress* does not increase much after the yield point has been reached, and also why there is so much *strain* in the postyield region. Each microcrack is fairly firmly anchored once it has traveled a small distance. It contributes its widow's mite to the overall compliance of the specimen at the moment of its formation, and will also contribute a little bit more as the load increases, not by lengthening, but by deforming. It is clear that to understand what goes on in the yield and postyield region of bone, a much greater understanding of the phenomenon of microcracking is needed than we have at present.

Zioupos (1998) gives an interesting and coherent discussion of these kinds of ideas. He points out that in some materials, including bone, there are situations in which it become *more* difficult for a major crack to travel as it gets longer. This is in contrast to the situation in truly brittle materials, in which the resistance to crack travel *decreases* as the crack gets longer. Peterlik and colleagues in Vienna (unfortunately not yet published) showed that in bovine fibrolamellar bone longitudinal/circumferential cracks, traveling longitudinally, behaved brittly and progressed easily once they had started. On the other hand, radial/circumferential cracks find it harder and harder to progress as they get longer. Such effects are likely to be particularly useful in fatigue loading, where the amount of energy available to push the crack forward is not as great as in impact loading, where there is lots of energy flying around.

It is not obvious how this *increased* resistance to crack travel comes about. The amount of microcracking increases as a crack elongates (Vashishth et al. 1994), but this in itself should not make further crack travel more difficult. In a very tough tissue like antler, as the crack travels many fibers remain bridging the gap, and as the crack opens up these fibers will be stretched and require more energy to extend them. In a brittle material the crack extends more easily as it gets longer because proportionally more material becomes relaxed as the crack extends, and the strain energy released can be fed into the crack front. However, in tough bone (not all bone is tough) the formation of microcracks uses up the strain energy and a long crack is not fed proportionately more energy. It may be that, in some types of bone, as the crack advances a larger volume of material becomes microcracked around the crack tip, and so the energy required becomes greater, not less, as the crack advances. This certainly seems to be the case in video pictures of antler specimens breaking, shot by Andy Sedman in my laboratory. In these pictures the reflected light forms a halo round the crack tip, and it gets bigger and bigger as the crack advances.

3.15 Strain Rate, Creep, and Fatigue: Pulling the Threads Together

One feature of the fracture of bone has been mentioned several times, and now it is time to try to pull the threads together. This feature is that fracture depends to some extent on the *rate* at which it loaded, or *how long* it is loaded, or *how often* it is loaded. The first feature shows itself as an increased stress at fracture at higher strain rates, the others as creep rupture and as fatigue. Do these phenomena have anything that holds them together? One idea is that (1) bone fractures when it undergoes a certain amount of damage, and (2) the rate at which this damage accumulates is a high power of the stress (or strain) level (Carter and Caler 1983). Suppose that these postulates are true. How would they explain creep rupture, strain rate dependence, and fatigue?

Creep rupture is the easiest to understand, and the other two follow from it to some extent. In creep rupture the bone does not break at once on being loaded and held at a constant load, but does break after a time. This accords with the postulates above, since damage will be accumulating during the load-holding period, and eventually enough will occur to make the bone break. Furthermore, the higher the load, the more rapidly the bone accumulates the damage and so the sooner it will fail.

Strain-rate dependence can also be explained according to the two postulates. Creep rupture shows that the longer the time at any stress level, the greater the amount of damage accumulated. If the loading rate is low, the bone will dwell at each stress for a longer time, accumulating more damage. Therefore, the stress by which the fatal amount of damage is accumulated will be lower at low strain rates than at high strain rates.

Cyclical loading, which causes fatigue, can be thought of as a continually interrupted creep rupture test. The damage accumulated during the peak loads remains during the time of unloading, but is added to in the next cycle. It ought to be possible to distinguish between the effects of cycling, as such, and creep rupture, by altering the frequency of the loading cycle. Different frequencies should, according to the two postulates above, have no effect on the fatigue life, because the total time the bone is stressed at any particular level is unaffected by the frequency.

Caler and Carter (1989) analyzed the relationship between creep rupture and fatigue. I shall consider this paper in some detail, because it is very important. Caler and Carter performed the following tests on wet human femoral cortical bone at 37°C:

- Creep rupture in tension
- Creep rupture in compression
- Fatigue in which the bone was loaded repeatedly from zero stress to a tensile stress; this experiment was done at 2 and 0.02 Hz
- Fatigue in which the bone was loaded repeatedly from zero stress to a compressive stress; this experiment was done at 2 and 0.02 Hz
- Fatigue in which the bone was loaded into both tension and compression, but with a mean in tension; this last experiment was carried out at 2 Hz only

In these experiments Caler and Carter used, as the measure of the intensity of loading, stress normalized by dividing by Young's modulus. In effect, before the specimen undergoes damage, this is equivalent to strain. After damage occurs, and the Young's modulus is reduced, there is no neat relationship applicable. It is generally found that normalizing by the initial Young's modulus makes the resulting data much cleaner and statistically more tractable (e.g., Mauch et al. 1992). (In fatigue tests Young's modulus decreases as the number of cycles increases and damage develops [fig. 3.3]; that is why the initial Young's modulus is used.)

The fatigue experiments are done under load control, so the maximum stress is the same in each cycle. If it were done under strain control, the stress would continually decrease as the compliance increased (because of increasing damage in the specimen), and eventually the stress imposed would become tiny.

The creep rupture experiments showed the relationship between applied stress and time to failure for these bones. Both tension and compression showed a straight line relationship between log normalized stress and log time to failure. The compression specimens took about 5 orders of magnitude longer than the tensile specimens to fracture at any particular normalized stress.

Caler and Carter (1989) proposed that the equation describing the curves is $t_f = A(\sigma/E^*)^{-B}$ where t_f is the time to failure, σ/E^* is the normalized stress level, and A and B are parameters. From this equation one can calculate the time to fracture if the bone is loaded at various different stresses, assuming that the effects of the damage accrued at different stresses are additive. One can do this by turning the equation into a damage rate equation, which assumes that the specimen fails when damage = 1, and then by integrating this damage equation over the proportion of time the specimen is loaded to different stresses.

The fatigue experiments in which the bone was loaded repeatedly from zero into tension showed that it took more *cycles* to load the bone to failure at the higher frequency (2 Hz) than at the lower frequency (0.02 Hz). However, the *time* taken to fracture the specimen was independent of loading frequency. (In my laboratory we have shown exactly the same phenomenon, loading being at 0.5 and 5 Hz, respectively [fig. 3.19].) Furthermore, the time taken to failure in Caler and Carter's specimens accorded very closely with the calculated time to failure if it was assumed that failure were simply caused by the accumulation of damage at the various rates appropriate for the varying stresses that the cyclical loading induces. These experiments, therefore, suggested that fatigue is simply the result of loading bone to particular stresses for particular lengths of time; the number of times the loading is changed is irrelevant. Zioupos et al. (2001) corroborated the finding of Caler and Carter for tensile loading.

In the compression experiments things were more difficult. Cycles to failure varied as a function of frequency, and again the specimens fell on the same line as a function of time to failure. However, the calculated time-dependent time to failure, derived from the creep–rupture experiments, was considerably greater than that observed. These experiments in compression remain a puzzle. Caler and Carter are forced to suppose that the number of cycles is important in compression, but not in tension. They are able to produce an equation for the effect of cycles that will allow the data to fit the experiment. However, this is really only an empirical fit. It suggests that tensile failure is a function of time alone, whereas compressive failure is a mixture of time and number of cycles, with the number of cycles being the dominant factor. It was possible to test these suggestions by loading cyclically into both tension and compression, and this Caler and Carter did. The resulting cycles/times to failure did not accord very well with the predictions, leading Caler and Carter to write gloomily: "These data lead us to believe that there is probably a complicated time-dependent and cycle-dependent damage accumulation associated with the complex load histories to which bone is exposed in vitro."

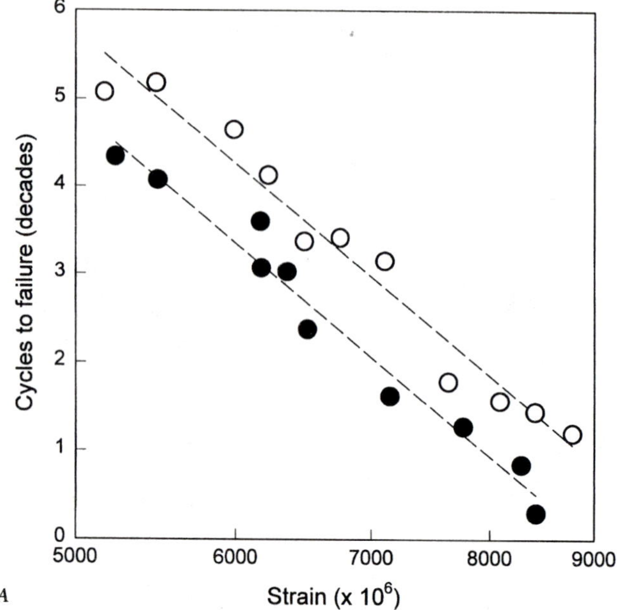

A

Wang and Ker (1995) and Wang et al. (1995) have analyzed creep and fatigue life in which the load is cycled at different frequencies in tendons of the wallaby *Macropus rufogriseus*, and have found that fatigue fracture is the result of a mixture of creep and also weakening due to reversals of load.

Carter and Caler's 1985 model proposed that the rate of damage accumulation occurred as a power law function of stress, and that the exponent was very high, about 17. I found a similar value (Currey 1988, 1989). What is likely to cause such a very strong dependency of damage accumulation on stress (or strain)? Rimnac et al. (1993) explored the possibility that damage in creep rupture (and also in other kinds of tests) is a thermally activated process. Such behavior would obey an Arrhenius relationship of the form $\dot{\varepsilon} = Ae^{-Q/RT}$, where $\dot{\varepsilon}$ is the creep rate, A is a constant, Q is the activation energy for whatever it is that controls the damage, R is the gas constant, and T is the absolute temperature. They performed tests at three different temperatures and found that, in an empirical model incorporating stress, the exponent for stress was 5.2, which is, of course, a good deal less than the values of 17 or so found by the other workers. Neverthless, an exponent of 5.2 is a very high power of stress, and it will be interesting to see whether a refined model will be produced.

Davy and his colleagues (Fondrk et al. 1999a,b) have performed

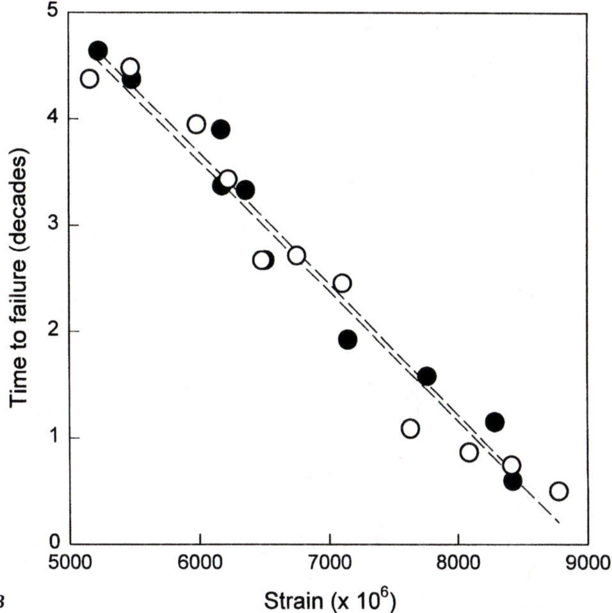

FIG. 3.19 Tensile fatigue behavior of bovine specimens, plotted in two ways. Strain is the strain to which the specimen was *originally* loaded in tension. As the specimen became damaged the strain needed to keep the load the same increased. The specimens were loaded in tension either at 5 Hz (open circles) or at 0.5 Hz (filled circles). (A) *opposite page:* Cycles to failure. Decades refers to powers of 10, so 2 = 100. The regression lines for the two frequencies are of virtually the same slope, but are of different heights. (B) *above:* Time to failure. The distributions for the two frequencies are barely distinguishable. (From Currey et al. [2001].)

many experiments in which they load bone into the region in which it is damaged and unload it rapidly. They do this under load control; that is, it is the load produced by the machine that is servo-controlled, rather than the deformation. They produce a model, based on the assumption that bone undergoes viscoelastic deformation and, when strongly loaded, also undergoes some damage. Using this model they explain three features of the observed behavior of bone:

- The slope of the load–deformation curve is initially straight, and then becomes flatter. (I have discussed this already.)
- In load control, after the load starts to decline, the deformation increases for a while, before declining.
- There is some residual deformation when the load declines to zero, leading also to a permanent increase in volume of the specimen.

Their model explains the behavior of the bone specimens they tested rather well, and does not have too many arbitrary parameters. (Any behavior of anything that can be described quantitatively can be described by a model, if the number of parameters used is great enough. Models with many parameters should be regarded with considerable suspicion.) Nevertheless, the parameters are arbitrary and do not, at the moment, elucidate the underlying mechanisms causing the bone's behavior.

3.16 Fracture in Bone: Conclusions

There are innumerable influences on the strength that bone will actually show in life. In interpreting studies made in the laboratory we must remember, for instance, that static tests, though useful, are different from the situation in which bone usually breaks in life: either in milliseconds or after thousands of cycles of loading and unloading. Strength increases with strain rate. On the other hand, evidence is beginning to appear that some of the variables concerning energy absorption decrease with strain rate. Certainly, the energy absorbed during crack travel decreases with increase in crack velocity. This energy reduction is associated with a much smoother fracture surface. Little work has been done on the effect of temperature on fracture behavior; indeed, most experiments are done at room temperature. Bearing such reservations in mind, we can nevertheless begin to see that bone has a number of design features reducing its likelihood of fracture.

Except in rare cases it is stronger in compression than in tension and, as we shall see in chapter 7, bones are rarely loaded in tension overall. However, if they are loaded in bending, then parts of them will be loaded in tension. Again, we shall see how the skeleton is designed to minimize bending, though it cannot be completely eliminated.

The microscopic structure of bone is such as to reduce the stress-concentrating effects of potential concentrators that must be present. Most bone, except woven bone, has a definite grain, produced by the cooriented cementing together of collagen fibrils and their mineral. This gives bone a microstructure equivalent to that of a fibrous composite with a very high volume fraction of fibers. The cement sheaths around Haversian systems and the layers containing blood channel networks in laminar bone are again structures that, though quite capable of holding the tissue together, form relatively weak interfaces at which cracks may be stopped, diverted, or made to split up, and the general level of energy input needed to keep the crack going increased.

Perhaps the most problematical aspect of the microstructure of bone in relation to fracture is the presence of Haversian systems. In no respect does Haversian bone seem to have mechanical properties superior to those of fibrolamellar bone. The penalty for having Haversian systems is high. A possible advantage is that it allows a reorientation of the grain of the bone if this orientation is wrong for some reason. This may be an important effect. If bone tissue within the substance of a bone is badly oriented, there is no way in which it can be altered except by Haversian remodeling. In the first chapter I discussed briefly how bone could become fossilized with the grain in the wrong direction. However, the work of Riggs et al. (1993a, b) does show that some kinds of Haversian systems are better than others in some loading modes. Unfortunately for the mechanical effectiveness of bone, Haversian systems develop, often extraordinarily densely, in bone that was suitably oriented already. This happens particularly in primates and carnivores. However, once bone has been remodeled, then there may well be an advantage in remodeling to have the best kind of Haversian system, with its fibrils aligned in the best possible way.

Another possible advantage of reconstruction is that it enables the repair of microcracks. It is very difficult to evaluate the importance of this. Aged humans show a considerable amount of microcracking, but the bone of most young animals seems to show very little or none (Riggs et al. 1993b), yet such young bone is able to undergo much microcracking. I deal with the role of microcracking in remodeling in chapter 11.

Indeed, the study of microcracking is one that has developed much in the last few years. Frost reported microcracking many years ago, but not much was done about it. Now, however, many laboratories are investigating it. One of the charms of microcracking, to me anyhow, is that when it occurs it diminishes the most important property of bone, its stiffness, and yet the ability to microcrack under extreme circumstances is immensely valuable, because it prevents bone breaking. Once again we see the conflicting demands for stiffness and toughness. However, bone, unlike a damaged crash helmet, can, if lucky, repair itself, be as stiff as before, and be ready to microcrack all over again if required.

Chapter Four

THE ADAPTATION OF MECHANICAL PROPERTIES TO DIFFERENT FUNCTIONS

IN GENERAL, the mechanical properties of bone material taken from limb bones of adult mammals and birds do not vary a great deal between different parts of the same bone, between bones, or between species. For instance, Biewener (1982) tested whole bones of animals as small as the mouse (0.04 kg body mass) and the painted quail (0.05 kg body mass). The strength of the bone *material* of these animals, and of somewhat larger ones, was about the same as the bone of the horse, the cow, and humans. However, as soon as one strays from testing the limb bone of adult, quadrupedal, terrestrial amniotes, it becomes apparent that bone has quite a range of mechanical properties and that much of this variation is adaptive. This chapter explores some of this variation. The chapter will show us how much bone is able to adapt to different requirements, and what changes in its structure and composition are necessary to produce this variation in the mechanical properties.

4.1 Properties of Bone with Different Functions

First, we look at the properties of three rather different types of bone of adult mammals: deer antlers, mammalian limb bones, and ear bones. The discussion in this section is based mainly on Currey (1979a).

Ordinary limb bones can be taken as "standard." They must be fairly stiff and strong, but also quite good at resisting impact loads. Compared with them, antlers have rather different requirements. In the red deer *Cervus elaphus*, as in all deer except the caribou (or reindeer) *Rangifer tarandus*, antlers are found in the males only. They are grown in the spring and summer, used in the rut in the fall, and are shed in the late winter. During the rutting season males compete to collect, maintain, and impregnate harems. The ability to maintain a harem depends on many factors; among these is the ability to outface an opponent male. For this, antlers are to some extent important insofar as they signal the age and physical state of the bearer. During the display between rivals the antlers' mechanical properties are irrelevant—waterproof cardboard would do as well. However, if two opposing males appear to each other to be closely matched, they may fight. Fighting involves smashing the antlers together, fencing with them, and attempting to make the opponent lose his footing. During the smashing together and fencing the impact properties of the antler are very important. Broken antlers are common; about 30% of deer on the Scottish island of Rhum had some fracture of the antlers by the end of the rut (Clutton-Brock et al. 1979). Broken antlers reduce fighting ability. During the pushing phase of the fight the antlers need to be reasonably stiff, but in this respect they are almost certainly well overdesigned. The stresses imposed during the pushing phase will be much lower than during impact and fencing, and so the static strength will be more than adequate.

Ear bones in mammals have markedly different functions from most other bones. By ear bones I mean the auditory ossicles, the otic bone around the inner ear, and the tympanic bulla. The function of the auditory ossicles is to transmit the vibration of the tympanum to the fenestra ovalis without distortion or loss of energy. In doing this, they usually increase the force of the vibrations and decrease their amplitude. However, in sea mammals, because of the nature of sound waves in water, the amplitude is increased and the force decreased. The ossicles will perform their function best if they are very stiff and if they have a low mass. If they are compliant, they will distort under the influence of the tympanum, rather than moving as a whole, and this will affect the movement of the oval window. If they are massive, it will be difficult for the tympanum to move them at high frequencies because of their inertia. Both these effects will degrade the sound energy reaching the fluid of the middle ear.

The otic bone, surrounding the cochlea, should also be stiff. As the oval window vibrates, the fluid in the cochlea must vibrate to stimulate the hair cells. It is important for the function of the cochlea that the energy of the vibrating window should remain in the cochlear fluid and not be dissipated into the otic bone. The amount of sound energy that is reflected from the interface between the fluid and the bone is determined by the impedance matching between the two media. For two media meeting at a flat surface the proportion of the total energy reflected is given by $[(Z_1 - Z_2)/(Z_1 + Z_2)]^2$, where Z_1 and Z_2 are the acoustic impedances of the two media. $Z = (\rho E)^{1/2}$ for solids and $Z = (\rho K)^{/2}$ for fluids, E, K, and ρ being Young's modulus of elasticity, the bulk modulus of elasticity, and density, respectively (Fraser and Purves 1960). (The bulk modulus has not been mentioned before. It is the ratio of the hydrostatic stress [the same on all faces of a cube] to the relative change in volume. For an isotropic solid it is given by $K = E/3(1 - 2\nu)$. Although bone has a bulk modulus, of course, it is not usually of much interest, and here the bulk modulus of relevance is that of the fluid.) Therefore, to maximize the reflection, the difference between Z_1 and Z_2 should be as great as possible. Of course, the situation in such a complicated structure as the cochlea is not like the planar meeting of two very large volumes of media implied by the formula above. Nevertheless, whatever the particular shape of the interface, the requirement to maximize the difference between Z_1 and Z_2 remains.

To function efficiently as a receptor capable of determining accurately the direction of a sound source, the ear should not be excited by vibration reaching it from the bones of the skull. The inner ears of many mammals, though not humans, have isolating fibrous pads between the otic bone and the rest of the skull, or are isolated in other ways. However this isolation is not complete, and it remains important that bone-borne energy should not pass into the otic bone.

The energy transferred can again be minimized by maximizing the impedance mismatch between the otic bone and the rest of the surrounding tissues. Unfortunately for our purposes, auditory ossicles and the otic bone are small and convoluted, so it is difficult to make test specimens from them. However, the tympanic bulla of the fin whale *Balaenoptera physalus* is large, and it has a mineral content, density, hardness, and general histology much like that of the other ear bones. It is reasonable to think that its mechanical properties will be similar.

The three types of bone—the antler of a red deer *Cervus elaphus*, the femur of a cow, and the tympanic bulla of a fin whale—were tested for a variety of properties. These were work of fracture (the work needed to drive a crack through the material, which gives a good idea of impact resistance), bending strength, Young's modulus of elasticity, mineral content, and dry density. The results are shown in table 4.1. The differences in mechanical properties are very large, and it is likely that they are produced mainly by differences in the amount of mineralization, though there could be some effect of histology; in particular, the bulla material was less regularly arranged than the other two bone types. The

TABLE 4.1
Some Properties of Three Bone Tissues

Property	*Antler*	*Femur*	*Bulla*
Work of fracture (J m^{-2})	6190	2800	20
Bending strength (MPa)	179	247	33
Young's modulus (GPa)	7.4	13.5	31.3
Mineral content (% Ash)	59	67	86
Density (10^3 kg m^{-3})	1.86	2.06	2.47

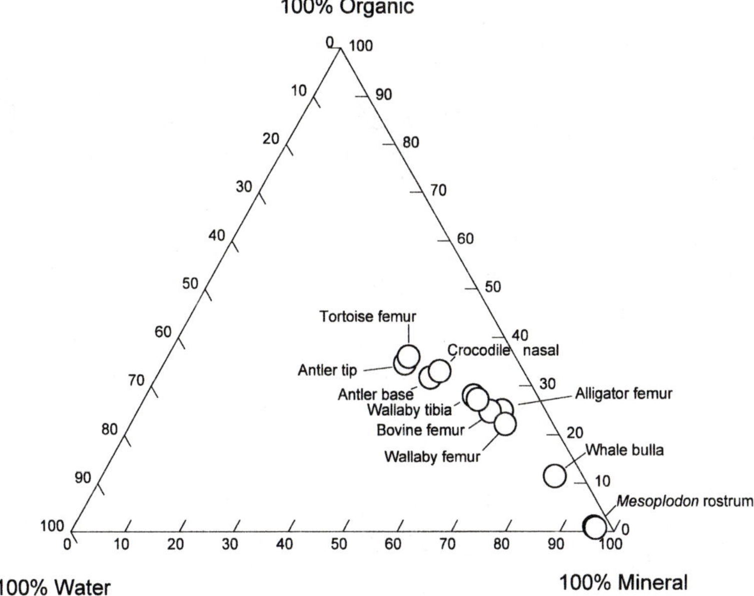

FIG. 4.1 Ternary diagram of the constituents of various bone types, showing the relative proportions, by weight, of the main constituents. (Derived from data in Zioupos et al. [2000].)

differences in the amount of mineralization also affect the amount of water in the bone. Figure 4.1 shows data from our laboratory concerning the relative amounts of water, organic material, and mineral in bones of different origins. In poorly mineralized bones, like antler, there is a high proportion of water. In bones with somewhat higher amounts of mineral, it would seem that the water is replaced by the mineral while organic material declines less (compare antler tip with alligator femur; the water has declined more than the amount of organic material). However, at higher levels of mineralization both water and organic material decline equally. Dry bone is much more brittle than wet bone (Sedlin and Hirsch 1966) even if the ratio of mineral to organic material remains the same. The water allows the collagen to deform more than it would if it were dry. (Dry tendon is brittle, but wet tendon is stretchy.) However, the details of this process have hardly been investigated at all.

The high mineralization of the bulla is the cause of its high modulus. However, it is just this great mineralization (and lack of water) that makes the bulla brittle as well as stiff. As we saw in the last chapter, "ordinary" bone has the ability to develop innumerable tiny cracks, which do not coalesce. This ability makes the bone tough. The bulla has so much mineral that it must behave almost as a single block, in which the thermodynamic criteria for crack extension are easily met. Tiny cracks of small length, which are stable and do not spread, cannot develop in the bulla because here cracks are unstable and spread easily. When seen under the scanning electron microscope, the fracture surface of broken bulla is smooth, quite unlike the pitted and ragged surface of a fracture in ordinary bone.

The antler has a rather low mineralization and a very high work of fracture. In fact, it is exceedingly difficult to break an antler specimen in impact if it is loaded across the grain; it usually deforms into a U-shape but does not fracture. The antler, therefore, has just the properties required of it. Compared with the "standard" bone, the femur, it is rather compliant, but any slight disadvantage this may produce in the pushing match is more than made up for by the very high impact resistance. On the other hand, the bulla, and presumably the other ear bones, have a much higher modulus even than standard bone. The question remains whether this is likely to be significant in hearing. In particular, is the impedance mismatch likely to be significantly enhanced by the high modulus and slightly higher density of auditory bones?

Table 4.2 shows the amount of sound energy reflected at the interface between different media. "Tendon" is as near as we can get to characterizing the collagenous connective tissue between the skull bones and the bone of the otic capsule. The table shows that the interface between air and *any* of the other media we are considering is very good at reflecting sound energy. Therefore, it would make effectively no difference whether the bone material was like femur or was more highly mineralized like the bulla. However, if the otic bone is partially isolated from the rest of the skull by fibrous connective tissue, as is often the case in mammals, the advantage of having high modulus bone is clear. Between connective tissue and ordinary bone there is 23% reflection; between connective tissue and high modulus bone the reflection is about 43%. Similarly, inside the cochlea, the amount reflected between fluid (which can be taken as water for our purposes) and the otic bone would be 31% if the bone were like femur, 50% if like tympanic bulla. Lees et al. (1996) have examined all the bones of the ear of the fin whale and show that they all have a very high acoustic impedance. These authors found that the density, and hence, they supposed, the modulus and impedance, of the otic bones of land mammals was also high, presumably for the same reason as in whales.

Clearly, the high modulus of the bone of the ear structures is important for their various functions. It is bought at great mechanical cost, however. Its work of fracture and bending strength are derisibly low.

TABLE 4.2
Proportion of Sound Energy Reflected Between Different Media

	Air	*Water*	*Tendon*	*Femoral bone*
Water	0.999	—	—	—
Tendon	0.999	0.010	—	—
Femoral bone	1.000	0.307	0.231	—
Bulla	1.000	0.504	0.431	0.066

Note. Calculated from Reflection = $[(Z_1 - Z_2)/(Z_1 + Z_2)]^2$.

However, the ear bones are not usually exposed to large forces, and so their inability to resist impact does not matter. They are so placed that if they were exposed to large loads it would almost certainly be from a blow to the skull that would be mortal. The otic bone in most mammals is almost surrounded by foramina, so that loads passing through the skull are forced to go around the otic bone. This is for good acoustical reasons, as we have seen, but it is very fortunate, to say the least, that the otic bone is difficult to load strongly. Nevertheless, in some mammals, the otic bone is an integral part of the floor of the skull. This is the case in humans and may be because our skull has altered so much in the last few million years that all the changes concomitant with an increase in brain size have not been completed. Whatever the reason, the otic bone is a weak region of the skull and is likely to fracture if the skull is subjected to a hard blow. I said above that the auditory ossicles should, ideally, be stiff yet light. The figures in table 4.1 show that, compared with the femur, a very considerable increase in stiffness (230%) is achieved by a very modest (20%) increase in density.

In summary, these three bones show that bone can exhibit considerable differences in mechanical properties, and where the reasons for these mechanical differences can be made out, the bone appears to be modified in ways suiting it for its various functions.

4.2 A General Survey of Properties

I have made a reasonably wide survey of the mechanical properties of a range of amniote bones (no fish or amphibia are included). These specimens were collected opportunistically, being dictated by what animals died or were culled in zoos. Except for the ossified tendon, all the bone specimens were of the same size and shape and the test was at the same strain rate, so the results are comparable. Apart from the values for the

TABLE 4.3
Mechanical Properties of Various Vertebrate Bony Tissues

Species and tissue	MVf	E	σ_{ult}	ε_{ult}	W
Red deer, immature antler	281	10.0	250	0.109	15.6
Red deer, mature antler	287	7.2	158	0.114	9.3
Reindeer, antler	300	8.1	95	0.051	3.2
Sarus crane, ossified tendon	322	17.7	271	0.062	12.1
Polar bear (3 months), femur	328	6.7	85	0.044	3.2
Narwhal, tusk cement	331	5.3	84	0.060	3.0
Narwhal, tusk dentine	340	10.3	120	0.037	3.7
Sarus crane, tarsometatarsus	341	23.1	218	0.018	2.0
Walrus, humerus	352	14.2	105	0.026	1.4
Fallow deer, radius	360	25.5	213	0.019	2.1
Human, adult, femur	362	16.7	166	0.029	2.8
Bovine, tibia	364	19.7	146	0.018	1.8
Polar bear (9 months), femur	366	11.2	137	0.042	3.3
Leopard, femur	375	21.5	215	0.034	3.4
Brown bear, femur	377	16.9	152	0.032	2.3
Donkey, radius	381	15.3	114	0.020	1.6
Sarus crane, tibiotarsus	382	23.5	254	0.031	4.1
Flamingo, tibiotarsus	382	28.2	212	0.013	1.4
Roe deer, femur	383	18.4	150	0.011	0.9
Polar bear (3.5 years), femur	386	18.5	154	0.022	2.2
King penguin, radius	394	22.1	195	0.010	0.8
Horse, femur	395	24.5	152	0.008	0.5
Polar bear (3 years), femur	397	16.5	142	0.028	2.5
Wallaby, tibia	402	25.4	184	0.010	1.1
Bovine, femur	410	26.1	148	0.004	0.3
Polar bear (7 years), femur	414	22.2	161	0.020	1.7
King penguin, ulna	421	22.9	193	0.011	1.2
Axis deer, femur	428	31.6	221	0.019	2.4
Fallow deer, tibia	430	26.8	131	0.006	0.4
Wallaby, femur	437	21.8	183	0.009	0.8
King penguin, humerus	453	22.8	175	0.008	0.7
Fin whale, bulla	560	34.1	27	0.002	0.02

Note. Testing conditions (cross-head speed, specimen shape, and so on) were essentially the same for all specimens, except ossified tendon, which necessarily had smaller cross sections than the rest. The figures are *median* values of various properties for tensile specimens from various bones from various species. *MVf*, Mineral volume fraction (volume of mineral in parts per thousand. This is probably underestimated in the case of the fin whale's ear bone.) The bones are listed in order of increasing mineral volume fraction. E, Young's modulus of elasticity (GPa); σ_{ult}, ultimate tensile stress (MPa); ε_{ult}, ultimate tensile strain; W, work under the stress–strain curve (MJ m^{-3}). The work under the stress–strain curve may be sensitive to specimen shape and, in brittle materials, to the specimen volume. Therefore, this property should be used only for comparative data, as here, derived from specimens all with the same dimensions.

mechanical properties themselves, table 4.3 gives the value for the mineral volume fraction.

YOUNG'S MODULUS

For vertebrates, it is the stiffness of their mineralized structures that is usually of overriding importance, and very often toughness is much less significant. Young's modulus is a rather structure-insensitive mechanical property, and, in general, there is a good relationship between mineral volume fraction and Young's modulus. For bone, which is essentially a composite of three components (hydroxyapatite, collagen, and water), the value of Young's modulus is determined to quite a high degree by two variables only: calcium content (or mineral volume fraction) and porosity. Figure 4.2 shows the relationship between Young's modulus and mineral volume fraction. Although it is clear, it is not a very clean relationship. However, porosity is also important, as might be expected, and if it is added as an explanatory variable the statistical fit becomes much better.

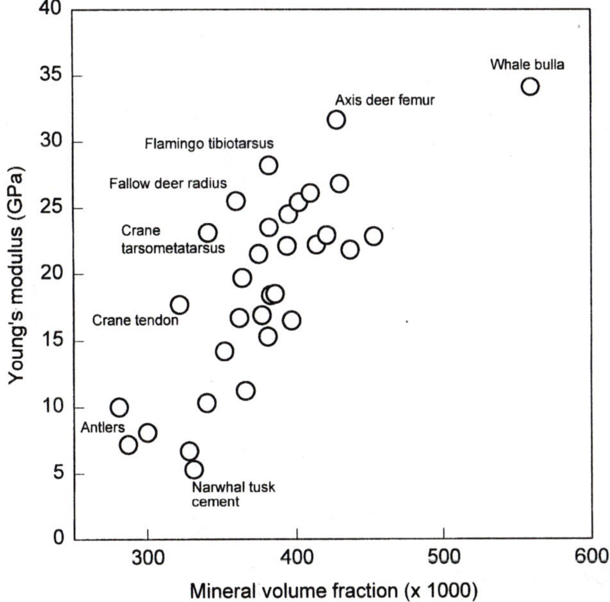

FIG. 4.2 Young's modulus as a function of mineral volume fraction. (Figures 4.2 to 4.6 are all based on the same set of specimens, taken from many species, loaded in tension. The circles refer to median values of a number of specimens from a particular bone. The specimens are a major subset of those shown in Table 4.3. Some outlying or otherwise interesting species are labeled.)

Figure 4.2 shows some outliers to the general trend. The flamingo tibiotarsus, fallow deer radius, axis deer radius, and the crane tarsometatarsus lie somewhat high. It is probable that the reason for the high stiffnesses is that these slender bones are extremely anisotropic, having a grain very highly oriented along the long axis of the bone. The ossified tendon of the Sarus crane is stiffer than one would expect from its mineral content. I shall consider it separately later. The low modulus narwhal tusk cement is made of very loosely organized bone, although it is more highly mineralized than the antlers.

STRENGTH

Tensile strength shows no clear relationship with mineral volume fraction (fig. 4.3). Excluding the antlers and the tendon there is perhaps a trend for increase in strength with increasing mineralization, but the very highly mineralized bulla is extremely weak, as I have discussed before. Some of the bones, like the immature antler, with a rather low mineral content are very strong. We have tested antlers that, though yielding at quite a low stress, show an enormous increase in load after

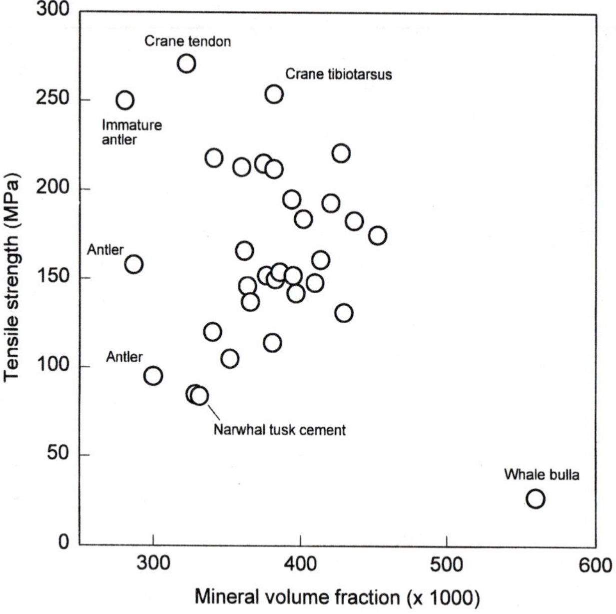

FIG. 4.3 Tensile strength as a function of mineral volume fraction. (For more detail, see legend to fig. 4.2.)

the yield point. Sarus crane ossified tendon is also very strong for its mineral content.

ULTIMATE STRAIN

Ultimate strain has an obvious relationship with mineral volume fraction (fig. 4.4A). The greater the mineral content, the less the strain. The relationship is even clearer if the ultimate strain is logged (fig. 4.4B). The large fracture strain associated with a low mineral volume fraction is produced by generalized strain occurring throughout the specimen, almost certainly through microcracking, and not just by processes occurring near to the fracture.

WORK UNDER THE TENSILE LOADING CURVE

Final strain is not in itself an important adaptive feature of bone. However, it does have a strong effect on the amount of *work* that can be done on a bone before it breaks. An indication of this is shown in figure 4.5, which is the relationship between the mineral volume fraction and the work under the stress–strain curve. This is not a proper fracture mechanics property, because it is rather dependent on specimen shape and volume. However, all the specimens in the table, except the crane tendon, had the same size, so the comparison is valid for these specimens. The eye should not be too seduced by the lower right-hand point, the whale's bulla, but even allowing for this there is a very strong relationship. Again, the relationship is even clearer if the work values are plotted on a log scale (fig. 4.5B). The range of values in the work under the curve is huge, a factor of 40 between greatest and least, even excluding the whale. It is interesting to compare the interrelationships of various properties. For instance, the crane tendon and narwhal tusk cement have very similar ultimate tensile strains (fig. 4.4B), but the crane tendon has a much higher work of fracture (fig. 4.5B). This is because the tensile strength of the cement is much lower (fig. 4.3). Figure 4.6 plots the interrelationships of ultimate strain, Young's modulus, and mineral volume fraction. It shows, as has been implicit in the previous graphs, that as mineral volume fraction increases, Young's modulus increases, but the toughness (as indicated by the strain at failure) decreases. This demonstrates again the feature of mechanics I have often mentioned, that bone cannot be both very stiff and very tough.

The results above relate to tensile loading. A somewhat similar pattern is shown in compression. Les et al. (1995, unfortunately not fully documented) examined the metacarpals of horses of various ages. The bones tended to become stiffer with age and there were considerable

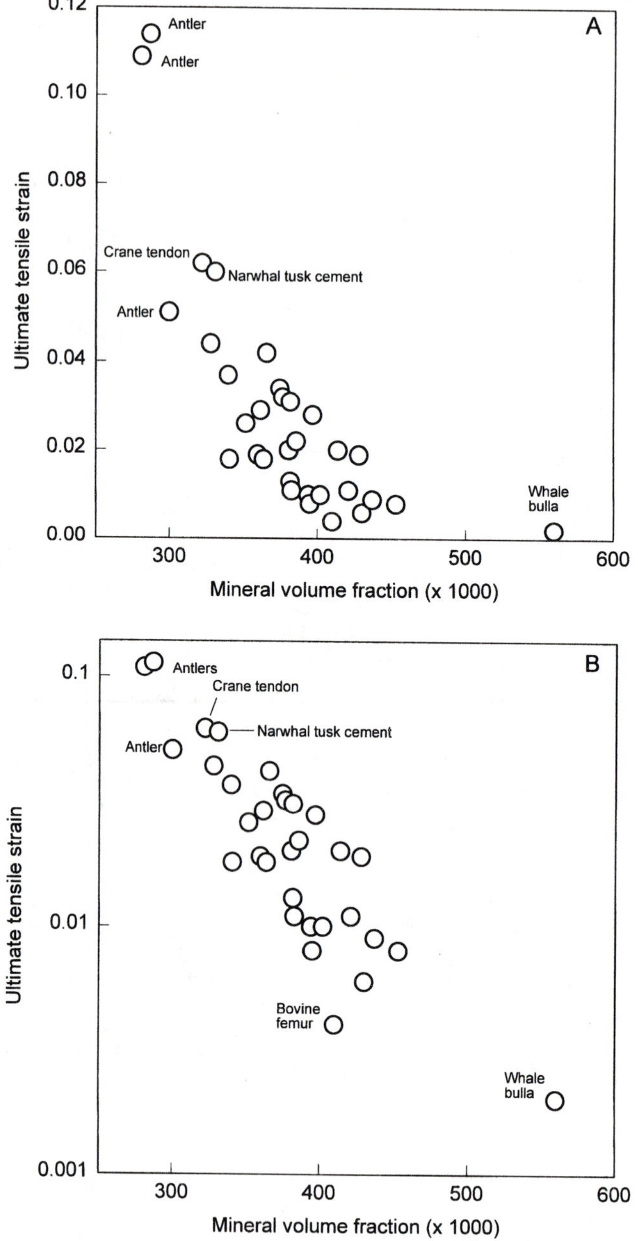

FIG. 4.4 Ultimate tensile strain as a function of mineral volume fraction. In figure 4.4B the ordinate is on a log scale. (For more detail, see legend to fig. 4.2.)

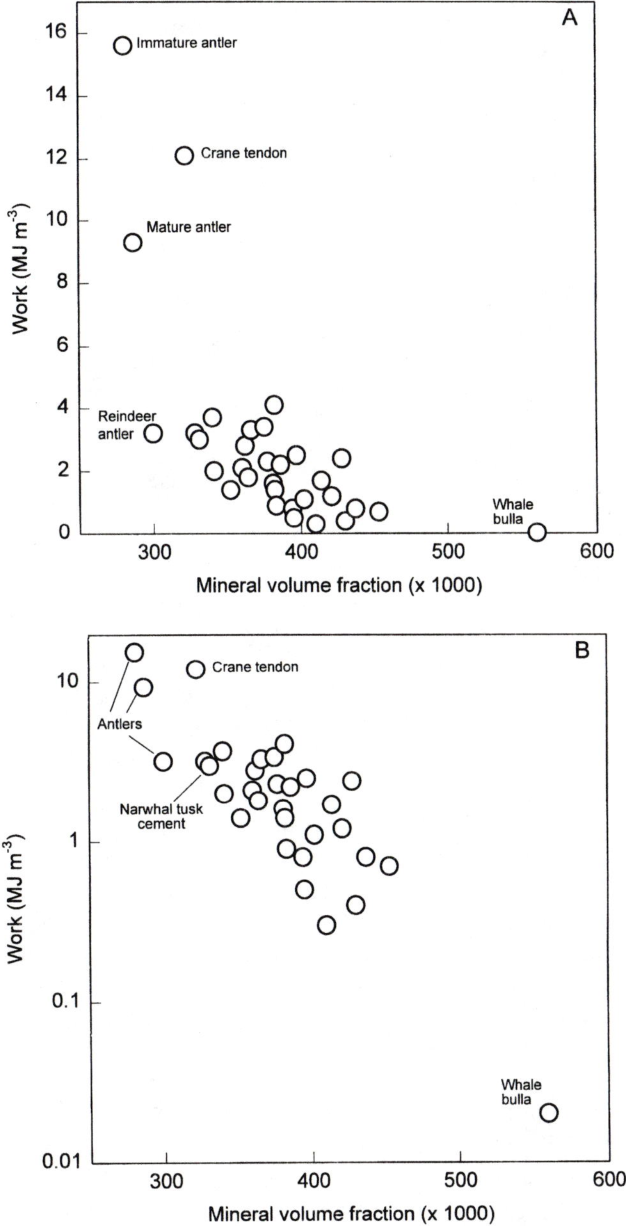

FIG. 4.5 Work as a function of mineral volume fraction. In figure 4.3B the ordinate is on a log scale. (For more detail, see legend to fig. 4.2.)

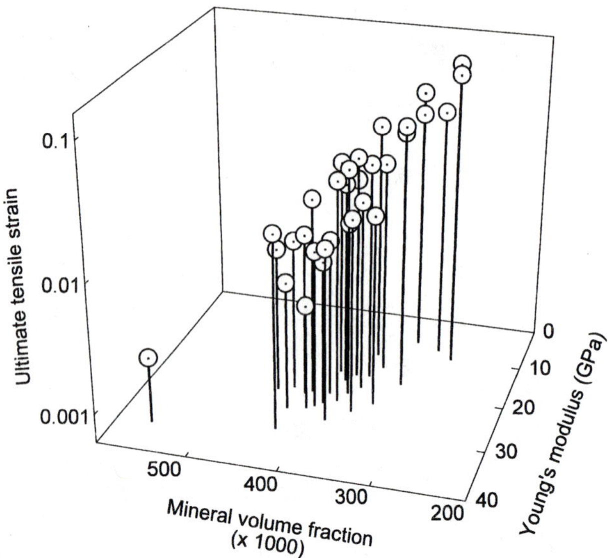

FIG. 4.6 The relationship between ultimate tensile strain, Young's modulus, and mineral volume fraction. (For more detail, see legend to fig. 4.2.)

variations in stiffness round the bone. As a result they were able to obtain specimens with very different values of Young's modulus (ranging from 4 to 20 GPa) from the same anatomical bone. They found that the postyield strain decreased with elastic modulus and that there was a threshold at about 14 GPa. No specimen with a Young's modulus of greater than 14 GPa showed a postyield strain of greater than about 0.004, whereas some specimens with a Young's modulus of 12 GPa had postyield strains of about 0.015. The greatest postyield strains, associated with Young's moduli of less than 8 GPa, were about 0.05. It is not easy to compare my tension results with the results of Les et al. in compression because postyield strain is such a different phenomenon in tension and compression.

All this shows, for a wide variety of species, what was shown for three special bones earlier: Bones cannot be both very tough and very stiff, though ossified tendon does come close to this ideal. Where a particular bone is placed on the relationship is determined by the mechanical requirements of the bone modified to some extent by such things as porosity, which may be a response to another pressure, that is the need to grow quickly.

Figure 3.12 shows the range of bending strengths and Young's modulus of elasticity for a large number of specimens. The reason for their covariation has already been discussed. Here the important thing to note is the extremely large range of both these variables that one can find in nonpathological bone.

Ossified tendon, which is not particularly outlandish (it is to be found in one's chicken or turkey dinner and, less approachably, in many dinosaurs), is a remarkable tissue. It starts off its development as standard tendon, but becomes progressively mineralized. It then usually undergoes secondary remodeling (Amprino 1948). It is extremely anisotropic in both its histology and its mechanical properties. The mineral crystals are very highly oriented along the length of the tendon (Wenk and Heidelbach 1999). The osteocytes (or tenocytes) are extremely elongated spindles, oriented along the axis of the tendon. If viewed normal to its long axis under crossed polaroids a tendon will all go dark at the same orientation. Its fracture strength is very high, and its mode of failure unusual. The transverse cracks that do develop almost immediately turn through a right angle and travel along the specimen to the grips. For a material of such a high stiffness it has a very high value for work under the curve because it has both a high Young's modulus and a large ultimate strain. Uniquely among bony tissues, ossified tendon is stiff, strong, and tough. However, it is not the wonder material that one might expect from this description, because it is also extremely anisotropic in strength. It is very weak in the transverse direction; despite its strength in the longitudinal direction it can be easily peeled apart with the fingers along its length. Although we have no measurements of its stiffnesses in the transverse direction, I suspect it is rather compliant. I shall explore, somewhat fruitlessly, the adaptive significance of the properties of ossified tendon in section 9.2. At the moment it is enough to say that ossified tendon achieves its very high strength by being composed of what are essentially isolated fibers, each of which is thin but strong. The connections between the fibers are sufficiently weak for cracks not to be able to pass from one fiber to the next.

Ossified tendon also occurs quite commonly as a pathological tissue (Fink and Corn 1982; Jecker and Hartwein 1992). In these situations its mechanical behavior is inappropriate for its position, and so is not of interest in relation to adaptations.

4.3 *Mesoplodon* Rostrum: A Puzzle

I must admit that sometimes it does not seem possible to provide any simple adaptive explanation for the mechanical properties of a bone. We have studied the rostrum of the skull of the toothed whale *Mesoplodon densirostris* (Zioupos et al. 1997). This bone is extraordinarily dense, having the highest mineral content and the highest Young's modulus of any bone recorded (46 GPa). Not surprisingly, it is also extremely brittle, and its strength and fatigue behavior are unimpressive (fig. 2.15). The histology of the bone is not totally bizarre in that it is full of blood channels and osteocytes with their interconnecting canaliculi, although there is a tendency for the blood channels eventually to have their lumen filled with mineral, thereby presumably leading to the death of the cells in the vicinity. The fine structure is peculiar; among other things, it appears that there is not the intimate relationship between the collagen and the mineral found in ordinary bone. Instead, the scanty collagen forms thin-walled tubes within which the mineral lies (Zylberberg et al. 1998). Furthermore, Rogers and Zioupos (1999) show by X-ray diffraction that the mineral is extremely well oriented along the length of the rostrum.

It is unlikely that the rostrum is used in fighting because, although it might make a rather efficient club, it is covered with soft tissues, which would damp out the effect of any blows. It might perhaps act as ballast, enabling the animals to sink more rapidly than they might otherwise. Males might possibly use it in some way in acoustic battles with other males. At the moment, its function, in this rarely found whale, is a mystery. It is annoying that this bone, which has a very extreme fine structure and which is mechanically the most extreme bone we have discovered, has an unknown function.

4.4 Property Changes in Ontogeny

The example of all these bones shows two ways in which structure and function vary: between different bones of the same or of different animals. We can also consider the changes that take place during development, because a bone does not necessarily have the same functions throughout life. Currey and Butler in 1975 and Currey in 1979b investigated the mechanical properties of the femur of humans aged from 3.5 to 93 years. Here I shall consider only the first 40 years or so, as later changes are senile and probably not adaptive. We tested the modulus of elasticity, the bending strength, and the impact strength. The results are shown in figure 4.7. The modulus of elasticity increases markedly with age, particularly in the earliest years and the bending strength increases slowly. The impact strength *decreases* very sharply. The changes are associated mainly with changes in the mineralization of the bone: greater mineralization leads to reduced impact strength but to an increase in other values. (Of course, the mineralization in these bones never approaches the very high values found in the ear bones, which are so high that they make the bones weak in bending as well as in impact.)

What do these results mean? Are the changes adaptive? A possible explanation is that the lower stiffness and higher impact strength of the young bones are merely a necessary correlate of their lower mineralization. Bone takes time to become fully mineralized. Natural selection produces a bone with a particular stiffness, and this requires some particular amount of mineralization. The immature bone is on its way to the fully mineralized state and meanwhile happens, fortuitously, to be stronger in impact than adult bone material.

A better way of looking at these results is to consider whether bone might at each age be well adapted. This view has two things to commend it: the differences in mechanical properties are so large that it is unreasonable to think that they will not be subjected to strong selection, and it is more profitable to think in terms of features of organisms being adaptive rather than accidental. To think in this way makes one set up testable hypotheses, whereas, by its nature, thinking that things are accidental prevents further analysis.

During the long time before humans became urbanized, efficiency in locomotion would have been of great importance. For this, high stiffness would be the feature most highly selected. It is, after all, the prime property of bone to be stiff, and the stiffer the bone material, the lighter the bones can be and still allow efficient locomotion. We shall see in chapter 7 that it is possible to quantify the optimum stiffness for bones, so that the mass of bone and muscle is minimized. The stiffness of the bone material turns out to be about right to minimize the mass. We know from our discussion of antlers and ear bones that bone is capable of having a wide range of values of mineralization. The whale's tympanic bulla achieves in a couple of years a mineralization that the femur of man never achieves. It would be *possible*, therefore, for the femur to be stiffer than it actually is by increasing its mineralization. The fact that the stiffness is not greater must mean that the tendency to increased mineralization is balanced, selectively, by some disadvantage. Since the resistance to impact decreases with increasing mineralization, it is probable that this is the counterbalancing disadvantage.

Children are acted on by rather different selective pressures from those acting on their parents. They neither trot for miles and miles after wounded prey nor, when really young, walk for miles and miles collecting berries or fetching water. For children, locomotory efficiency is not very important. On the other hand, they are extremely exploratory and have an alarming tendency to fall over, off, and out of things. For children, bones that do not break easily are of great selective value. The mechanical properties of their bones fit very well with their require-

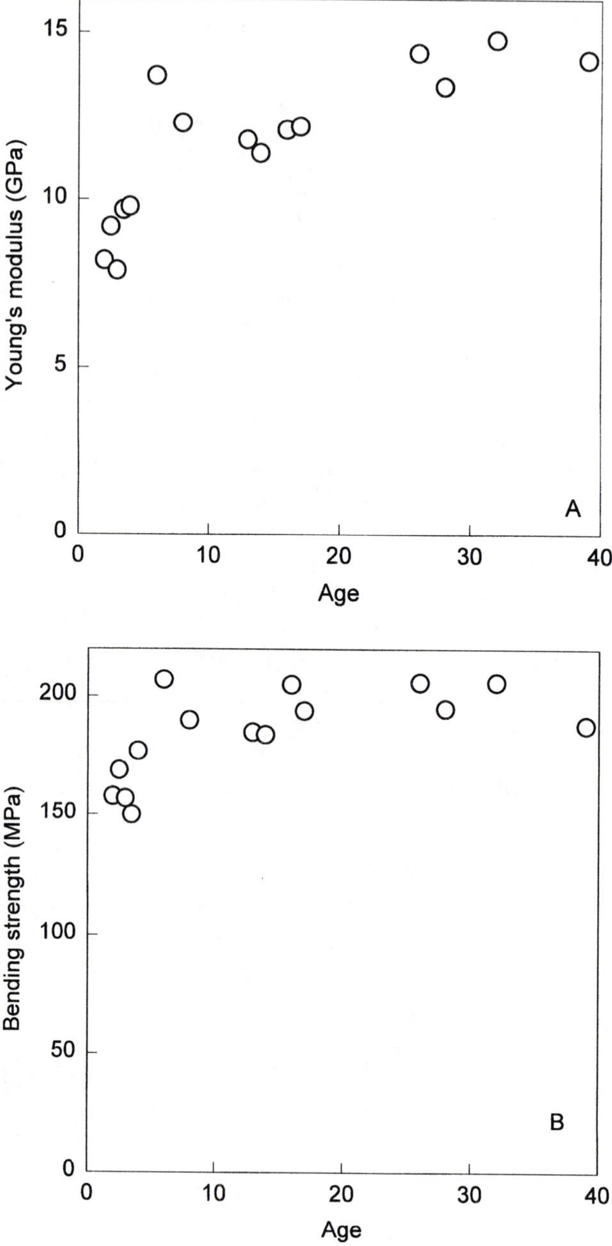

FIG. 4.7 Mean values for three mechanical properties of human femoral bone of various ages. The Young's modulus and bending strength values are taken from the same specimens. The impact energy absorption values are sometimes from the same bones as the specimens for the other two properties, sometimes different. (A) Young's modulus as a function of age. The 6-year-old is anomalously high. (B) Bending strength as a function of

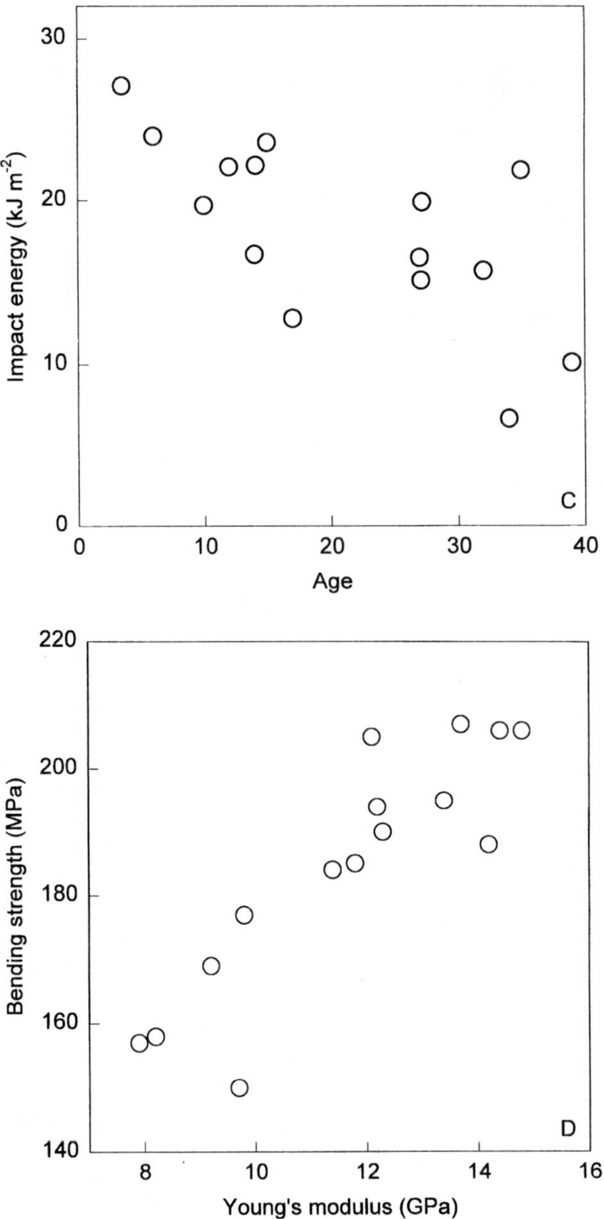

age. The change with age is less marked than was that of Young's modulus, but the young bones are less strong. (C) Impact energy as a function of age. There is a very marked decline. (D) Bending strength as a function of Young's modulus. This shows the usual relationship between these two properties. See Section 3.10 for a discussion of this phenomenon.

ments. Although they are not very stiff, they are remarkably resistant to impact. When children's bones break, they do not break cleanly; often the crack runs out of energy, and a greenstick fracture results, which, even in societies devoid of orthopedic surgeons, may heal pretty well.

This explanation of the ontogeny of the mechanical properties of bone is reasonable, but it is hardly conclusive. However, it is possible to perform a natural experiment by looking at bones that have a different set of selective pressures during life. Ungulates such as sheep and deer must be able to stand very soon after birth, and run within a few hours or so. Life is too serious and predator-ridden for the years of clownish gamboling that so develops the intelligence of the higher primates. Therefore, we would expect the time during which the bones are resistant to impact to be very short.

A striking example of ontogenetic changes is the case of the full-term fetus of the axis deer and its mother (Currey and Pond 1989). The mother, who was aged 2 years and 7 months, had to be destroyed just before she would have given birth, and so we were able to compare genetically closely related animals. The mother's bone had an extremely high Young's modulus (32 GPa) and, correspondingly, the fetus's bone had a Young's modulus that might be found in a 6-year-old human child. The mother's bone had a high tensile strength; this and the high Young's modulus probably being aided by the extremely high orientation of the bone material along the length of the bone. The energy absorption of the deer's bones was very low, however. Deer are even more leggy than sheep and so their extreme properties are not surprising.

In section 7.8, I discuss three other species, the polar bear, the musk ox, and the Californian gull, and show how changes in the material properties during ontogeny are matched by changes in the architecture of the bone as the functional requirements alter. It might still be argued that these apparent differences are the result of the human bone's *actual* age being much less than its *apparent* age, but that the same is not true for sheep. If a long bone is increasing in diameter quickly and keeps a reasonably constant wall thickness, then the bone will be added to on the outside at the same time as it will be eroded on the inside. As a result it will be continually changing its constitution (fig. 4.8). Similar things might be true of sheep's metacarpals, but, in fact, they achieve nearly adult diameter after only a few months' growth. However, studies on the growth of the human femur show that although the external diameter increases rapidly in the earlier years of life, much of the bone in a specimen from the middle of the cortex of a 7-year-old child would have been in existence for several years. Therefore, the marked difference in impact strength between the bones of a child and of a 6-month-old sheep cannot be attributed to immaturity of the child's bone.

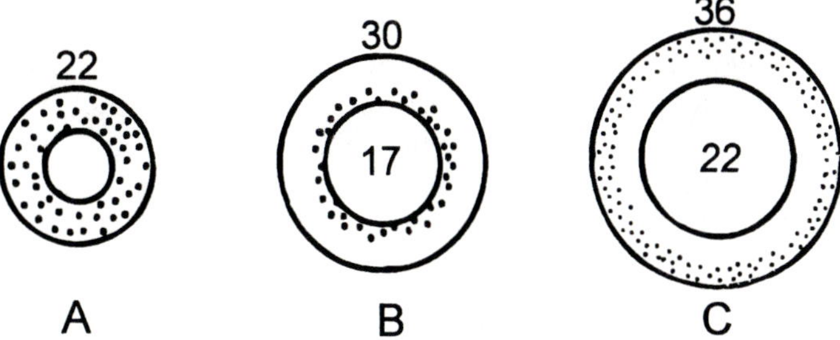

FIG. 4.8 How the age of bone tissue changes with the animal's age. Numbers by the sections show external and internal diameters. (A) The external diameter of the bone is 22 mm. All bone present at this age is shown heavily stippled. At (B) the external diameter is 30 mm; the internal surface is eroded away, leaving less of the original bone (heavily stippled). At (C) none of the original bone is left. New bone laid down between (B) and (C) is shown lightly stippled. As secondary remodeling takes place, there will be islands of young bone in the middle of older bone.

A study by Papadimitriou et al. (1996) on an insectivorous bat *Tadarida brasiliensis* shows not only ontogenetic increases in mineralization but also a decreasing amount of mineralization in the adult along the arm skeleton from proximal to distal. The humerus is the most highly mineralized, the metacarpals less so, and the proximal and middle phalanges even less so. Remarkably, the distal phalanges are completely unmineralized. Papadimitriou et al. produce reasonably convincing reasons why there should be this gradation in the wing skeleton, in which the distal parts can be rather flexible without disadvantage. This is unlike the situation on *Coelurosuravus*, the Permian reptile I mention in section 7.3.2, which has well-mineralized bony elements all the way along its wings.

The low Young's modulus of young bone is carried to an extreme in the fetus. McPherson and Kriewall (1980a,b) show that the bone in the calvaria of the fetal skull has a Young's modulus of about 4 GPa at birth. This value is about one-third of the value for femoral cortical bone of a 12-year-old. Fetal calvarial bone is unlike adult calvarial bone in that it does not consist of two layers of compact bone separated by cancellous bone; rather it is uniform, though somewhat porous. Its very low modulus allows the skull to be molded without cracking as it passes down the birth canal. The father of Tristram Shandy (the hero of one of the most remarkable experimental novels ever written in English) says "that the lax and pliable state of the child's head in parturition, the bones of the cranium having no sutures at that time, was such. . . . it so happened in 49 instances out of 50, the said head was compressed and moulded into the shape of an oblong piece of dough, such as a pastry-cook generally rolls up in order to make a pie of" (Sterne 1767). The fetal skull grows very quickly, and so most of the bone in it must be recently formed. One would expect it, therefore, to be rather lightly mineralized and inevitably to have a low Young's modulus. However, Kriewall et al. (1981) claim that the mineral content at term is not significantly different from that of an adult. This is most improbable. Kriewall and his colleagues did not measure the mineral content of adult bone but took my values (Currey 1969). Since the methods we used were slightly different, it would be best to suspend judgment on this until age comparisons have been carried out using the same specimen size and experimental procedures for all ages. I should be surprised if such a study did not show a clear difference.

Any relationship between age and mechanical properties is complicated by the fact that bone age is not the same as the age of the animal in which the bone sits, because bone is, in humans anyhow, always being remodeled. Once osteoid is laid down and mineralized, bone tissue can be considered to have formed. However, although mineralization initially proceeds very rapidly, almost 70% of the final amount being reached extremely quickly (Ascenzi et al. 1965), it inevitably slows down as the presence of mineral itself impedes the flow of ions though the bone (Wong et al. 1995). The resulting small differences in neighboring parts of the tissue might seem unimportant, but are not, because the mechanical properties of bone vary considerably as a function of mineralization over just the range of mineralizations in which the process becomes slow. Not only may contiguous parts of the bone, such as each side of a boundary between a secondary osteon and interstitial lamellae, be of different ages and therefore have different amounts of mineralization, but parts that are contiguous and of almost the same age may also differ considerably in their mineralization. The age of the bone at a particular level in a secondary osteon is almost the same to within a few weeks. However, mineralization proceeds from the nearest blood vessel, and as the bone becomes more mineralized, the process becomes slower and slower. This results in a density gradient of mineralization in secondary osteons, with the bone nearer the central canal being more highly mineralized than the bone near the outer limit of the osteon (Crofts et al. 1994). As a result there are differences in the mechanical properties within the secondary osteon (Rho et al. 1999). The amount of remodeling differs in different bones, and therefore some may show the effect of aging more than others. Kingsmill et al. (1998) show that the human mandible is more highly mineralized than many other bones in the body, and becomes very highly mineralized in old people.

I hope that some people will have been surprised by this chapter, in discovering what a large range of mechanical properties bone can have. Most of this variation is produced by variation in mineralization and in anisotropy. Perhaps the most important generality to come out of this exploration is that it seems not to be possible to have bone that is both stiff and tough. Increasing stiffness involves the tighter packing of the mineral, which in turn makes the passage of cracks easier. I was taught in the artillery that it was inadvisable, prior to a battle, to set up an observation point in a high but conspicuous church tower: "You pay for the best seats." In biology, compromise is pervasive.

Chapter Five

CANCELLOUS BONE

THIS BOOK HAS until now dealt almost entirely with compact bone. In a trivial sense cancellous bone can be thought of as being compact bone with a large number of large interconnecting holes in it. However, this way of thinking does not take us very far and is often misleading. It turns out that cancellous bone is a highly specialized tissue, extremely different in its mechanical properties from compact bone. Its study takes us to that uneasy, quaking conceptual ground between material and structure where definitions are useless, and where we still have a great amount to learn.

Three major questions can be asked about cancellous bone:

- What are the mechanical properties of cancellous bone *material*?
- What are the mechanical properties of cancellous bone *tissue*, and how are they affected by its density and the arrangement of its bony struts and plates?
- What are the *functions* of cancellous bone?

5.1 Mechanical Properties of Cancellous Bone Material

I distinguish the properties of the *material*, which is the solid bone making up the trabeculae, from the *tissue*, which is the behavior of the whole structure, trabeculae holes and all. There is no reason a priori why the mechanical properties of cancellous bone material should differ much from those of compact bone. It is made of the same constituents, in roughly the same proportions. However, several studies suggest that cancellous bone material is less stiff and, therefore, probably weaker than compact material. This is an important suggestion, because most analyses of the mechanical properties of cancellous bone tissue have to assume some properties for the bone material. What is the evidence that cancellous bone material differs in its mechanical properties from compact bone? Five main methods are used to evaluate these properties: buckling studies, "standard" mechanical tests, nanoindentation, acoustic methods, and back-calculations from finite element analysis.

BUCKLING STUDIES

Cancellous bone trabeculae are small and thin, and difficult to test. However, Runkle and Pugh (1975) and Townsend et al. (1975b) measured the loads required to cause the trabeculae to buckle, when loaded in compression end to end and, using the Euler buckling formula, calculated the Young's modulus. The values these workers found were 9 to 14 GPa. (Euler buckling is the buckling that occurs when a slender rod is loaded in compression. I deal with it in some detail in section 7.7.)

"STANDARD" MECHANICAL TESTS

It is possible, though with great difficulty, to machine test specimens from trabeculae and perform standard mechanical tests on them (Choi et al. 1990; Kuhn et al. 1989; Ryan and Williams 1989). The difficulty with this approach is that in testing very small specimens (of diameter about 200 μm) one is entering a whole new world, in which all sorts of size artifacts may appear. Values from such tests vary between about 1 and 8 GPa, markedly to extremely lower than values obtained from larger specimens of compact bone. Rho et al. (1993) attempted to overcome this problem by testing tiny specimens machined from compact bone, as well as from cancellous bone, and found the cancellous specimens to be considerably less stiff than the compact specimens.

NANOINDENTATION

This method examines the Young's modulus of tiny volumes of bone using a very small indentor, and can be applied to polished trabeculae and cortical bone. Again, the results are varied (Turner et al. 1999; Zysset et al. 1999).

ACOUSTIC METHODS

These were briefly described in section 3.1. The values obtained for Young's modulus are usually higher than those found using mechanical test specimens, and the values for cancellous bone are sometimes considerably less than those for cortical bone (Rho et al. 1993) and sometimes similar (Ashman and Rho 1988; Turner et al. 1999).

BACK-CALCULATIONS FROM FINITE ELEMENT ANALYSIS

Finite element analysis (FEA) has been employed with increasing sophistication (Mente and Lewis 1989; van Rietbergen et al. 1995; van Lenthe et al. 2001). (I say a little about the FEA technique in section 7.10.) If one can characterize the detailed morphology of a block of cancellous bone, one can predict, using FEA, how it should behave on loading, if one also knew its material Young's modulus of elasticity. If the real block is loaded, one can calculate what value of Young's modulus of the material will produce the observed deflections under load. The values obtained from such back-calculations are generally low.

Various other methods that have been used, and the results to date are summarized in table 5.1. The results are a real dog's breakfast, with different workers suggesting very different values. At the moment, perhaps all that can be said is that the consensus is that the Young's modulus for the material of cancellous bone is lower than that of neighboring compact bone, but it is not at all clear by how much. In general, the more recent studies have suggested the higher values for Young's modulus. Guo and Goldstein (1997), in a comprehensive study of this matter suggest that the difference in modulus is about 20–30%. I have a considerable scepticism about the values that are less than about 6 GPa, if only because in my laboratory we have obtained values as great as this for *whole blocks* of cancellous bone, completely discounting whatever effects the porosity itself may have, and these are considerable (Hodgskinson and Currey 1992).

If cancellous bone material has a considerably lower Young's modulus than that of compact bone there is a question as to why this should be so. There are at least two possible reasons. One is that the mineral content of cancellous bone is less than that of neighboring compact bone. There is some evidence for this (Hodgskinson et al. 1989). Although the difference is not large, cancellous bone having about 90–95% of the mineral content of the compact bone, this could have a noticeable effect on the strength and stiffness (section 4.2). Another possible reason is that although the lamellae and the general grain of cancellous bone is roughly along the line of the trabeculae, in detail it is not as well organized as it is in compact bone. This explanation is not convincing, because compact bone loaded at about 90° to its predominant orientation, the worst possible orientation, has a Young's modulus about half that of bone loaded along its grain (Reilly and Burstein 1975). Even very highly oriented lamellar bone, produced in dogs by suppressing remodeling, has an anisotropy of at most about 2:1 (Weiner et al. 1999). Cancellous bone is much better oriented along the trabeculae than compact bone loaded at right angles to its grain. Indeed, Rinnerthaler et al. (1999) show that the mineral crystals in cancellous bone are really very well oriented in relation to the orientation of the trabecular struts.

TABLE 5.1
Various Estimates of the Elastic Properties of Cancellous Bone Material

Authors	*Bone*	*Method*	*Young's modulus (GPa)*
Runkle and Pugh (1975)	Human distal femur	Buckling	8.7 (dry)
Townsend et al. (1975b)	Human proximal femur	Buckling	14.1 (dry) 11.4 (wt)
Ashman and Rho (1988)	Bovine femur Human femur	Ultrasonic	10.9 (wet) 12.7 (wet)
Choi et al. (1990)	Human tibia	3-point bending	4.6 (wet)
	Human tibia: cortical	3-point bending	4.4 (wet)
Kuhn et al. (1989)	Human ilium	3-point bending	3.7 (wet)
	Human ilium: cortical	3point bending	4.8 (wet)
Mente and Lewis (1989)	Human femur	Cantilever + FEA	6.2 (wet)
Ryan and Williams (1989)	Bovine femur	Tension	0.8 (drying)
Jensen et al. (1990)	Human vertebra	Structural analysis	3.8
Rho et al. (1993)	Human tibia	Tension Cortical	10.4 (dry) 18.6 (dry)
Rho et al. (1993)	Human tibia Human tibia	Ultrasound	14.8 (wet) 20.7 (wet)
van Rietbergen et al. (1995)	Human tibia	3-D FEA	6 (wet)
Turner et al. (1999)	Human femur Human femur: cortical Human femur Human femur: cortical	Ultrasound Ultrasound Nanoindentation Nanoindentation	17.5 (wet) 17.7 (wet) 18.1 (dry) 20.0 (dry)
Zysset et al. (1999)	Human femur Human femur: cortical	Nanoindentation Nanoindentation	11.4 (wet) 16.7 (wet)
van Lenthe et al. (2001)	Bovine femur	FEA and ultrasound	4.5 (wet)

Note. All values are for cancellous bone unless stated.

5.2 Mechanical Properties of Cancellous Bone Tissue

The ideas about stiffness and strength applicable to compact bone have to be severely modified for cancellous bone. This is because the arrangement of the struts and little beams (trabeculae) of cancellous bones in space mean that it is behaving as a structure as well as a material. The loads in cancellous bone can be transferred from place to place by bending moments, and compressive loads may cause individual trabeculae to buckle. This buckling may be Euler buckling or, at least in older human bone, by a crease traveling right across the trabecula, as happens in wood loaded in compression. Therefore, an analytical account of cancellous bone would have to explain not only the reaction of very small volumes of bone to force, just as in compact bone, but also how it was that those particular small volumes came to be subjected to those forces. This latter is not a problem with carefully machined test pieces of cortical bone.

There is a serious problem in performing mechanical tests on cancellous bone, and that is that *end effects* are far more important than in compact bone. It is very difficult to perform tensile tests, because of the problems associated with gripping a foamy structure, and superficially it seems rather easy to perform compressive tests. One can machine cubes or similar shapes, and load the specimen between the flat platens of a testing machine. Unfortunately, as Odgaard and his co-workers have demonstrated, the ends of cancellous specimens deform much more than the middle (Odgaard et al. 1989; Odgaard and Linde 1991). This is mainly because the part of the specimen butting up against the platens consists of many isolated struts that are not supported sideways and so deform rather easily. (This is called the *damage* artifact by Keaveny et al. [1993].) Under these conditions a calculation of Young's modulus that assumes that the whole specimen is deforming uniformly produces an underestimate of the stiffness of the bone in the middle of the specimen, which is behaving much more like bone in the bulk of a real bone. There is another end effect, the *friction* artifact. This is caused by friction between the specimen and the platens. The Poisson effect will make a compression specimen tend to spread out laterally. If this spreading is prevented by friction at the ends then, for reasons outlined at the end of section 2.2.2, this will produce an overestimate of the stiffness.

As is always the way, these two effects, although having opposing results on the estimates of stiffness, do not cancel out neatly. Calculations of the degree of underestimation by Keaveny et al. (1993) suggest that it can vary considerably according to the Poisson's ratio of the bone (the greater the Poisson's ratio the greater the friction artifact), the aspect ratio of the specimen (longer thinner specimens have a lower friction and a lower damage artifact, but may buckle), and the size of the specimen (in general, the bigger the better, as long as it is still homogeneous). The average size of the underestimation of Young's modulus for a reasonable-shaped specimen is about 40%. This is a considerable effect and needs to be borne in mind in what follows below. If we are comparing specimens that have roughly the same Poisson's ratio and are not too much concerned about absolute values, then compressive tests are probably sufficient. If absolute values are of prime concern, then some means must be found of measuring the actual deflection of the middle of the specimen. At the moment this is extremely difficult.

A word of warning here. Many authors (e.g., Mente 2000) use the term *bulk modulus* to mean something quite different from the *bulk modulus of elasticity* that I mentioned in section 4.1. The usage is made clear in this quotation from Mente: "*tissue modulus* will be used to refer to the modulus of the bony material that makes up individual trabeculae and *bulk modulus* will refer to the modulus calculated from larger samples (including the effects of architecture and porosity)." A little later in the same book, Rho (2000) uses *bulk modulus* in the sense that I used it in chapter 4. Since both Rho and Mente were discussing the estimation of the stiffness of cancellous bone, it is obvious that the possibility of confusion arises, but the context should usually make things clear. One hopes.

It is possible to measure the elastic properties of cancellous bone using ultrasonic methods. However, the theory of this is not as soundly based as it is with compact bone, because it is not really clear what the velocity of the sound waves depends on (Haire and Langton 1999). The modes of vibration of a mass of interconnecting struts like cancellous bone are very complex, and the rules that apply for a compact mass may not apply to such a structure. However, it does seem that the kinds of answers obtained by ultrasonic techniques and direct mechanical techniques are, fortunately, reasonably similar. The papers mentioned below all used ordinary compression.

In three important papers (Carter and Hayes 1976b, 1977a; Carter et al. 1980) Carter and his co-workers put forward the idea that "all bone can be mechanically viewed as a single material" (Carter and Hayes 1976b). These workers isolated specimens of human and bovine cancellous bone and determined their elastic and strength properties in relation to their apparent density. *Apparent density* is the dry mass of the bone, with marrow removed, divided by the volume of the specimen determined from its outside dimensions. They found that both the strength and the modulus fitted power law equations very well, and that

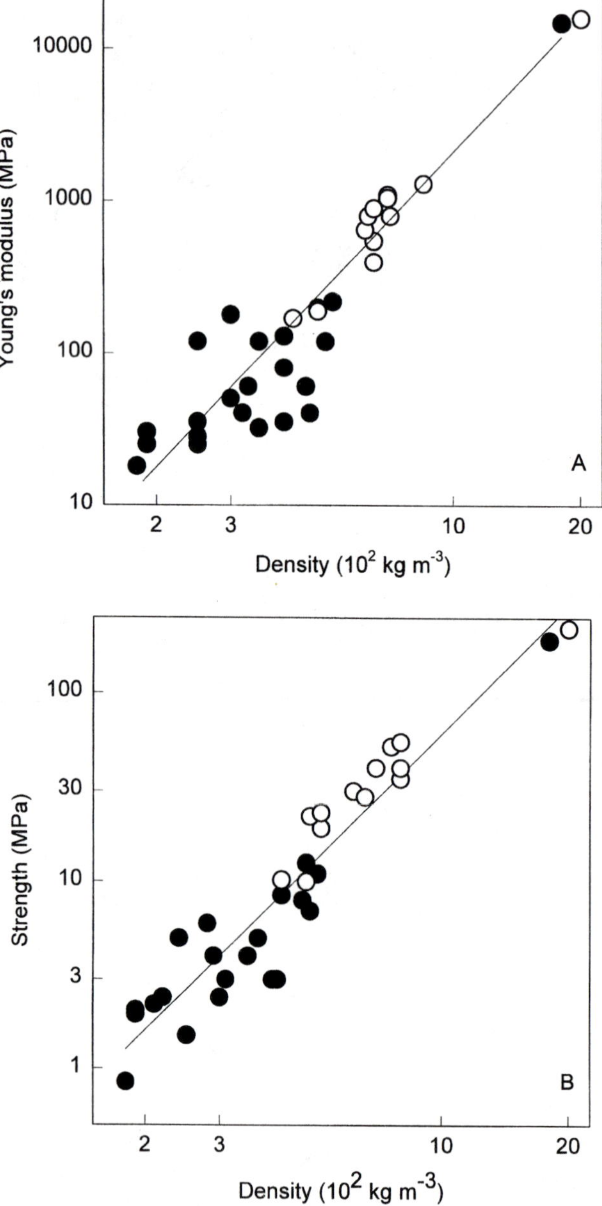

FIG. 5.1 Relationship between the density of bone and its mechanical properties. (Note logarithmic scales on both axes.) Abscissa: density. Open circles: bovines; solid circles: humans. (Derived from Carter and Hayes [1977a].) The symbols at the top right (which are solid bone) have been confirmed by many investigators, and should be heavily weighted in the mind's eye.

compact bone fell on a continuation of the cancellous distribution (fig. 5.1). The relationships were Young's modulus = kD^3, compressive strength = $k'D^2$, where D is the apparent density. This was an empirical observation and did not explain why the mechanical properties varied with density in the way they did.

Since that paper appeared, there have been many studies in which people have tried to relate strength and stiffness to density. Many of the earlier studies were discussed by Rice et al. (1988). This paper is an example of a metaanalysis, in which work by different authors is gathered together, placed on the same footing as best as can be, and reanalyzed. It is a splendid work and shows that the consensus in the literature is that the exponent for both strength and modulus is about 2. Carter and his colleagues, remember, suggested that the exponent for modulus was about 3. Later work (e.g., Hodgskinson and Currey 1992; Yang et al. 1999) has confirmed that the exponent for stiffness is nearer 2 than 3. Keller (1994), who dealt with compressive properties, found that if very high density cancellous bone was included in the anlysis, then the exponent was nearer 3 than 2. That paper has been reanalyzed by Hernandez et al. (2001). They considered cancellous and cortical bone together, a procedure about which I have some reservations, and produced a two-parameter model including the effects of both apparent density and the mineral content of the bone material. Their model equations for both compressive strength and compressive modulus are

$$\text{Compressive strength (MPa)} = 794\ \text{apparent density}^{1.92} \times \text{mineral}^{2.79}$$

$$\text{Young's modulus} = 84\ \text{apparent density}^{2.58} \times \text{mineral}^{2.74}$$

An influential paper (certainly one frequently cited [168 times by August 2001]) concerned with this matter is that of Gibson (1985). Gibson supposed that cancellous bone was constructed in various different ways (cubes made of struts, cubes made of plates, and so on) and then explored, using beam theory, the effects of changing the density on these different structures. The results are shown in table 5.2. She developed theoretical relationships between density and mechanical properties that suggest that different scaling (power) laws apply at different ranges of density. For instance, she proposed that cells will change from being open to being closed at a relative density greater than about 0.2, and that, therefore, the exponent for the density dependence of Young's modulus will change, as predicted by table 5.2, from a quadratic power law to a cubic one. I find Gibson's work entrancing, but the paper as a whole dubious because, I think, there is not enough evidence to show that the changes between the different types of structures occur at the densities she claims. Nevertheless, the ideas Gibson produces are in-

TABLE 5.2
Density Dependence of Young's Modulus and Failure in Cancellous Bone

Behavior	*Deformation mechanism*	*Cubic open cell*	*Cubic closed cell*	*Hexagonal open cell*
Young's modulus	Cell wall bending	$(\rho/\rho_s)^2$	$(\rho/\rho_s)^3$	—
	Axial deformation	—	—	ρ/ρ_s
Elastic collapse	Elastic buckling	$(\rho/\rho_s)^2$	$(\rho/\rho_s)^3$	$(\rho/\rho_s)^2$
Plastic collapse	Plastic yielding	$(\rho/\rho_s)^{1.5}$	$(\rho/\rho_s)^2$	ρ/ρ_s

Source. Taken from Gibson (1985).
Note. The entries in the body of the table give the power dependence of the mechanical property on the ratio of the actual density of the cancellous material to the density of solid bone (ρ/ρ_s). ρ, density; ρ_s; density of solid bone.

tensely interesting, and anyone concerned with the mechanical properties of cancellous bone should be familiar with this paper.

Carter et al. (1980) also found that the Young's modulus and tensile strength obey the same power law as the equivalent compressive properties, and, indeed, the constants k and k' were nearly the same. This is to be expected of Young's modulus; one would not expect the modulus measured in tension and in compression to be different. However, the similarity of the compressive and tensile strengths is not found in fully compact bone, of course. Zysset and Curnier (1996) found effectively no difference between the tensile and compressive Young's moduli measured on the same specimen. However, Røhl et al. (1991) found that although Young's moduli were the same in tension and compression, the strengths were different, being about 27% greater in tension than in compression (once allowance had been made for the statistical differences in density between the specimens tested in different modes). I can only lament this lack of agreement in the literature.

Keaveny et al. (1994a), using a method that eliminated end effect artifacts, found essentially no difference between compressive and tensile moduli of bovine cancellous bone as a function of density. As was to be expected the modulus increases with the density of the cancellous bone, but the distributions of the values of modulus against density were indistinguishable. However, the strengths were different (Keaveny et al. 1994b). Both yield stress and ultimate stress were the same in tension and compression at low moduli and density, but as the modulus (and density) increased, the compressive yield and ultimate stresses became larger than the tensile ones, and at highest moduli compressive loading produced about twice the values compared with tensile loading. They also found that the tensile yield strain was about 70% of the compressive yield strains. Unlike the stresses, yield strain was constant with modulus and density, being about 1.1% in compression and 0.8% in tension.

It is curious that Carter and co-workers, in their earlier papers, found that Young's modulus was roughly proportional to the cube of density, whereas practically everyone else found a near-quadratic relationship. These earlier values got everyone off on the wrong foot; for instance, anyone who (improbably) cares to compare this book with its first edition will note that some theoretical calculations, and some of the conclusions, about the behavior of cancellous bone have had to be changed because of the different exponent that I have now used.

The theory of the relationship between porosity and mechanical properties of porous and foamy materials is not well worked out. There are various empirical studies, usually dealing with rather low porosities. The best, very readable, introduction to the subject is by Gibson and Ashby (1997).

The mechanical properties of cancellous bone are a function mainly of variation in the apparent density and also of the arrangement of the bony trabeculae with respect to the loading direction. Despite the irregularity of the lamellae in cancellous bone being one of the features put forward to explain its apparently feeble mechanical properties, in fact, the grain of the individual lamellae in cancellous bone usually lies more or less along the length of the struts or sheets. Therefore, because the struts are loaded only from their ends (they cannot be loaded significantly via the marrow), the bone material in the struts will be loaded fairly well in the direction of its grain. It will obviously be advantageous that the material should be loaded in the direction of its grain, because it is stiffest and strongest in that direction. Anisotropy in cancellous bone will be produced, not so much by the micron-scale structure of the material, but by the arrangement of the trabeculae themselves. Figure 5.2 shows diagrammatically a small block of rather anisotropic cancellous bone. Obviously, it will be stiffer when loaded in the A-A direction than when loaded in the B-B direction. When the block is loaded in the latter direction, the finely hatched regions will be subjected to larger bending moments than when the block is loaded in the A-A direction. They will therefore deform more and there are more of them in this direction, six as opposed to two in the line of action of the force in this figure, and so the block as a whole will be more compliant.

Cancellous bone can be quite anisotropic. Townsend et al. (1975a) examined the anisotropic elasticity of 18 cubes of cancellous bone from the human patella. The median of the ratio of the greatest to the least value for modulus in each of three orthogonal directions was 1.7. The greatest ratio was 4.8. Linde et al. (1990) testing cubes from the proximal human tibia found a mean ratio for stiffness of 3.7, with a few

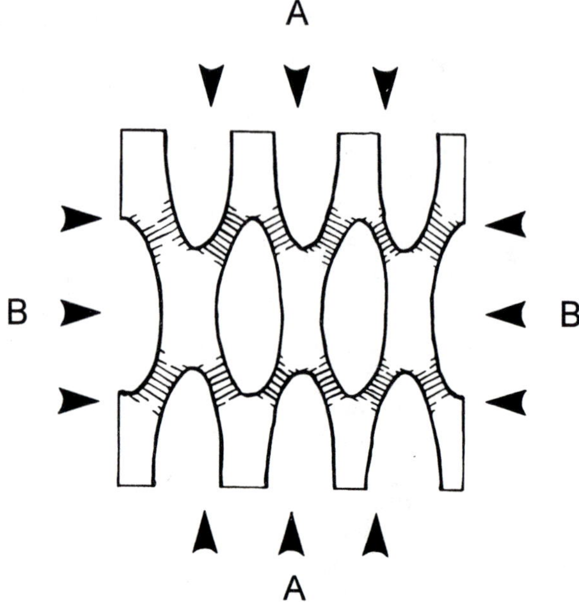

FIG. 5.2 Anisotropy of cancellous bone. The hatched regions are particularly subjected to bending.

cubes having a ratio greater than 8. Hodgskinson and Currey (1990a,b) using cancellous bone from various nonhuman animals found values up to about 7. (It should be remembered that these values are based on testing of cubes in different directions. Poisson's ratio is likely to be different in different directions, and therefore the end-effect artifacts are likely to be different as well, so the differences observed may be actually slightly greater or smaller than those reported.)

Measured values for Young's modulus vary somewhat from study to study, because different types of cancellous bone are being measured. For example, Dalstra et al. (1993) record a maximum value of 280 MPa for human pelvic trabecular bone. Hodgskinson and Currey (1992) reported minimum and maximum values of Young's modulus from 4 to 350 MPa in human material and from 35 to 7000 MPa in nonhuman material, the highest values coming from very dense cancellous bone in the horse. The lower values do not have much meaning because, as one moves further and further away from the load-bearing region of a bone, the cancellous bone becomes so spindly and wispy that the concept of the Young's modulus of a cube made of this bone ceases to be valid.

One can capture much of the mechanically important features of the architecture of cancellous bone with two properties: its apparent density and its fabric. *Fabric* is a measure of the degree of anisotropy of the tissue, and also a measure of the direction of that anisotropy. One can think of fabric as a three-dimensional ellipsoid. The ellipsoid will have three major axes, orthogonal to each other. How these are arranged in space will show the direction of the anisotropy, and the relative lengths of the major axes will show the amount of the anisotropy. The ellipsoid will be long and thin if the trabeculae are oriented mainly in one direction, and will be almost spherical if the trabeculae are arranged randomly.

The relationship between structural density and stiffness has been explored at some length in this chapter and, though not totally straightforward, it is fairly well understood. The question of how one determines fabric or other architectural properties and how one then uses that knowledge to estimate their effects on the mechanical properties, particularly its elastic coefficients, is one of considerable difficulty, both conceptually and computationally. The short introduction by Odgaard (2001) to determining the architecture of cancellous bone is accessible and has a useful reference list. The discussion by van Rietbergen and Huiskes (2001) in the same volume, on relating elastic properties to fabric and other measures of architecture is a good lead-in to this difficult field. In general (Hodgskinson and Currey 1990a,b; Linde et al. 1992), the Young's modulus and the compressive strength of cancellous bone are directly proportional to each other, the strength being about one-hundredth the Young's modulus (fig. 5.3).

There is a property of cancellous bone that attracts a different amount of interest according to the background of the investigator. This is *connectivity*. Pathologists and endocrinologists are much more taken with the importance of this property than are engineers (Odgaard et al. 1999). (Odgaard is a mechanically knowledgeable orthopedic surgeon!) The concept of connectivity is a tricky one and difficult to apply in practice to a structure like cancellous bone. The general idea is that it is a measure of the density with which parts of the structure are connected to each other. If a trabecular strut is cut through, the connectivity of the structure is decreased. Similarly, if a solid body has a hole made right through, it its connectivity is increased. Odgaard and Gunderson (1993) give a good account of the concept of connectivity and show how attempts to infer the connectivity of a specimen of cancellous bone from two-dimentional slices are invalid. In one sense connectivity is a question of scale, because a small part of a highly connected specimen will not look very connected. Indeed, the question of the mechanical importance of connectivity may be simplified to the question of whether it is better to have a few large trabeculae or many thin ones (see section 5.3.8). It has been found in general (e.g., Goulet et al. 1994) that once

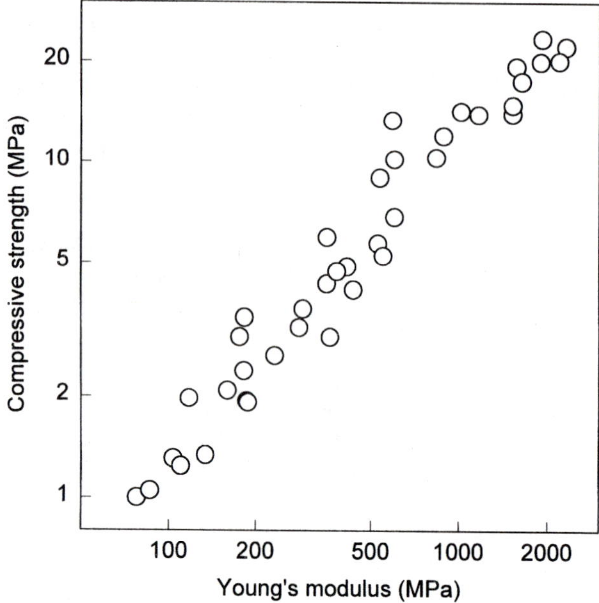

FIG. 5.3. Relationship between Young's modulus and compressive strength of bovine cancellous cubes. Both axes are on log scales. Young's modulus is about 100 times greater than the strength, and this relationship is maintained over the whole range of values. (Derived from data in Hodgskinson and Currey [1990a].)

the amount of bone in specimens has been factored out, differences in connectivity between specimens do not have any important effect on stiffness or strength.

5.3 Functions of Cancellous Bone

Cancellous bone is found in many places in bone. It useful to distinguish five locations:

- At the ends of long bones, under synovial joints
- Where it completely fills short bones
- Where it acts as a filling in flattened bones
- Under protuberances to which tendons attach
- In the medullary cavity of *some* long bones

A glance at any section of cancellous bone shows it to be far from randomly organized. Indeed, it was the apparent relationship between the cancellous bone in the head of a femur, and a crane-shaped bar, that indirectly led Wolff to produce his famous (though unhelpful) "law of

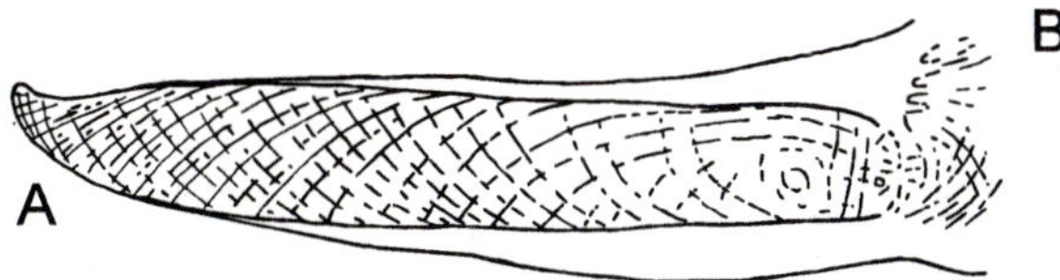

FIG. 5.4 Cross section of the ischium of a horse. This is a fairly flat plate loaded by adductor muscles at its free end (A). The predominant orientation of the trabeculae is sketched in a rather idealized way. In region B the loading system is quite different.

bone transformation" (Wolff 1892). The subject of the intellectual origins of Wolff's law is dealt with most interestingly by Roesler (1981), and a pleasing demolition of its validity is by Cowin (2001c).

Figures 5.4 and 5.5 show two bones in which cancellous bone occurs densely. Two things are shown. First, there is a fairly clear distinction between compact and cancellous bone, there being little bone that it is difficult to classify as one or the other. Second, the trabeculae, as they fan out from the cortical bone, appear to cross each other roughly at right angles. This latter feature is not universal in cancellous bone, but does occur often enough to need explanation.

5.3.1 Principal Stresses

Think of a piece of bone with an arbitrarily complicated set of normal and shear forces acting on it. Inside the bone we can imagine a very small cube, sufficiently small for the forces in it to be uniform throughout. As was mentioned in section 2.2.1, it is always possible to rotate the cube in such a way that there are no shear stresses acting on its faces. There will be three normal stresses (any of which can be zero, of course) acting on each of the opposite pairs of faces, and therefore acting mutually at right angles. These stresses are called the principal stresses.

To simplify the discussion, suppose that we are considering stresses only in a plane, ignoring the third dimension. The two-dimensional sheet we now have has many small areas, each of which could, in the imagination, be rotated so that only principal stresses act on it. It is possible to follow stresses through the sheet; they will change direction in various ways, but at any point there will always be two at right angles. If the bony sheet is solid, then the stresses are not constrained to go in any particular place by the bone itself, but only by the forces acting on the bone. However, suppose that the bone were filled with finely spaced small holes, set sufficiently closely that the general direc-

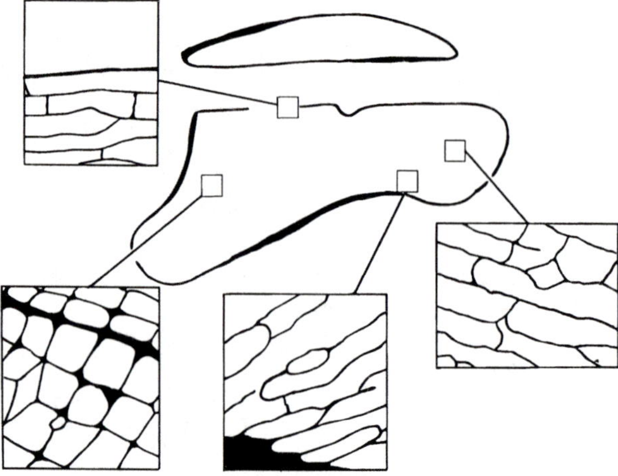

FIG. 5.5 Sagittal section of a cervical vertebra of a horse. The boxes show regions enlarged. Note that, in general, the trabeculae meet at right angles.

tion the lines of stress follow are effectively unaltered from the situation in the solid sheet. Are there any particularly advantageous or disadvantageous ways the holes could be placed? Yes: the holes should be set so that the solid bone remaining lies in the direction of the principal stresses.

As an example, figure 5.6 shows a small portion of a cancellous network consisting of struts left between holes. Each strut joins at its end to other struts at a node. Suppose the length of each strut is 10 units and that each has a square section of side 1 unit. Suppose the nodes A, B, and A*, B* have to carry a load F between them. What will be the maximum stress in the struts connecting the two nodes? In the case of A, B it will be F. In the case of A*, B* it will depend on the value of φ. Take the node B* as being held rigid. The force F can then be decomposed into two forces: $F \cos \varphi$ acting along the strut, and $F \sin \varphi$ acting to produce a bending moment of $10F \sin \varphi$ at the other end of the strut, at the node B*. The maximum stress caused by a bending moment M in a beam is Mc/I, where c is half the depth of the section, and I is the second moment of area of the cross section. Here c and I have values of 0.5 and $1/12$, respectively. Therefore, the greatest tensile stress caused by the off-axis component of F is $60F \sin \varphi$. This will be added to the tensile force to give a total of $F \cos \varphi + 60F \sin \varphi$. When the strut is at right angles to the load's line of action it will, notionally, be bearing a maximum stress of $60F$, which is 60 times greater than the maximum

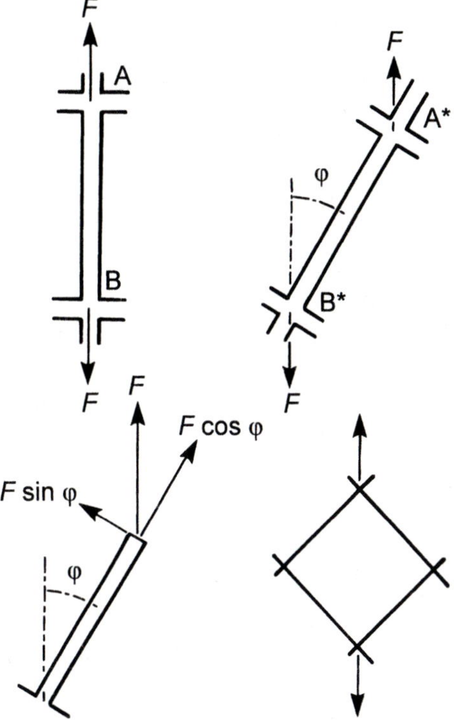

FIG. 5.6 Bending forces in trabecular networks. Explanation in text.

stress when it is aligned with the load. However, the situation is not as bad as this because, so far, I have ignored all the other struts in the meshwork. If φ were 90°, another strut would be in line with *F* and would be bearing the load. In fact, the worst orientation is at $\varphi = 45°$, when two struts are sharing the load borne by one strut when φ is zero (fig. 5.6 lower right). Even so, given the quite arbitrary ratio of length to breadth we have set up, the maximum tensile stress will be 22 times as great as in the strut oriented in line with the force. In fact, things rapidly become quite complex as we consider the other principal stress, which may be the same as, or different from, the first one in both magnitude and sign. However, despite these complications, it remains true that if the struts are oriented in the direction of the principal stresses, they will experience no bending moments, and this will minimize the stress they experience. It is easy to show, by extension, for the three-dimensional case, that all these struts ought to be mutually perpendicular.

5.3.2 Arrangement of Trabeculae in Cancellous Bone

The question of whether the trabeculae in cancellous bone are arranged in relation to the principal stresses has been the subject of much rather desultory argument (Murray 1936; Roesler 1981; Heřt 1994; Biewener et al. 1996; Cowin 1997). One problem is that it is usually very difficult to determine what forces are acting on a bone, so it is even less clear what stresses will be acting inside it. Another, related, problem is that bone with cancellous bone inside it is no longer a homogeneous body, and the distribution of stress is itself constrained by the presence or absence of bony tissue at particular points. Nevertheless, in some bones the distribution of stress is intuitively clear, and we can see whether the cancellous struts are well arranged.

The ischium of the horse has a posterior flange, which is loaded in bending by the action of adductor muscles. Its mechanical situation is like that of a cantilevered beam loaded at its free end. The principal stresses acting in a cantilever are shown in figure 5.7, and the trabecular arrangement in the ischium in figure 5.4.

The spinous process of a thoracic vertebra of a horse is shown in figure 5.8. This spine measured 185 mm dorsoventrally, 12 mm from side to side, and 33 mm anteroposteriorly. The main function of the spine is to bear muscle pulls in the anteroposterior direction. However, because it is about three times longer in this direction than it is laterally, it is likely, quite often, to bend in the direction shown by the small arrows. The trabeculae seem to be arranged to resist such bending.

Vertebral centra are arranged this way, because of the stress-distributing property of the intervertebral disk, which is usually loaded almost axially, even if the spine is flexed. A principal compressive stress will run, therefore, from one end of the centrum to the other. Although the compressive forces are to some extent resisted by the cortical shell, there is no doubt that the cancellous bone will bear the major part of the load (Silva et al. 1997). Possibly, because the body of the centrum is concave along its length, compressive forces between the two ends will cause the walls of the centrum to bow inward, causing radially directed compressive stresses (Frost 1964). This effect may not be very marked, but, however large it is, it will be less than the effect of the end loading in producing axial compressive stresses, and so the expected direction of the principal stresses will be longitudinal and transverse. This is the orientation of the trabeculae. Galante et al. (1970) showed that trabecular plugs from the human centrum were about 2.5 times stronger when loaded in the longitudinal direction than when loaded dorsoventrally or laterally, so the transverse fibers are certainly weak but do have significant strength.

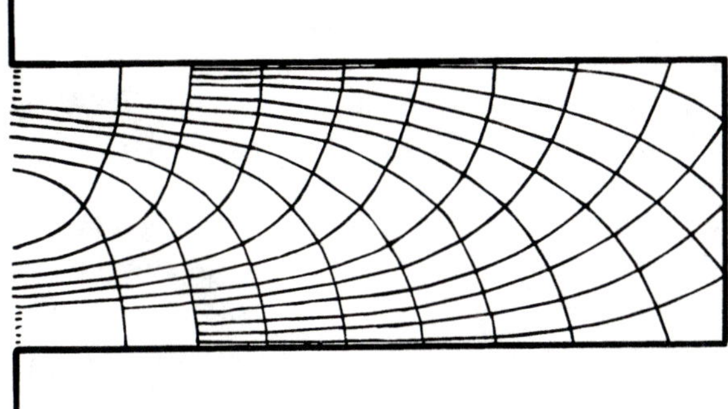

FIG. 5.7 Principal stress trajectories in a cantilever loaded at its free end. The more crowded the trajectories, the greater the stress. They have been omitted for clarity near the root of the cantilever, where the stresses are greatest.

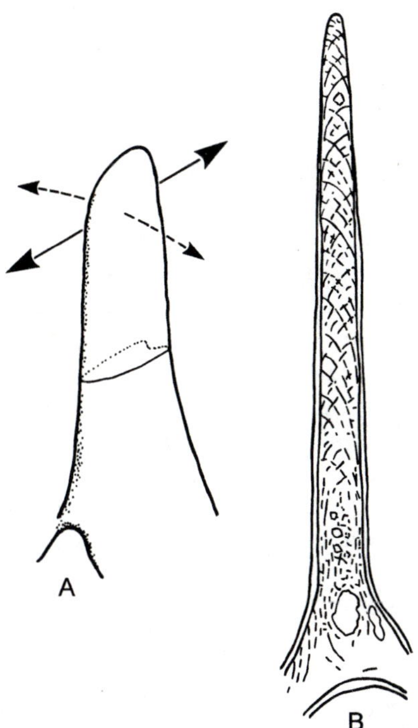

FIG. 5.8 Spinous process of a thoracic vertebra of a horse. An indication of the cross-sectional shape is shown halfway down. (A) The direction of loading. (B) Transverse section showing trabeculae.

In all these bones the trabeculae are oriented in the direction one would expect if they were aligned with the direction of the stresses in a solid body. They seem to be a bony incarnation of the principal stresses, arranged so that they will not be subjected to significant bending moments. However, the trabeculae are not mere embodiments of mathematical abstractions; they have their own lives to lead. If there were no radial stresses in the centrum, and it is not really certain that there are any, it would still be bad design to have only longitudinal struts. These longitudinal trabeculae would have a very high ratio of length to second moment of area, and therefore would undergo Euler buckling at a very low load (see section 7.7). The radial trabeculae may be resisting radial stresses, but they also are pinning the longitudinal trabeculae and preventing buckling. Unfortunately, therefore, the distribution of trabeculae in the vertebral centra has two very different types of explanation.

Until a decade or so ago there were few convincing attempts to model the mechanical effects of different trabecular architectures, because a set of interconnecting struts and sheets is very difficult to analyze. There has, however, recently been an explosion of interest in the possibility of modeling the mechanical behavior of cancellous bone using FEA (section 7.10). The interest is caused by the advent of microCT (microcomputed tomography). Computer tomography allows one to determine the *three-dimensional* structure of solid objects noninvasively. The resolution of the method has now increased to the point where it can resolve individual trabeculae reasonably precisely (Jacobs et al. 1999). The anatomy of the cubes can therefore now be stored on computers, which, with their increased power and speed, can perform calculations that were impossible a few years ago. At recent meetings of biomechanics societies there have been many papers on the subject, and within very few years the relationship between structure and mechanical properties will be much more precisely worked out.

Lanyon (1974) showed a clear similarity between the principal strains, as measured with strain gauges in vivo, on the cortex of a bone and the arrangement of the trabeculae in the cancellous bone beneath the surface. The bone was the calcaneus of the sheep, which is a suitable bone for this experiment because it is loaded symmetrically about its midline, so the flat lateral cortices, to which the strain gauges are stuck, are likely to have the same direction of strain as the trabeculae. Although the directions of the principal strains vary somewhat during the locomotory cycle, the direction of the trabeculae conforms quite closely to that of the principal strains when these are greatest. Biewener et al. (1996) found essentially the same thing in the calcaneus of a marsupial, the small kangaroo-like potoroo *Potorous tridactylus*. Both sets of tra-

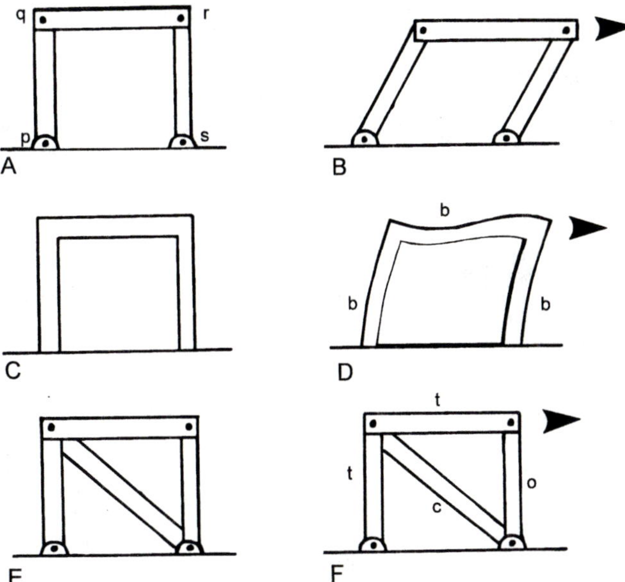

FIG. 5.9 Braced frameworks. The structure in (A) has no sideways rigidity (B). If one or more joints are made rigid (C) the structure resists bending (D) by means of bending stresses set up in its members (b). If one extra diagonal element is put in (E) the structure is rigid (F), resisting the load by tension (t) or compression (c) stresses. The right-hand member (o) is in fact not loaded in this situation.

beculae, which crossed nearly at right angles, were very closely aligned with the direction of the greatest principal compressive and tensile stresses.

So far I have discussed the apparent orthogonality of trabeculae as being related to the directions of the principal strains in a solid bone of which, as it were, the cancellous bone is a skeleton. However, a superficially completely different approach also suggests why the observed structure may be adaptive. This is the consideration of cancellous bone structures as minimum weight braced frameworks.

Look at the structure shown in figure 5.9. In part A it is pin-jointed at p, q, r, and s; that is, it is free to rotate at these points. A sideways load (B) will obviously make it collapse. This can be prevented if one or more of the pin joints are replaced by a rigid joint (C). The sideways load is now resisted by *bending* moments set up in the structure (D). An alternative way of making the pin-jointed structure stable would be to add an extra member (E). The structure now bears the loads as axial stresses, and deforms little under load (F). A structure in which load is resisted by changes in length of its members, rather than by bending, is

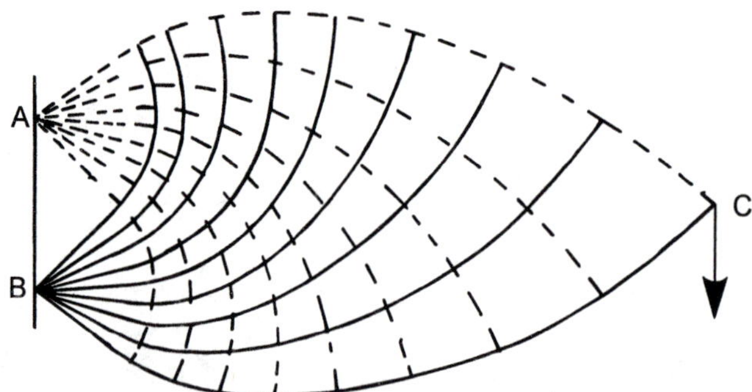

FIG. 5.10 A minimum weight cantilever framework.

known as a braced framework. A clear though old introduction to the subject of braced frameworks is by Parkes (1974). The fact that braced frameworks do not bear bending loads shows the probable relationship they will have with cancellous bone, in which I have shown it will be advantageous not to have bending moments, even though the struts might be capable of bearing them.

The method of deriving braced frameworks of least weight is not simple. What is interesting is the type of structure that results. They are orthogonal nets. Depending on the loading system they may be rectangular or, in systems with concentrated loads, they are often developments of equiangular spirals. Figure 5.10 shows the minimum weight framework for a cantilever, pinned at A and B and loaded at C. The elements that carry tension are shown interrupted; the compression members are shown continuous. The resemblance of this general pattern to, for instance, the pattern of trabeculae in the ischium of the horse is striking (fig. 5.4). Just because it is striking we should remember that there are great differences in the two situations. The cancellous bone is enclosed, as always, between two thin sheets of compact bone, and the loading system is more complicated in the bone. However, the resemblance is such that it is very likely that cancellous bone is often a distorted version of a minimum weight braced framework. Unfortunately, deriving minimum weight frameworks for complex loading systems is very difficult.

Heřt and his co-workers will have none of the idea that cancellous struts in long bones are organized in the direction of the principal stresses (Fiala and Heřt 1993; Heřt 1994). They suggest that the struts are oriented to resist the compressive stresses at the extremes of joint motion. They produce some interesting geometrical arguments showing that if there were struts in all directions, the struts under the greatest load on average will be those oriented at the greatest angle to the long axis of the bone. The reason for this, roughly, is that when the load is along the long axis of the bone, the longitudinal struts are also supported by the obliquely arranged struts in both directions, but that at the extremes the oblique struts are supported only by the longitudinal struts, and barely by the other oblique struts. Therefore, in the process of remodeling, the more lightly loaded longitudinal struts will disappear. Of course, Heřt and co-workers do not deny the existence of struts loaded in tension in such extreme cases as the patella and apophyses, where tendons attach for bone. They do deny it for the generality of situations, however. Pidaparti and Turner (1997) quantify some advantages, related to the reduction of large shear strains, in having nonorthogonal struts when loading comes in various directions over the gait cycle.

Even if the trabeculae are arranged in the direction of the principal stresses, this does not necessarily imply that they are optimally arranged. An instructive example is the distal condyle of the horse third metacarpal. This, the most proximal bone of the "toe" of the horse, is loaded in various directions in the anterior–posterior plane, but hardly at all laterally. Accordingly, the cancellous material is arranged in sheets, oriented along the length of the bone, and in the anterior–posterior direction, with just a few cross-struts holding the sheets in their proper spacing and also preventing lateral Euler buckling. This appears adaptive. However, horses are, in fact, very prone to fractures of this bone. The fracture involves a crack that travels *between* two sets of sheets. Once the cortical shell has split, there are only a few feeble cross-struts to prevent the crack traveling; they usually fail to do so, and a catastrophic fracture results (Boyde et al. 1999). The orientation of the sheets makes the bone stiff in the direction of loading, but the lack of many cross-struts is a potentially fatal weakness.

5.3.3 Joins Between Trabeculae

An obvious feature of cancellous bone is that the joins between trabeculae are well radiused; that is, there is a smooth curve between one strut and the next. The straightforward mechanical adaptationist explanation of this is that it reduces stress concentrations, and so it does. However, there is another somewhat more subtle adaptation. If one considers cancellous bone to be a structure of minimum weight, then one should consider how the weight should be best distributed. It turns out that even if such a thing as a stress concentration did not exist, it would be efficient to have more of the mass of the bone near the joins than in the middle of the lengths of the trabecular struts themselves. The reason for this is that the trabeculae are bent on loading, and the bending moments are greatest where the trabeculae join each other. In the same way that an efficient cantilever has more weight near its root, so, too, should an efficient trabecula. This has been worked out for honeycomblike structures by Simone and Gibson (1998) and for open-foam structures (but not emphasizing the weight-saving features of this design) by Warren and Kraynik (1988). It so happens that in many circumstances in foams a surface called a *Plateau border* develops, driven by surface tension. This produces a curve of constant radius of curvature between the struts or sheets that are joined. Of course, the radiusing of trabeculae is not produced by the same mechanism (although D'Arcy Thompson would no doubt argue that it was), but the mechanical result is similar. The increase in stiffness for the same weight can be considerable. However, it would require detailed histomorphometry to work out how important this effect might be in cancellous bone.

5.3.4 Energy Absorption of Cancellous Bone

The discussion above deals with how the struts of cancellous bone may be related to the loads falling on the bone. However, there is another feature of cancellous bone that may be important in extremis. This is its energy-absorbing ability. The load–deformation curve of cancellous bone has a characteristic shape (Fyhrie and Schaffler 1994). In compressive loading there is an initial linear region, then there is a drop in the curve, and then a long plateau region that is fairly flat (fig. 5.11). When the curve dips, trabeculae have started to buckle. The flattish plateau region (not to be confused with the Plateau border!) is where more and more of the length of the specimen collapses. Eventually, the trabeculae collapse so far that the bone becomes compacted in that region and so resists further collapse, but the bone is hopelessly damaged. In tensile loading, on the other hand, the linear region is much the same as in compression, but once the specimen starts to fail, the trabeculae tear across at one level, and the load–deformation curve is roughly triangular. The energy, shown by the area under the load–deformation curve, is much greater in compression than in tension.

It is often assumed that cancellous bone is good at absorbing energy. The findings of Carter and Hayes and other workers concerning the relationship between density and Young's modulus and strength show that this is not true if we concern ourselves with elastic, preyield behavior. Consider a volume of linearly elastic material loaded to its yield

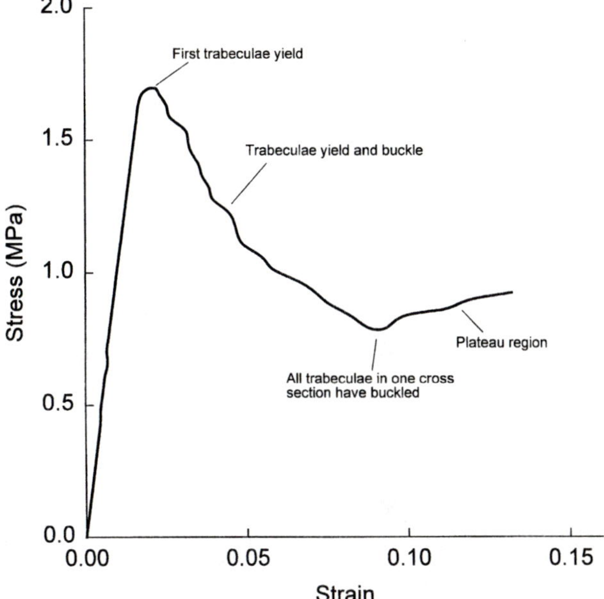

FIG. 5.11 A stress–strain curve for cancellous bone loaded compression. The test was stopped at the right-hand end. The values are characteristic for cancellous bone, but there is great variation in life. (Derived from Fyhrie and Schaffler [1994].)

stress S. If it has a Young's modulus E, then the strain at yield will be S/E. The area under the stress–strain curve, which is equivalent to the work done on the specimen or the energy it absorbs, is (stress × strain)/2 = $S^2/2E$. Since, as we have for cancellous bone, in general, yield stress $\propto \rho^2$, where ρ is density, and Young's modulus also $\propto \rho^2$, so $S^2/2E \propto (\rho^2)^2/\rho^2 \propto \rho^2$. The area under the stress–strain curve will be proportional, then, to the apparent density squared, and denser bone absorbs more energy per unit volume. If selection were favoring a material of minimum mass to absorb energy, it would select for a material of maximum value of $S^2/\rho E \propto \rho^4/\rho\rho^2 = \rho$ (table 7.3). In other words, for the criterion of minimum mass, the denser the bone the better. If mass of fat is taken into account, then cancellous bone is at a further disadvantage, because it will add to the mass of the bone without improving its mechanical properties.

These calculations apply, however, only when elastic deformations are being considered. If the trabeculae buckle in compression, then the cancellous load–deformation curve will enter the long plastic zone and will absorb considerably more energy than would compact bone. Furthermore, cancellous bone often has hematopoietic red marrow, in which case the marrow is useful, not mere packing, and should not be considered simply as useless mass in the bone.

5.3.5 Cancellous Bone in Sandwiches and in Short Bones

Sandwiches and short bones are classic places where cancellous bone occurs, and it is important to analyze what it is doing there. However, discussion of these adaptations will have to wait until after we have considered the role of marrow fat in affecting the optimum placement of bony material. I discuss them in sections 7.3 and 7.4.

5.3.6 Cancellous Bone in Tuberosities

The cancellous bone in protuberances and tuberosities onto which muscles and tendons attach does not have a shock-absorbing function. These bumps and lumps are not loaded in impact by muscles; this is ensured by the compliance of the tendons of the muscles themselves. The cancellous bone merely has the job of leading the forces from the area under the tendon insertion into the nearby compact bone. The struts of the cancellous bone very often show a clear orthogonal arrangement, as in figure 5.12, a section of the greater trochanter of the femur of a cow. The similarity of this arrangement to that of minimum weight frameworks is again striking. Because the line of action of the tendons is often easily determined and often does not vary much through the cycle of movement, the cancellous bone lying under muscle insertions would seem to be useful for testing the idea that cancellous bone is often arranged optimally for weight saving. This has not been done, as far as I know.

5.3.7 Medullary Bone

There is one type of cancellous bone that probably has no mechanical function at all: the medullary bone of many egg-laying birds (Simkiss 1967). It appears to act solely as a reservoir of calcium to be put into eggshells before they are laid (Dacke et al. 1993). It forms a loose, extremely irregular meshwork, packing the medullary cavity of long bones. This bone is considerably more highly mineralized than ordinary bone, and the apatite plates in it are much more irregularly arranged (Ascenzi et al. 1963; Bonucci and Gherardi 1975). Medullary bone is

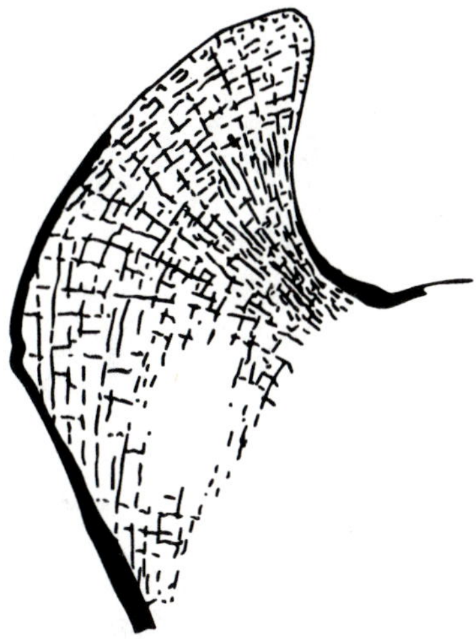

FIG. 5.12 Stylized drawing of a frontal section of the greater trochanter of the femur of a calf, to indicate the main directions and density of the cancellous struts. The blank region has very few trabeculae. At the top, where the adductor tendons insert, the cortex is very thin.

destroyed at the beginning of the egg-laying period, but it is almost simultaneously laid down. The newly formed bone is initially very poorly mineralized, so that there is a net loss of calcium from the bone to the body, which is used for the eggshells (van der Velde et al. 1984). The bone feels extremely soft and mushy to the touch, and apart from having some slight effect in preventing buckling of the thin cortex walls of the bones of the laying chicken, can have no useful mechanical effect. It is clear that, probably uniquely for a bony tissue, its overwhelmingly important function is metabolic not mechanical.

The medullary cavity of many large bones often contains a rather sparse isotropic meshwork of cancellous bone. This arrangement is found, for instance, in horses and dinosaurs. It is very unlikely that the bone is used as a temporary store of mineral. Alan Boyde suggested to me that the trabecular meshwork acts merely as a support for the marrow, which, in the very large space available, might be damaged by the pounding the bone receives if not supported in some way.

5.3.8 The Size of Trabeculae

In all the discussion so far, I have made effectively no comment about the actual size of the trabeculae, except to mention in section 1.6 that their diameter is about 0.1 mm, and their length before they make connections with other trabeculae is about 1 mm. It is as if the volume fraction of the trabeculae, and their orientation, is all that is important. However, it would be possible to have the same amount of trabecular material disposed in fewer, longer, and fatter trabeculae, or more, shorter, and thinner trabeculae. In fact, in nonpathological bone, trabecular shape is remarkably invariant with body size. Swartz et al. (1998) have investigated this matter. Excluding bats, which behave rather differently, they found that in the equations

Trabecular diameter $\propto$ Body mass^a, and
Trabecular length $\propto$Body mass^b

the values for exponents a and b were about 0.06 and 0.05, respectively. The range of body masses they investigated was 4 g to 4×10^7 g. These exponents imply that over the whole ten millionfold range of body masses the diameter increased by a factor of only 3 and the length by a factor of 2.5, resulting in a mass increase of the average trabecula of only about 25.

The Gibson and Ashby book (1997) seems not to make mention of the size (as opposed to the shape) of struts as an important variable in cellular structures, and Lorna Gibson tells me that she sees no reason why, except in a few very special circumstances, it should be important mechanically. However, *very* small struts would probably be disadvantageous, because remodeling would have to be extremely precisely carried out by the osteoclasts if the trabeculae were not to be cut in two. This very characteristic feature of bone structure would repay further analysis.

5.3.9 Cancellous Bone with No Compact Bone

Normally, cancellous bone is covered with at least a thin sheet of compact bone. I have mentioned the mechanical reason for this already and will develop the matter in section 7.4.3. However, one does occasionally find cancellous bone that is not so covered. This condition is seen in some bones of some whales (de Buffrénil and Schoevaert 1988). The periosteal bone in the long bones starts as compact bone, but as the bone grows in width, a wave of endosteal erosion follows it, eventually reaching it so that there is no compact bone left in the adult. Another feature of these bones is that they have no endosteal medullary cavity, and the cross section of the bone is of continuous cancellous bone, although the trabeculae are sparsest in the center and denser toward the periosteal surface. The mechanical significance of this arrangement is (to me) obscure. A very similar histology is also found in fossil ichthyosaurs, which were fast-swimming reptiles, whose general habits were much like those of porpoises (de Buffrénil and Mazin 1990). It is interesting to compare this histology with that of the Sirenia (the manatees and dugongs), another mammalian group whose members spend their whole life at sea. These animals again often have bones without a medullary cavity, but in this case the bones are solidly compact and have no cancellous region. The whale and ichthyosaur situation is no doubt related to their being buoyed up in water, and therefore weight being of little consequence, yet also diving deep and therefore not requiring to be neutrally buoyant at the surface. (Whales' lungs collapse at depth, and therefore neutral buoyancy at the surface would imply considerable negative buoyancy at depth.)

5.4 Conclusion

I hope I have shown that to consider cancellous bone simply as compact bone with a great deal of porosity is so simplistic as to be really misleading. The arrangement of the trabeculae in space often has a satisfying look to it that makes its function seem obvious. Unfortunately, this obviousness is difficult to back up with numbers or real physical insight. The trabeculae are rarely arranged so regularly that analytical solutions of their function are possible. However, with the coming of powerful finite element analysis, we are beginning to obtain estimates of strains in particular parts of the trabecular lattice. When such analyses are married to the interaction of the cancellous bone with the surrounding shell of compact bone, we shall begin to make real progress.

In chapter 7, I shall return to the matter of cancellous bone, particularly from the point of view of minimum mass in whole bones.

Chapter Six

THE PROPERTIES OF ALLIED TISSUES

VERTEBRATES have evolved a number of tissues that, although they are not all called bone, are clearly very closely related to it, being made of collagen mineralized with hydroxyapatite. The tissues I shall deal with in this chapter are calcified cartilage, cement, dentin, and fish scales. I shall also consider enamel, which is very different from bone, having about 97% mineral by weight (Boyde 1989) and more or less lacking collagen. Although nonbony, it does have some extremely interesting structure–function relationships with instructive similarities to what is found in bone and some shell materials in mollusks. The early evolution of vertebrate mineralized tissues is discussed, rather dauntingly, by Smith and Hall (1990).

6.1 Calcified Cartilage

Calcified cartilage is derived from cartilage. Cartilage, existing in a variety of intergrading forms, is a connective tissue made principally of collagen (type II collagen, not type I as is found in bone), proteoglycans, and water. The mechanical properties of cartilage arise from the interplay of these three materials. The proteoglycans consist of a core of protein with floppy keratan sulfate and chondroitin sulfate side chains. These side chains interact weakly with water molecules. Three quarters of the water in cartilage can be bound by these side chains, making the cartilage gellike. The collagen fibers provide a more coherent resistance to loading. Cartilage has no vasculature and the embedded cells, chondrocytes, are nourished by diffusion through the watery matrix. The mechanics of cartilage are seriously difficult, because it undergoes large strains and is very viscoelastic. A flavor of the difficulties can be gained from part II (Cartilage Biomechanics) of volume I of Mow et al. (1990), although this deals only with joint cartilage. I deal a little more with cartilage mechanics in chapter 8.

Cartilage can become mineralized, and the mechanical properties of mineralized (calcified) cartilage are of some interest, but are hardly known. Calcified cartilage appears in various places in vertebrates but two places are of particular importance: at the ends of growing long bones, and in the general skeleton of many sharks.

In so-called endochondral ossification, characteristically seen in the epiphyseal plates of long bones, the bone is preceded in time by cartilage. After a while the cartilage cells hypertrophy, and the cartilaginous matrix becomes calcified. The mineral involved in the calcification is a very poorly crystalline version of apatite (Rey et al. 1991). The crystals are randomly oriented in relation to the collagen fibers. The calcified cartilage is then invaded by cells that destroy most of it, and by other cells that produce bone instead. So cartilage is replaced by bone. While the calcified cartilage is present it acts to some extent as a skeletal material. An attempt to measure its elastic modulus was made by Mente and Lewis (1994), using little beams of calcified cartilage and of subchondral bone, the bone lying immediately beneath the cartilage. They used beam theory to account for the differences in modulus observed when the calcified cartilage occupied a greater or lesser proportion of the depth of the beam. They found that the calcified cartilage was considerably less stiff than the subchondral bone, their Young's moduli being 0.35 and 5.7 GPa, respectively. A modulus of 0.35 GPa is extremely low; tendon has a Young's modulus in tension of about 1 GPa. Mente and Lewis suggest that in this way it acts, in synovial joints, as a transitional zone of intermediate stiffness.

Calcified cartilage is not usually mechanically a very high-quality material. For instance, the costal cartilages of dogs are initially cartilaginous. They then become mineralized, and then finally are replaced by bone. Fukuda (1988) showed that as soon as the cartilage becomes mineralized, it starts to develop cracks. This certainly suggests that this particular type of cartilage is not mechanically very well adapted to anything, though presumably the fact that it is stiffer than the cartilage it replaces is advantageous as the dogs become larger. This lack of mechanical excellence in this not very critically placed tissue is probably compensated for, in terms of natural selection, by some other property, such as rapidity of construction.

Of more general interest, because it is not a transient structure, is the calcified cartilage found in many sharks. The Chondrichthyes, the cartilaginous fish, have skeletons usually made of cartilage. Sometimes, however, the cartilage becomes calcified, and therefore much stiffer. The calcified cartilage contains layers of prismatic calcium phosphate, probably apatite. The cells of calcified cartilage often remain alive, unlike those in the cartilage of developing bone. Typically, the layers appear only on the outer surface of the cartilage, not in the interior. Moss (1977) writes, "the hydroxyapatite deposits characteristically form a series of subsurface plaques, which seem to serve as points of attachment of uncalcified perichondral collagen fiber bundles. It is almost as if these calcified plaques served as 'staples.'" Dingerkus et al. (1991) showed that in large sharks, such as the great white shark *Carcharodon car-*

charias and the tiger shark *Galeocerdo cuvieri*, smaller individuals had totally cartilaginous jaws, but that as the animals grew larger they developed first one and then many layers of mineral. Presumably, this is an adaptation to the need for stiff jaws in large animals that exert large forces. It is not merely an aging change, because smaller species in the same genera never develop calcified cartilage, even when totally mature.

Sometimes the loads on the skeleton are very large, and in one case the sharks have developed an extraordinary convergence with what happens in trabecular bone. The cownose ray *Rhinoptera bonasus* crushes hard-shelled prey between mineralized tooth plates. Summers (2000) shows that not only do the tooth plates have many layers of mineral on the surface, but they have internal mineralized trabeculae that look remarkably like those found in the ends of long bones or in flat bones. The trabeculae are surrounded by the cartilage, and seem well arranged to resist the loads imposed on them. One particularly interesting thing about these trabeculae is that they are often hollow; that is, they are hollow cylinders. This is an efficient way of constructing trabeculae, because it reduces the risk of buckling. Indeed, it is somewhat surprising that the trabeculae of cancellous bone seem never to be hollow.

6.2 Collagenous Tissues of Teeth

Teeth in mammals, and many other vertebrates, are made of three main components: enamel, dentin, and cement. The enamel usually forms a cap on the outside of the visible part of the tooth, and is underlain by a base of dentin. Enamel is laid down by cells called ameloblasts, and dentin by odontoblasts. During the development of teeth these cells initially form two apposed layers. As they lay down their respective tissues the two layers move apart from each other. As the tooth erupts the exposed ameloblasts disappear, and no more enamel can be laid down. The odontoblasts, as they lay down dentin, move toward the center of the tooth. A mature tooth consists of an outer cell-free layer of enamel, and a hollow base of dentin in which there are no cell bodies. This base extends well below the cap of enamel, as the root, and lies in a bony socket, or alveolus, in the jaw The central lumen is called the pulp cavity.

6.2.1 *Cement*

There needs to be a tissue attached to the dentin in the alveolus that can be remodeled during growth and movement of the teeth. This is provided by the cement, which is attached to the outside of the dentin, and serves to hold the collagen fibers of the periodontal ligament, which unites the teeth to the bone in which they lie.

Cement is essentially a modified bone, with osteocyte-like cells (cementocytes) connecting via canaliculi, to the periodontal ligament, which spans the gap between the tooth socket and the tooth. (There would be no point in cementocytes connecting to the dentinal surface, because there are no life-giving cells there.) The collagen in cement is partially produced by the cementocytes themselves, but a fairly large proportion is derived from the periodontal ligament. Electron microscope pictures show the cementocyte-derived fibers wrapping round the periodontal fibers. Both become fully mineralized. Often a very thin layer of cement, adjacent to the dentin, is acellular. The complexities of cement development are described by Bosshardt and Selvig (1997). Yamamoto et al. (2000) show very clearly how some human cement has a plywood-like structure very like that of normal lamellar bone.

Cement usually occurs just as a thin film, and therefore little is known about its mechanical properties. However, in my laboratory we have been able to examine it in narwhal tusk, where it is thick enough to make test specimens. I shall consider it along with narwhal dentin. (Cement should not be confused with the "cement line" or sheath, which surrounds Haversian systems.)

6.2.2 Dentin

Dentin is a major constituent of teeth and has a constitution very like bone, so its similarities to and differences from ordinary bone are of great interest. Possibly dentin was the first kind of bonelike vertebrate mineralized tissue to be developed; it had certainly been developed by the Ordovician, about 450 million years ago (Smith et al. 1996). A good general description of dentin and of teeth in general is by Ten Cate (1989).

Dentin is a mineralized tissue whose overall composition is roughly the same as that of bone, human dentin having about 20% by weight of collagen and other organic material, 70% mineral, mainly in the form of hydroxyapatite, and 10% water (Frank and Nalbandian 1989). However, the *histology* of dentin is somewhat different from that of bone. There are no cells in dentin. Instead, the bodies of the odontoblasts, which lay down the tissue, remain outside it (in the pulp cavity), but leave long, rather straight "dentinal" processes containing living tissue, which permeate the tissue. The orientation of the collagen fibrils, and that of their associated hydroxyapatite crystals, is normal to that of the dentinal tubules. Very often they are not only normal but, at

TABLE 6.1
Young's Modulus of Various Dentins, and of Cement Where Stated

Tissue and loading mode	*Value (GPa)*	*Authors*	*Tissue and loading mode*	*Value (GPa)*	*Authors*
Tensile			**Indentation**		
Human	19.3	1	Human molar	8.7–11.2	15
Human	11.0	2	Human (crown) peritubular	29 dry	14
Bovine	13.1	13	Human (crown) intertubular	20 dry	14
Human	16.6	5	Human (crown) intertubular	15 wet	14
Human	13.7	11	**Compression**		
Human	12.5	9	Human (crown)	12.6	10
Human (crown)	12.5	6	Human (root)	12.8	10
Human (root)	7.8	6	**Bending**		
Narwhal	8.9	7	Human molar	10.2	16
Elephant tusk	9.6	18	Human	16.7	3
Human cement	2.4	20	Elephant tusk	7.7	18
Narwhal cement	6.9	7	*Bovine bone*	*24.0*	12
Nano-indentation					
Human molar	19.9	19			

Note. Authors: 1, Bowen and Rodriguez (1962); 2, Lehman (1967); 3, Renson and Braden (1975), 6, Stanford et al. (1960); 7, Brear et al. (1993); 10, Huang et al. (1992); 11, Sano et al. (1995), 12, Reilly and Burstein (1975); 13, Sano et al. (1994); 14, Kinney et al. (1999); 15, Meredith et al. (1996); 16, Jameson et al. (1993); 18, unpublished observations in the author's laboratory; 19, Mahoney et al.(2000); 20, Yu et al.(1999) [large data set, but experimental details extremely scanty].

any particular point through the thickness of the dentinal wall, extremely uniformly oriented in one direction. The dentin immediately surrounding the tubules, the peritubular dentin, has a high mineral content, lacks collagen, and is much harder and stiffer than the rest of the dentin, the intertubular dentin (Kodaka et al. 1991; Kinney et al. 1999).

Dentin grows from the outside in, reducing the size of the pulp cavity as it increases in thickness. The mineralizing front does not advance uniformly, as is the case in, say, lamellar bone (Francillon-Vieillot et al. 1990). Instead, mineral is deposited initially in the form of *calcospherites* (Ten Cate 1989). At first these are roughly spherical heaps of crystals, pierced by dentinal tubules in one direction, and by collagen fibrils in another direction. The calcospherites grow to about 10 μm

TABLE 6.2
Fracture Properties of Various Dentins and of Narwhal Cement

Tissue		*Authors*	*Tissue*		*Authors*
Tensile strength (MPa)			**Compressive strength (MPa)**		
Human molar	37	16	Human (crown)	295	10
Human	52	1	Human (root)	251	10
Bovine	59	1	*Mature bovine*		
Human	41	2	*bone*	294	12
Human	106	11	**Work of fracture (kJ m^{-2})**		
Narwhal	120	7	Human, paral-		
Narwhal cement	140	7	lel to tubules	0.55	8
Bovine	129	13	Human, normal		
Mature bovine			to tubules	0.27	8
bone	160	12	Narwhal	1.7	7
Bending strength (MPa)			Narwhal		
Human molar	166	16	cement	3.1	7
Human	131	3	*Mature bovine*		
Elephant tusk	115	18	*bone*	0.9	18

Note. Authors: 1, Bowen and Rodriguez (1962); 2, Lehman (1967); 3, Renson and Braden (1975); 7, Brear et al. (1993); 9, Watts et al. (1987); 10, Huang et al. (1992); 11, Sano et al. (1995); 12, Reilly and Burstein (1975); 13, Sano et al. (1994); 16, Jameson et al. (1993); 18, unpublished observations in the author's laboratory.

diameter, abut, and often fuse. When they abut they tend, in general, to meet along rather flat planes. However, sometimes they do not fuse completely, but leave a hypomineralized space between them. These hypomineralized regions are called *interglobular dentin*. In humans this lack of fusion is probably to some extent pathological, being found particularly in people exposed to high levels of fluoride or suffering from vitamin D deficiency (Ten Cate 1989).

The mechanical properties of dentin and cement are not well known. Tables 6.1 and 6.2 give many of the values that are to be found in the literature. Dental scientists are mainly interested in human teeth. These are rather small, and it is difficult to obtain reasonable-sized test specimens from them. We have, however, carried out some extensive studies of narwhal tusk, and, as a result, it figures more prominently in the table than it does it the great scheme of things.

These data show that, compared with mature bovine bone, or other bone that is discussed in chapter 3, dentin has a somewhat lower Young's modulus, markedly lower in the case of the narwhal tusk tissues. The values for strength are very variable, the highest values approaching those of bovine bone. The work of fracture of human dentin is lower than that of the narwhal dentin but higher than that of bovine bone. The human values should be regarded with care, because the specimens were so small, and it is difficult to know how comparable these values are to those produced from larger specimens.

Kinney et al. (1999), using an atomic-force microscope, obtained values for the modulus of the intertubular and peritubular dentin (table 6.2). They then made a finite element model to examine how much the stiff peritubular dentin affected the overall value of the modulus. Their model produced a value for the dentin of 16 GPa, which is close to the intertubular value (15 GPa), suggesting that, over the range of volume fractions of peritubular dentin observed in teeth, these stiff tubes have little effect on the modulus of dentin. It would be interesting to know whether they affect strength. Although it goes against the general thesis of this book to say so, it may be that these high modulus tubes are not adaptive at all, but are merely the consequence of some inevitable excess mineralization round the dentinal tubules. This is unlikely, however, because the apparent lack of collagen in the peritubular dentin and the clear histological boundary between the two types of dentin suggests that it is a specialized structure.

6.2.3 *Narwhal Dentin*

The mechanical properties of narwhal dentin have been described in three papers (Brear et al. 1990b, 1993; Currey et al. 1994). The stress–strain trace has no sharp yield point, and the postyield stress increases fairly steadily until failure. This is unlike the situation in well-mineralized bone, in which the stress barely increases after yield, but is similar to that in antler bone (section 4.2). The mechanical properties in tension are roughly as follows (the figures in parentheses are for well-mineralized bovine bone, for comparison) (Brear et al. 1993): Young's modulus, 9 GPa (24 GPa); ultimate tensile stress, 120 MPa (140 MPa), ultimate strain, 9% (2%). These values are strongly influenced by the orientation of the specimen. Narwhal tusk is layered, and the different layers have different orientations. The values given are for specimens in which the preferred orientation is along the length of the specimen.

The Young's modulus of dentin is unremarkable for a material of its mineral content. However, it is extremely good at resisting impact, being a classic notch-insensitive material. A notch-insensitive material is one in which a stress-concentrating notch on the tensile surface of an impact specimen has little effect on the amount of energy it can absorb.

TABLE 6.3
The Impact Energy Absorption of Same-Size Specimens of Narwhal Cement, Dentin, and Bovine Bone

Tissue	*Condition*	*Impact energy absorption (kJ m^{-2})*
Cement	Unnotched	31.4
Cement	Notched	22.0
Dentin	Unnotched	36.3
Dentin	Notched	30.8
Bone	Unnotched	12.6
Bone	Notched	1.5

Source. From Brear et al. (1993).

Note. The bone is less able to absorb energy than the tooth tissues, but is particularly badly affected by the presence of a notch, which affects the tooth tissues much less.

This is the trademark of toughness (table 6.3). The cement is also good at resisting impact.

We gained some idea of how this toughness is achieved by examining it during loading. To understand dentin's toughness one must understand its histological structure. In narwhal tusk, interglobular dentin (which is mainly pathological in human dentin) is pervasive and normal. Figure 6.1 is a scanning electron microscope picture of the inner surface of developing Narwhal tusk dentin. The specimen has been chemically treated so that no free-standing organic material remains. On the left, the calcospherites look like little stalkless mushrooms, pierced by the cavities that contained the dentinal tubules in life. In the middle of the picture the calcospherites are more or less fused, and on the right there are hardly traces of them. The right-hand side is at a later stage of development. The mature surface shows little trace of the globules; the topography of the fracture surface is determined to a large extent by the predominant direction of the collagen fibrils, which is up and down the picture in this view. Although the calcospherites are no longer visible, the boundaries between them are hypomineralized and can be seen in backscattered electron images.

We tested specimens in tension at various orientations with respect to the predominant direction of the collagen fibrils. Along the $\approx 0°$ direction, dentin was very tough and it was difficult to produce a clear break. In these tests ($\approx 0°$), when the ultimate stress point was reached, a crack developed that split the specimens longitudinally, not transversely, which, in an isotropic material, is the direction in which the crack would travel. Narwhal dentin achieves a remarkably similar and tough behavior over a large span of angular variation (10–70°). The

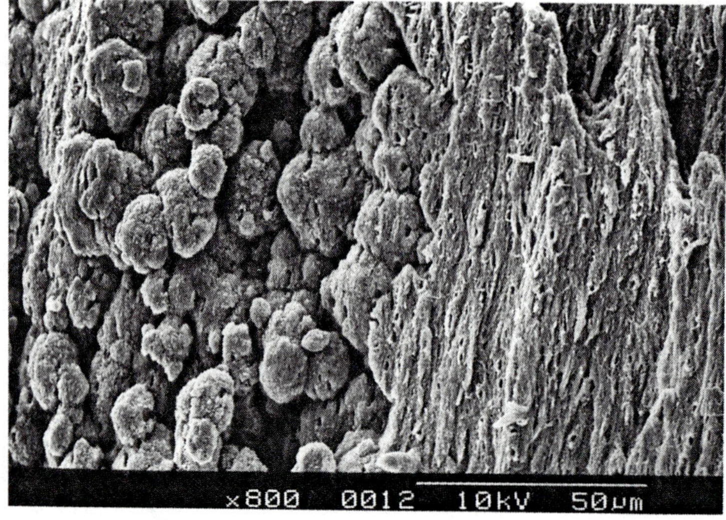

FIG. 6.1 Scanning electron microscope picture of the anorganic inner fracture surface of developing narwhal tusk dentin. Width of field of view is 150 µm.

ultimate strain stayed at very high values even when loading was at 70° to the fiber angle, resulting in a tough behavior that is insensitive to the direction of loading.

These puzzling results were to a considerable extent explained by prefailure images of microcracking obtained by the use of LSCM. In the *unloaded* state, the abutting regions of the calcospherites show up rather clearly in the LSC microscope, being more prone to absorb the stain than the spherites themselves. Because of the way the spherites form and are closely in touch with each other in all directions (as in a close-packed hexagonal system), these regions of abutment have no particular orientation with respect to the collagen fiber architecture.

When the specimen is stretched by a uniaxial load, lines of increased staining appear at about 45° to the direction of loading. These observations suggest that the regions between the spherites are relatively more compliant, and undergo irreversible changes of some kind along a direction of maximum shear, and these changes, whatever they are, allow more stain to settle there than previously. However, these 45° regions (of what look like incipient failure) actually bear no particular relation to the direction in which a second stage of microcracking develops in which the tiny microcracks form "clouds" at a lesser angle to the loading direction. At a third stage of fracture, the critical level for the initiation of a fatal macrocrack is reached, which travels through the structure at an angle close to the fiber direction as explained earlier.

It appears, therefore, that there are overall four phases in the behavior of narwhal dentin when it is loaded in tension: (1) linear elastic behavior, not associated with any permanent changes; (2) damage in the direction of maximum shear (this damage involves some kind of microcracking or loosening of the structure and is associated with irreversible changes that are presumably associated with nonlinear stress/strain behavior and absorb energy); (3) the appearance of dense clouds of microcracks, not now confined to the boundaries between the calcospherites (microcracking now occurs in between the fibers inside the spherites as well as outside them); and (4) the development of a fatal macrocrack, orientated almost along the grain of the dentin.

These studies suggest that narwhal dentin has three quite different mechanisms for toughening. One makes use of the difference in properties of fully mineralized dentin and the hypomineralized, interglobular dentin. The second mechanism makes it easier for a crack to form along the fibrillar direction. A third, more classical, mechanism makes use of the fact that dentin is arranged in layers, coaxial to the long axis of the tusk but with a different helical angle at different levels through the thickness (Brear et al. 1993). The latter two mechanisms act so as to deflect a macrocrack off its main route.

I have dealt with narwhal dentin at length because it shows well how, in a mineralized tissue that has a particularly uniform development and structure, it is possible to arrange things so that a high toughness can be achieved. Ordinary bone has to grow in complex ways, and usually undergoes reconstruction, and is less precisely arranged, making it difficult to achieve such toughness. The adaptive significance of the toughness of the unicornlike tusk of the narwhal is unknown. A rather wild suggestion about its function is put forward by Brear et al. (1993).

6.3 Enamel

Enamel's function is to provide a hard surface for the crushing, slicing, and trituration of food and, in some species, for wounding enemies. Enamel is about 97% by weight mineral, essentially apatite, 1% organic material, mostly protein, which is not collagen, and 2% water (Waters 1980). There are no cells or cell bodies in enamel. Enamel mineral crystals are about 25 nm thick, 100 nm wide, and 500 nm, or possibly much more, even up to tens of microns, in length, giving them the aspect ratio of spaghetti (Lowenstam and Weiner 1989). The enamel crystals are, therefore, much larger than the apatite crystals found in bone and dentin. Many of the crystals are bound together in bundles called prisms. There is little protein within the prisms; what protein there is in

TABLE 6.4
Mechanical Properties of Human Enamel

Tooth	*Prism orientation w.r.t. loading axis*	*Young's modulus (GPa)*	*Strength (MPa)*	*Authors*
		Compression		
?	Across	30.3	194	6
?	Along	9.0	134	6
Canine	Across	33.0	253	6
Molar	Across	32.4	250.0	6
Molar	Along	10.0	103.3	6
Molar	Along	80.5	376.9	2
		Tension		
?	?	?	10.3	1
		Bending		
?	?	131.0	75.8	7
		Ultrasonic		
?	?	76.5		5
?	?	73.0		4
?	?	74.0		3
		Nanoindentation		
Molar	Along	87.5		8
Molar	Across	72.7		8
Primary Molar	Along	80.4		19

Note. Authors: 1, Bowen and Rodriguez (1962); 2, Craig et al. (1961); 3, Gilmore et al. (1969); 4, Lees (1968); 5, Reich et al. (1967); 6, Stanford et al. (1960); 7, Tyldesley (1959); 8, Habelitz et al. (2001); 19, Mahoney et al. (2000).

enamel is concentrated at the interprism boundaries. These prisms are the characteristic feature of mammalian enamel, and how they are arranged has considerable effects on the enamel's mechanical properties. Unfortunately, because enamel comes in rather small lumps it is not easy to obtain very reliable values for the mechanical properties. Waters (1980) provided an overview of the mechanical literature up to 1979, and surprisingly little work seems to have been done since. Table 6.4 is derived mainly from Waters, with a few more recent additions. There is a rather alarming variation in reported values of Young's modulus. Spears (1997) has produced a finite element model that takes into account the variation in the major orientations of the crystals through the tooth. The results essentially agree with the larger values for Young's modulus shown in table 6.4.

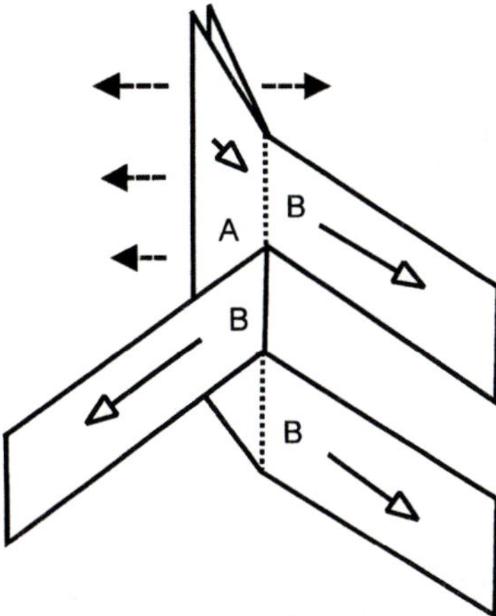

FIG. 6.2 Diagram of the way in which the crossed sets of prisms, appearing as the Hunter–Schreger bands, cause difficulties for crack travel. If the tensile load is in the direction of the black arrows, a crack, coming out of the page, will travel easily in region A, which it is shown here as having opened up. When the crack reaches the decussation, shown by the dotted line, it will split up and travel off in different directions in region B. Travel in region B is difficult. The crack cannot move far like this, and will tend to come to a halt. This system is also shown in figure 3.14B.

Enamel, being 97% or so mineral, is likely to be very brittle. The structure of enamel is described, with many illustrations, by Boyde (1989). In mammals there is usually a thin superficial layer in which the enamel crystals lie normal to the surface. However, deep to this, things become more complicated. Here the enamel is frequently arranged so that it is difficult for cracks to travel very far. The prisms are often surrounded by an interprismatic matrix. This consists of apatite crystals that are not part of the prisms. In earlier groups the crystals are often in roughly the same orientation as those in the prism, but in more derived groups the matrix may be at a considerable angle to the prisms. The enamel prisms themselves are arranged in a decussating, plywoodlike structure. This is like the arrangement that is frequently seen in lamellar bone, but in enamel the arrangement is on a somewhat larger scale, and is easier to visualize. There is an exactly comparable arrangement of calcium carbonate layers in the so-called crossed-lamellar structure of many mollusks (Currey and Kohn 1976). There is an "easy" direction for the cracks to travel, in which the cracks are just separating layers of mineral prisms. There is also a "difficult" direction in which the crack, as it advances, is continually having to break across the prisms if the crack front is to remain coherent (fig. 6.2). These crossed sets of prisms appear microscopically as so-called Hunter–Schreger bands. In the earliest mammals the prisms are aligned almost all in one direction, Hunter–Schreger bands developing when mammals became larger and developed carnivorous or herbivorous habits, as opposed to being insectivorous. The suggestion is that the larger size and more abrasive mode of loading made the development of crack-resistant enamel more important (von Koenigswald et al. 1987). Sometimes the "easy" direction is horizontal with respect to the long axis of the tooth, sometimes vertical, and sometimes it changes direction partway through the thickness of the tooth. Pfretzschner (1986) is a clear introduction to these matters.

Much work has been done on the way in which the predominant direction of decussation is related to the local state of stress. For instance, in many early herbivores the tooth cusps are rather domed. Finite element analysis shows that such domes are likely to suffer tensile stresses acting in the horizontal direction (hoop stresses), and therefore cracks are likely to run vertically. In later herbivores the teeth tend to become taller and the maximum tensile stresses, as the crowns of the teeth shear past each other producing bending, are more vertically directed, so cracks will tend to run more horizontally. It is satisfying to see that as the shape of the teeth changes, so the plane of decussation, shown by the direction of the Hunter–Schreger bands, changes from being predominantly horizontal to being predominantly vertical, thereby ensuring that the cracks have to run in the "difficult" direction.

The Hunter–Schreger bands show variations appropriate to details of the kinds of loading to which the teeth are exposed. Flat or wavy bands, oriented in a more or less horizontal plane, will make it difficult for vertical cracks to travel through the enamel. However, as Stefen (1997) shows in mammalian carnivores, those species that have the habit of crushing bone, as opposed to tearing flesh, have Hunter–Schreger bands that are even more complex in their orientation, becoming zigzaggy, with some extra connections between next-but-one neighbors, making it yet more difficult for the cracks to travel. One wonders why all teeth are not triply reinforced like this, since a strong tooth must be an advantage. Perhaps it is a difficult trick to bring off. A similar, and particularly beautiful adaptation, is shown in many rodents' incisors, in which alternating layers of prisms, and the interprismatic matrix, together form an interpenetrating lattice work oriented in three orthogonal directions, producing a stiff but tough material (fig. 6.3). These and other adaptations of enamel structure to loading are discussed in Rensberger (1992, 2000), Rensberger and Pfretzschner (1992), von Koenigswald (2000), and several chapters in von Koenigswald and Sander (1997) and in Teaford et al. (2000).

The different mechanical properties of the three major constituents of

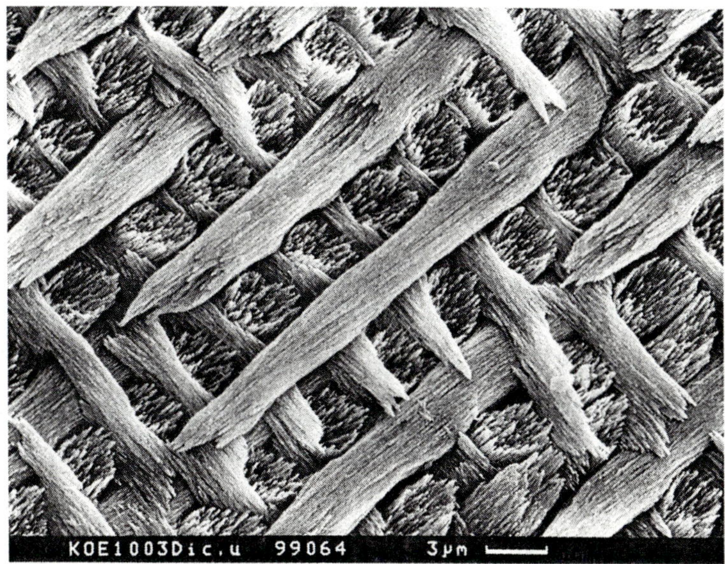

FIG. 6.3 Scanning electron micrograph of the enamel of the incisor of a marmot *Dicrostonyx torquatus.* Alternating layers of prisms are oriented in orthogonal sheets (NE–SW and NW–SE in this photograph). The interprismatic matrix forms what are in effect prisms oriented orthogonally to the true prisms. There is no direction in which a crack can travel easily. Scale bar 3 µm. (Photograph given by Wighart von Koenigswald.)

teeth—enamel, dentin, and cement—are famously made use of in the teeth of many mammalian herbivores and rodents. Although the business part of the tooth when it first erupts has a continuous coating of enamel, this sooner or later wears away, exposing successive layers of enamel, dentin, and cement. These three tissues wear at different rates, and the resulting tooth has a basically flat surface but with sharp ridges and valleys caused by the enamel standing proud of the faster-wearing cement and dentin. This arrangement produces an excellent surface for grinding up the cellulose walls of plant cells, which are often rendered gritty by silica inclusions (Janis and Fortelius 1988) (fig. 6.4A). The relationship in the grinding cheek of rodents between the direction of movement of the teeth and the direction in which the enamel prisms come to the surface is beautifully adapted to the different loads that are placed on the teeth in different points of the chewing cycle (von Koenigswald and Sander 1997). This relationship has also been found in a marsupial, the koala (Rensberger 1997). Rodents also make use of the different wear properties of enamel and dentin. The outer layer of their curved, ever-growing incisors is covered with enamel, while the inner

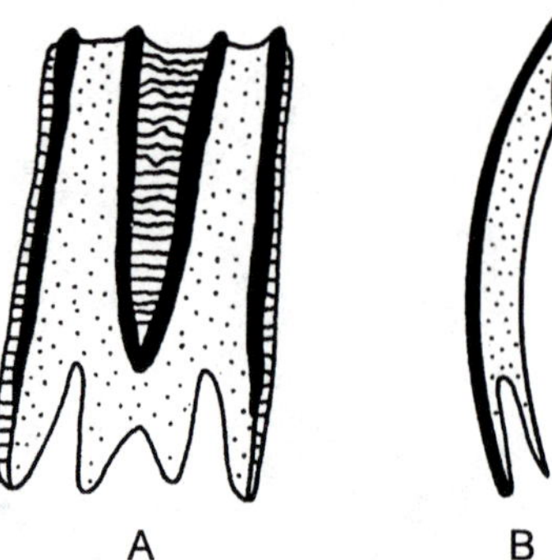

FIG. 6.4 Diagrams of how the differential wear of different tissue types is used to produce sharp grinding or gnawing tooth surfaces. Black areas, enamel; dotted areas, dentin; striped areas, cement. (A) Longitudinal section through the tooth of a herbivore, showing how after a while the enamel stands proud of the cement and dentin. (B) Longitudinal section of an incisor of a rodent, showing how the enamel and dentin interact to produce a self-sharpening chisel.

layer is made of dentin, which wears faster. The combination produces a self-sharpened tip to the tooth (fig 6.4B).

Some rodents and insectivores have iron salts incorporated in their enamel, which colors it yellow or brown. It is most probable that these salts increase the hardness of the enamel, though the evidence is not as clear-cut as one would like. Granados (1986) found that in hamsters with a mutation that made some incisors colored and others not, the pigmented incisors were almost invariably longer than the unpigmented ones; that is, they wore less quickly. On the other hand, Söderlund et al. (1992) found little difference in shrew's teeth between the colored and less colored parts and, if anything, the unpigmented teeth were slightly harder.

Sharks are perhaps the most toothy of vertebrates, and Preuschoft et al. (1974) show that their enamel is beautifully adapted to various different functions. Sharks' teeth are designed either for crushing or for slashing. In crushing teeth of *Hybodus*, *Heterodontus*, and *Ptychodus* there is a thin layer of enamel lying on a bed of dentin. Typically, the enamel layer is 0.2 mm thick, the underlying dentin 15 mm. The function of this enamel is to act as a very hard and wear-resistant cap to the dentin. Being so thin, it will hardly contribute to the stiffness of the tooth as a whole, which will be determined by the dentin. The dentin has a considerably lower Young's modulus than the enamel. The thinness of the enamel layer prevents it from undergoing large strains when it is compressed locally and so loaded in bending. (For a discussion of this mechanism see section 2.2.4.) The hydroxyapatite prisms, in general, lie roughly normally to the outer surface of the tooth, and may therefore be good at resisting compression. The mechanical properties of this enamel are unknown, because it is always so thin.

In slashing teeth found in *Odontaspis*, *Isurus*, and *Carcharodon*, the enamel layer is thicker, averaging 0.6 mm in the species examined. There are three layers of highly mineralized enamel sheathing the dentin. On the outside is an extremely thin layer of *shiny* enamel. This is only about 4 µm thick. It consists of mineral fibers 2.5 µm long and about 0.15 μm thick. These form a random mat of fibers, lying in the plane of the layer. There is no evidence of an organic matrix. Below is a layer of *parallel-fibered* enamel (not to be confused with parallel-fibered bone). Here the mineral has a very high, spaghetti-type aspect ratio (about 2.5 μm long by 0.03 μm across) embedded in a matrix of organic fibers. The fibers and their associated needles lie either normal to the surface of the tooth or parallel to it. In any fairly small region the fibers parallel to the surface are also parallel to each other, but this general direction may change over a few millimeters. Underneath the layer of parallel-fibered enamel is a *tangle-fibered* layer. There are many fibers that run predominantly in the radial direction, but there are many fibers that run in all directions normal to them.

Preuschoft et al. (1974) suggest that these parallel-fibered enamels are principally designed to resist tension. The parallel-fibered enamel will be highly anisotropic in tensile strength, because there will be an easy and a difficult way for the cracks to travel. They found a tensile strength of 140 MPa in the difficult direction and 35 MPa in the easy direction. They also suggest that that the shiny layer on the outside, with no detectable organic material, acts as a very hard surface, and that any cracks produced in it will not travel through the parallel-fibered layer.

Teeth are usually used for cutting or crushing, and the mechanical properties of the material of enamel and dentin would not seem suitable for function as bristles. However, this is what the teeth of the pterosaur *Pterodaustro guiñazui* did. The long lower jaws had many hundreds of filamentlike teeth forming a comb, presumably to allow filter feeding like the baleen of whales or the ridged roof of the mouth of flamingos. The teeth of *Pterodaustro* are true teeth, nevertheless, having a sheath of enamel on one side, and dentin surrounding a central pulp cavity (Chiappe and Chinsamy 1996). The teeth are very long in relation to the transverse dimensions (about 30 mm $\times$ 0.3 mm $\times$ by 0.2 mm) and so, as was explained in section 2.2.4, could have undergone quite large deflections without dangerous strains being imposed on the enamel or dentin.

The fact that enamel and dentin have such different stiffnesses pro-

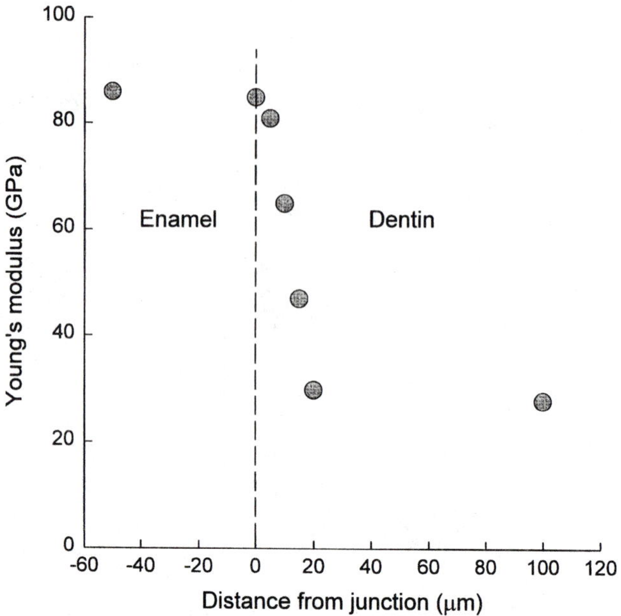

FIG. 6.5 Young's modulus of the tissues at the dentino-enamel junction in a human incisor tooth. These values were determined by nanoindentation. This technique uses the unloading curve of a nanohardness indentor to determine the Young's modulus of a very small piece of material, less than one micron across (Oliver and Pharr 1992). The Young's modulus is reasonably constant away from the junction, but changes over about 20 μm. (This picture was derived from data in Fong et al. [2000].)

duces a problem. When the tooth is loaded there is likely to be a mismatch in the stresses at the boundary between the two tissues. Because the tissues are not able to move relative to each other, there will be stress concentrations here, and as a result the two tissues will be prone to spring apart. Two recent studies (Fong et al. 2000; White et al. 2000) have shown that there is, in fact, not a totally sharp division between the two tissues, but instead there is a zone, about 20 μm or more wide, in which the mechanical properties change from being characteristic of pure enamel to that of pure dentin (fig. 6.5). This mechanical transition is paralleled by an interpenetration of the two tissues. In this way, the transfer of load from the brittle enamel to the more compliant and tougher dentin can proceed fairly smoothly.

Almost as if to show that mineralized tissues can do anything, occasionally dentin can be so highly mineralized as to become like enamel. This so-called "petrodentine" is found in the teeth of lungfish (Smith 1984). Although it does not have dentinal tubules, it is certainly dentin, not enamel, because its sparse protein is collagen, it connects with the normal dentin at its boundaries, and it is produced by cells that are clearly derived from ordinary odontoblasts (Ishiyama and Teraki 2000). It acts, in conjunction with more normal dentin, to produce hard ridges flanked by softer shoulders, as in herbivore and rodent teeth, where the harder and softer parts are enamel and dentin. Microhardness of minerals is a good measure of their relative abrasion resistance. I have found that the microhardnesses of the petrodentine and dentin of the lungfish *Lepidosiren* are very similar to those of conventional enamel and dentin.

6.4 Fish scales

Fish scales show an enormous variety of structures. However, because most of the variation is in fossil fish, virtually nothing is known about their mechanical properties. Many teleost fish have bone without included bone cells, and the range of structures seen in scales of different types forms almost a complete spectrum, from what is obviously "typical" bone to what is obviously "typical" enamel (Meunier and Huysseune 1992). There are some structures that are unique to the fish. Enameloid is a structure that looks like enamel, with very long crystals, but has collagen as its organic matrix; isopedin in the basal plates of some scales is said to be bone that has failed to mineralize, and chondroid has features intermediate between cartilage and bone. This situation is often found in biology, of course, since nature is not concerned with categorization, but with producing effective results.

Many fish scales frequently have a regular "plywood" structure. Each ply is a sheet of ossified fibrils running in one direction and appears, at least in herring, to be only one fibril thick $\approx 1\ \mu m$ (fig. 6.6). The change of fiber direction between plies can be a regular 90° or can be more subtle. Meunier (1987) characterizes the orientation of successive bony layers in the scales of 25 fish, and shows that almost all kinds of arrangement exist, including strict orthogonality and helicoids, but also more complex arrangements. The precise mechanical significance of these different arrangements is not known, but clearly any plywood arrangement is going to make the scale more isotropic in mechanical behavior, and stronger. We have found some fish scales to be so tough they are difficult to tear even after immersion in liquid nitrogen!

6.5 Dentin vs. Bone

Although in the early evolution of vertebrates, and in present-day fish scales, typical dentin and typical bone were the extremes of a complete

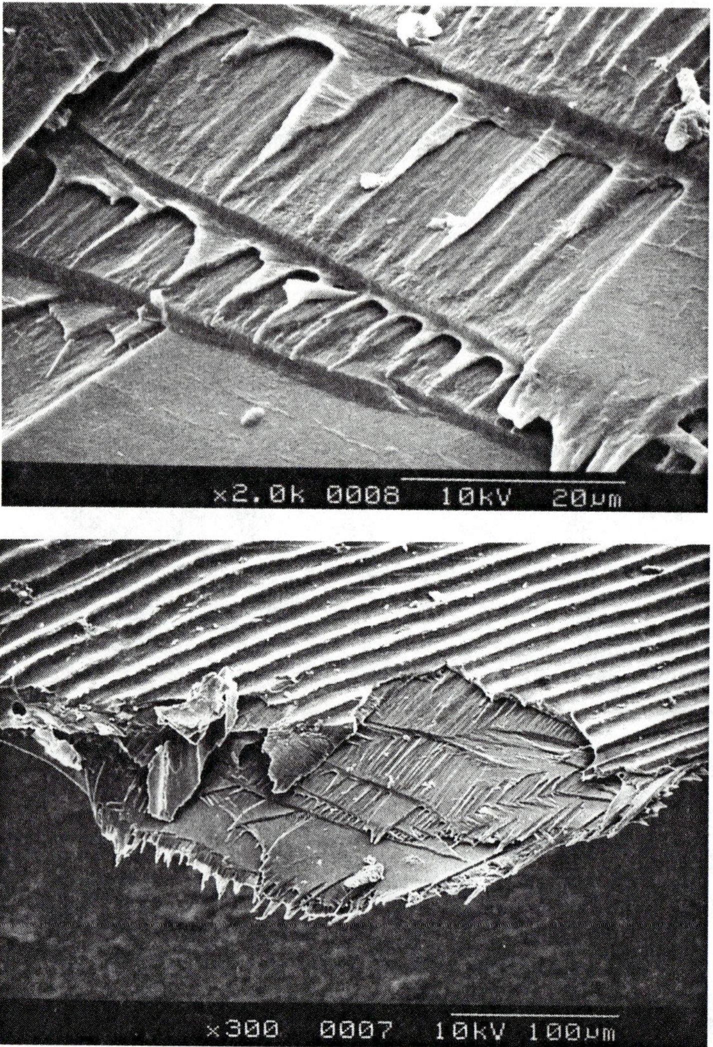

FIG. 6.6 Scanning electron micrograph of the bony scale of a herring, showing the orthogonally arranged plywood structure. The sculpturing on the surface in the lower picture probably has some hydrodynamic function. Width of field: upper, 60 µm; lower, 440 µm.

spectrum of types, in modern mammals there is a clear distinction between the two. This seems a little strange if their mechanical properties are considered, because bone has a range of properties that includes those of dentin, so it is not obvious that mechanically there is much to choose between them. Two main features distinguish dentin from bone.

First, the cells bodies in dentin send processes out throughout the dentin, whereas in bone the osteocytes (if the bone is cellular) have rather complex connections through a series of intermediates from the blood vessels to the furthermost osteocytes. Second, although dentin can under some circumstances undergo a form of remodeling in the formation of *reactionary* or *reparative* dentin (Smith 2000), in general, once dentin is laid down it is a fixture and is not remodeled.

It is possible that the sensory requirements of teeth, lying as they are exposed to the outside world, are different from those of the more deeply buried bone. Teeth are exquisitely sensitive to *local* stress and trauma and we are conscious of our teeth in a way quite unlike our sense of what our bones are up to. Perhaps the odontoblast processes are designed to carry local information to the pulp cavity in as direct a way as possible, and the information produced by bone cells, which will be dealt with in chapter 11, needs to refer to longer-acting situations, and to some extent integrate information over larger volumes, than is the case in teeth.

The modeling and remodeling that take place in bone is associated with an intimate blood supply and, as we shall see, is probably often initiated by bone undergoing microdamage. The conditions in teeth are different, since dentin can undergo very little modeling, and the enamel, being totally lifeless, can undergo none at all. This leads, as anyone over the age of about 40 will know, to a gradual degradation in the appearance and mechanical integrity of the enamel. It would be extremely interesting to know the extent to which dentin undergoes damage in normal day-to-day living, and whether this is less than is suffered by bone.

One could build a series of hypotheses about why dentin *needs* to remodel less than bone, but it would go well beyond the evidence, so I shall not. There is a view that the histology of dentin is constrained by developmental and phylogenetic constraints. I find such arguments unconvincing, but they are well displayed in an interesting article by Smith and Sansom (2000). People interested in teeth should certainly read this work and others in the same book.

It is unfortunate that we have so little idea of the mechanical properties of many of the interesting materials described in this chapter. This is mainly because of the small size and convoluted shape of many of the structures in which they are found. Their histological structure shows that they are often very well designed mechanically, but, unfortunately, we usually cannot quantify how well.

r Seven

..APES OF BONES

7.1 Shapes of Whole Bones

THE SHAPES of whole bones are marvelously varied and are, of course, intimately related to their functions. It would be futile to try to explain all the different shapes we see in mechanical terms, although it might be possible, in theory. But to do so even in a single case one would need to have a good idea of how the bone was loaded, and this is often very difficult to find out. This chapter will, therefore, deal principally with the adaptations of long bones, because they are relatively simple in shape and we have a reasonable idea of the loads falling on them and what they have to do.

Despite the variation in the shape of bones, a large proportion of them fall into a few groups. I shall describe these briefly, mainly as a reminder of facts the reader is no doubt aware of. Tubular bones, usually called long bones, such as the humerus, radius, ulna, femur, tibia, fibula, metacarpals, and metatarsals (fig. 7.1A), are elongated in one direction and the section is often roughly circular. These bones are hollow; in their midsection the wall thickness is characteristically about one-fifth of the overall diameter. They are expanded at their ends and capped with a layer of synovial cartilage, forming part of the synovial joint. Near the ends the central lumen is filled with cancellous bone. In places the plain tubular shape may be distorted where the bone is drawn out into flanges and tubercles for the attachment of muscles and ligaments. Tubular bones are designed to carry compressive loads and bending moments over reasonably long distances. These bending moments are often large, and the bones must withstand them without breaking or deforming too much.

Short bones are bones or parts of bones that are roughly the same size in all directions. Examples are wrist and ankle bones and the centra of vertebrae (fig. 7.1B). Many phalanges are intermediate between short bones and tubular bones in their structure. Short bones tend to have very thin cortices and to be completely filled with cancellous bone, the trabeculae running from one end to the other. They are designed to carry loads, usually compressive loads, over short distances. They are not usually subjected to much bending.

Tabular (flat) bones are bones or, often, parts of bones that are rather

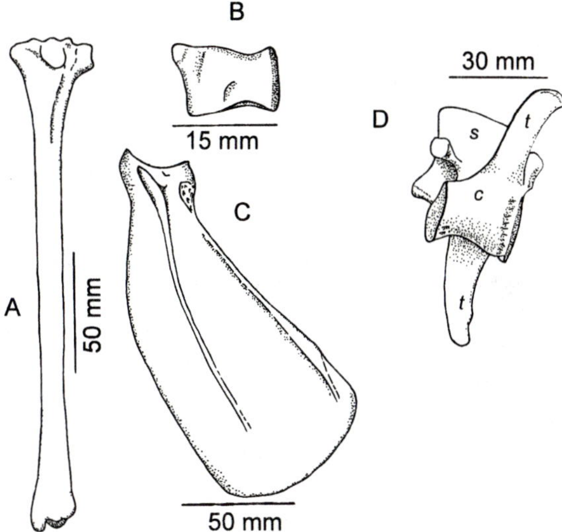

FIG. 7.1 Variously shaped bones. (A) Dog's tibia. The bone is straight, considerably laterally expanded at the proximal end, slightly expanded at the distal end. (B) Dog's tarsal bone. This has a complex but rather boxlike shape. It has thin walls and is filled with cancellous bone. (C) Dog's scapula. This is greatly flattened in the plane of the paper. (D) Lumbar vertebra of a fallow deer *Dama dama* seen from below and slightly to one side. The centrum (*c*) is thin-walled and filled with cancellous bone. The transverse (*t*) and spinous, (*s*) processes are greatly flattened.

flat, so that one dimension is much less than the other two. Examples are the scapula (fig. 7.1C), the iliac blades, many bones in the vault of the skull, the carapace of chelonians, and many bones of fish. They usually consist of two thin sheets of cortical bone separated by some cancellous bone, but sometimes, as in some scapulae and the vault of the skull of many small mammals, the two cortices are not separate. Fish seem not to have true cancellous bone; instead, their bone is pierced by many, many tubes all running roughly parallel to each other. Ribs are intermediate between tubular and tabular bones. Flat bones are designed either for protection, as in the vault of the skull, or to provide a base for the origin of extensive muscles, as in the scapula or the iliac blades.

Often the extensions to bones for muscle and ligament insertions are so large that they take on the anatomical and mechanical characteristics of tabular bones For instance, the vertebrae are in two parts: the centra are like the carpals, but the projections around the spinal cord are like the flat bones (fig. 7.1D). The turbinals in the nose are very thin and scroll-like. They are fragile and are designed not to bear significant loads, but merely to support the nasal mucosa. There are some other bones, particularly in the skull and parts of the pelvis, that have shapes that do not fit into this classification, but, in general, most bones fit somewhere within the types listed above.

7.2 Designing for Minimum Mass

I have labored the point, perhaps too much, that bone has as its primary requirement the need to be stiff. However, the point must be made again: the long bones of vertebrates act primarily to exert forces on the environment, and in so doing they must withstand large bending moments. These bending moments will, of course, tend to distort them. The bones must not distort too much. Natural selection will favor animals that can perform particular locomotor functions with the greatest efficiency. By efficiency I mean here merely that the function can be performed, and that the cost to the animal in metabolic and other terms, such as time for production, is minimized. It is obviously *possible* to perform mechanical functions with materials or structures that seem, even to layman, to be wildly unsuitable. The Incas made fish hooks of gold. Gold is unsuitable, not because it is rare and expensive (besides copper and tin it was the only metal the Incas had), but because it has such a low modulus of elasticity and low yield stress.

For locomotor structures, the feature that is likely to be subjected to particularly stringent selection is the mass of the material that will perform a particular function. The reason for this is perhaps obvious, but deserves a little comment. Any animal traveling with some velocity has some kinetic energy. This energy has two components: the *external* and the *internal* kinetic energy (Alexander 1975). The external kinetic energy is half the mass of the animal times the square of the velocity of its center of mass ($E = \frac{1}{2}MV^2$). Also, individual bits of the animal may have velocities relative to the center of mass, and these will also have kinetic energies associated with them. Energy associated with the relative movement of separate bits of the animal is called the internal kinetic energy. A flying bird has external kinetic energy from its movement through the air, but it also has internal kinetic energy associated with the beating up and down of its wings. All the kinetic energy of the bird must have come from its muscles. If the bird has some mass and needs to travel at a particular velocity, its external kinetic energy is essentially fixed. However, the energy associated with the beating of the wings will depend to a considerable extent on the mass of the wing bones. Furthermore, unless it has a means of storing the internal kinetic energy of the up- or downstroke, for making it do useful work, this energy will be lost, and the muscles must do work to recreate it during each cycle. Therefore, although at any moment the external kinetic energy may be considerably greater than the internal energy, the internal energy is more likely to be wasted. It has been calculated that in a man running at 6 m s^{-1} about 40% of the total power output provides external kinetic energy and 32% internal kinetic energy, the remainder of the power output being used for changes in potential energy (Cavagna et al. 1964). Alexander and his co-workers have written a fine series of papers about the way in which animals minimize the energy lost in locomotion (Alexander and Jayes 1978; Alexander 1982, 1988, 1993).

In all discussions of the design of bones there should always be, at the back of our minds, the question, How can the mass of material necessary to do the job be minimized?

7.3 Long Bones

Let us start with long bones. (In fact, by far the greater part of this chapter is concerned with long bones.) The most obvious features of long bones that need to be explained are that they are usually *thick-walled*, *hollow* tubes, *expanded* at the ends, and having *cancellous bone* rather than compact bone under these expanded ends.

7.3.1 Why Are Long Bones Hollow?

Finding out how to build a structure that will perform a particular function with minimum mass is called minimum mass (or weight) analysis. *Weight* and *mass* are often used interchangeably, yet they are very different. A body always has the same mass, but its weight depends on the local effect of gravity. A neutrally buoyant fish of mass 100 g has this mass wherever it is; its weight is zero in water, but when caught and flopping about on the river bank it has a weight of 100 grams force, or about one newton. Note that kinetic energy, discussed above, is a function of mass, not weight. However, everyone knows roughly what is meant by an animal "weighing" 100 g so the problems of ambiguity are small.

We need to set up some not too arbitrary task that a bone must carry out and then see what kind of structure will best do it. Take a simple example: a limb bone loaded as a beam (fig. 7.2). It is required to support a pair of bending moments M along a distance L, and in so doing the middle of the length may deform by an amount Z only. Note

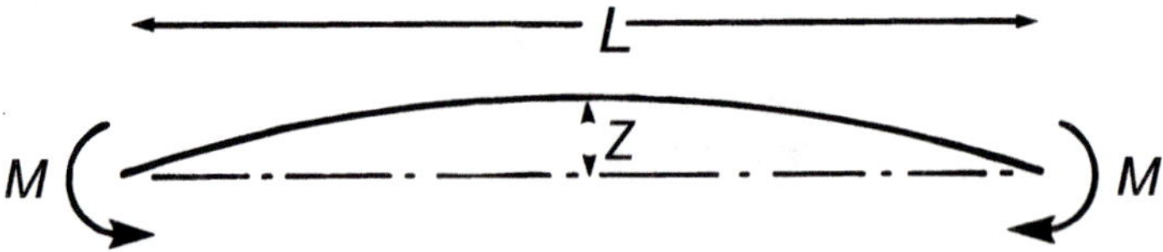

FIG. 7.2 Possible design criterion for a bone. A limb bone is initially straight and has a length L. It must not distort in the middle by more than an amount Z when subjected to a pair of bending moments M.

that this says nothing about how *strong* the bone has to be. Our criterion of failure is excessive deformation, not rupture. This is not because strength is not important, but merely that we here assume that if the bone is stiff enough it will be strong enough.

The beam is loaded in so-called pure bending; that is, it is exposed to the same bending moment all along its length. Although limb bones are not often loaded in pure bending, this is handy for present purposes because it means that we can ignore shear stresses. This uniform bending moment immediately implies that the beam should be of uniform cross section. The formula giving the maximum deflection of a beam of uniform shape all along its length in pure bending is $Z = ML^2/8ET$. The terms Z, L, and M are already fixed for us so that we can alter only E, Young's modulus of elasticity, and I, the second moment of area. Let us for the moment assume that E is also fixed. How should the build of the bone be arranged to minimize the mass of the bone? If a beam is bent, as here, it will get shorter on one side and longer on the other (fig. 7.3). The amount of deformation will decrease toward the middle of the depth of the beam and there will be a neutral plane, which will remain the same length (though it will become curved) when the bone is bent. Seen in section, this neutral plane is a neutral axis. The second moment of area is defined as $I = \Sigma y^2 \delta_{area}$ (fig. 7.3C). As shown above, because Z is inversely proportional to I, to minimize Z, I should be as large as possible. What effect will this have on the mass of the beam? If the density of the bone is ρ, then the mass of the whole bone is $\rho LA = \rho L \Sigma_i^N \delta_{area}$. Therefore, the ratio of mass to second moment of area is $\rho LA/I = \rho L \Sigma \delta_{area} / \Sigma y^2 \delta_{area}$. This implies that to minimize mass for a fixed value of I, L, and ρ the whole of the area should be as far as possible from the neutral axis, because this will minimize the ratio $\Sigma \delta_{area} / \Sigma y^2 \delta_{area}$. The larger the second moment of area, the less the area, and therefore the less the mass, necessary to limit the deflection.

In fact, the mass can notionally be made as small as one likes by making I sufficiently large. The ideal shape for doing this would be two very thin sheets a long way from the neutral plane. But there would be

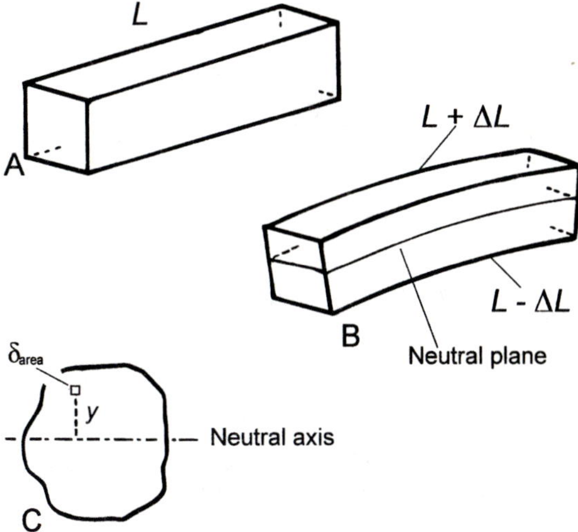

FIG. 7.3 (A) A beam, originally of length L is bent by moments as in figure 7.2, (B) changing the length of all fibers except those lying in the neutral plane. (C) The cross section has a second moment of area $I = \Sigma y^2 \delta_{area}$ about the neutral axis. If the bone were loaded about a different axis, the second moment of area would be different.

nothing to stop the two sheets collapsing onto each other, and so a web would have to be inserted to keep the flanges apart (fig. 7.4A,B). This arrangement is the theoretical ideal, but would work only if the orientation of the neutral axis were fixed. Nevertheless, this is apparently the ideal solution for the problem as initially set up.

If the neutral axis were *m-m*, rather than *M-M* (fig. 7.4B), then there would be a great deal of bone close to the neutral axis, which would make the value of I much less, and the shape therefore less efficient. For bending moments acting in two directions at right angles the square box is the best sectional shape (fig. 7.4C). The sides act as flanges or webs according to the orientation of the neutral axis, and in either case there is little bone close to the neutral axis. In real long bones the direction of loading is quite likely to come, occasionally, from any direction and a hollow cylinder is the best solution. It is also the least mass solution for torsional stiffness (fig. 7.4D).

7.3.2 How Hollow Should Bones Be?

So far it would seem that an indefinitely fat bone with vanishingly thin walls would be the least mass solution to the problem set by natural

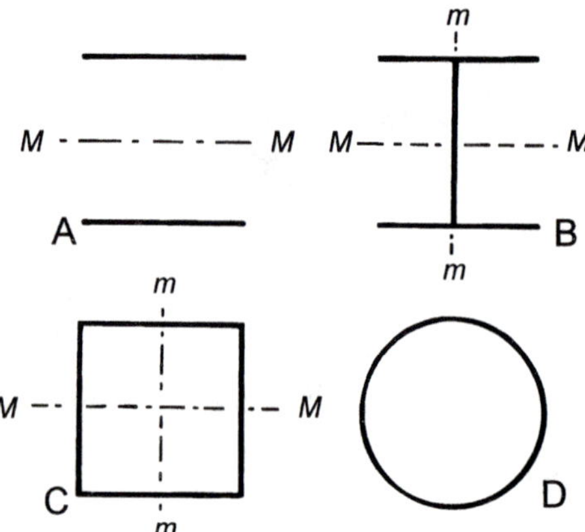

FIG. 7.4 Possible mass-minimizing shapes for the cross sections of beams. (A) Theoretically good arrangement when the plane of bending is about *M–M*. (B) However, a web is needed to keep the flanges apart and to bear the shear stresses. Even so, if loading changed so that the neutral axis became *m–m*, mass would be *concentrated* near the neutral axis, a bad arrangement. (C) The best arrangement if there are two mutually perpendicular neutral axes. (D) The hollow cylinder—often the best arrangement. The neutral axis can be at any angle through the center of the cylinder.

selection. In fact, most bones, though hollow, are quite thick-walled. Why is this? One limitation on the thinness of walls is the possibility of local buckling. Local buckling is the form of buckling seen when a thin-walled plastic drinking cup or an aluminum beer can is crushed. The buckling starts as a little wrinkle, which spreads initially at its two ends and then usually in many directions, leading to a general collapse. (It is instructive, and fun, to load an empty beer can slowly in compression in a testing machine.) Local buckling occurs when the walls are so thin relative to the overall size of the structure that the shape of the structure does not support the wall sufficiently to prevent it from bending in an easy direction (fig. 7.5). The analysis of local buckling is very complex (Young 1989) and for many situations is not properly worked out. Such theoretical results as exist are likely to be considerably in error unless the structures and loading system are perfect: any stress concentrations or little local sideways forces are likely to be very injurious. Local buckling does not depend on the *strength* of the bone material at all, but solely on its stiffness.

Brazier (1927) showed that in the ideal case the bending moment causing a thin-walled cylindrical tube to buckle is $(1.1)ERt^2$, where R is the midradius of the wall, and t is its thickness. For any bending moment, the cross-sectional area necessary to prevent failure *by buckling*

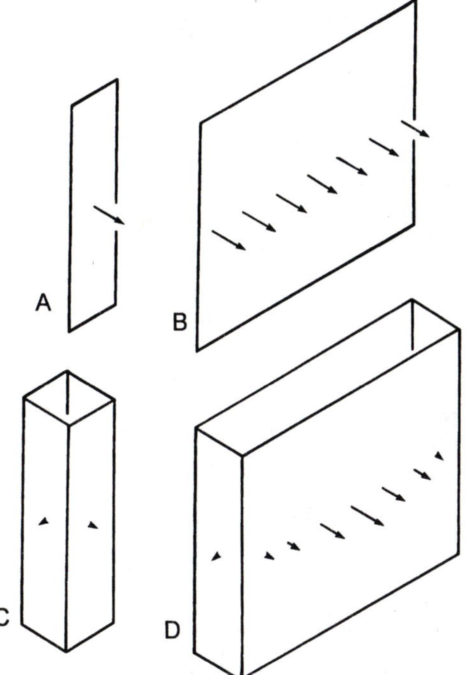

FIG. 7.5 Buckling. All four structures are imagined as being loaded in compression at the top and bottom ends. The resulting displacement of the middle is shown by arrows. (A, B) A narrow and a wide sheet will deflect by the same amount. (C) A square-sectioned hollow box subjected to the same stresses in the walls as (A) and (B) will deflect little because the neighboring walls constrain each other. (D) If the walls of the box are wide, the constraining effect of the neighboring walls is lost, progressively, away from the corners. The complex stress systems set up can produce wrinkling.

can be shown to be $\propto (R/t)^{1/3}$ (Alexander 1983). This implies that, as the radius increases and the wall thickness becomes relatively less, the mass must *increase*. The mass required when R/t is 20 is 70% greater than that required when R/t is 4. However, when the value of R/t is small, local buckling is not a realistic possibility. So, in a bone without any marrow, the cross-sectional area, and therefore the mass, necessary to provide a particular resistance to buckling must *increase* as the diameter of the bone increases.

Suppose that a bone can fail in one of two modes, by buckling or by breaking. The question then is How should things be arranged so that the bone can carry out its functions and yet be as light as possible? For a particular cross-sectional area, and therefore constant mass, as the value of *R/t* increases the resistance to failure by breaking increases, while the resistance to buckling decreases. In these circumstances it would be best to arrange things so that the bone fails by both modes at the same time. (If, for instance, it were so thin-walled that it failed by buckling *before* it failed by breaking, then a reduction of *R/t* would increase the buckling resistance but would, for a limited reduction, not decrease the strength to below the buckling resistance.) Calculations by Alexander (1982) show that, for reasonable values of strength and stiffness of bone, the point in a bone of constant cross section where both modes of failure should happen at the same load is when *R/t* is about 14.

If, as here, the minimum mass for a particular *stiffness* is being found, we can say nothing a priori because the stresses in the bone may be low, indeed, well below the critical stress. However, as will be shown in chapter 10, skeletons are designed so that they are not greatly understressed; that is, they are likely to be designed so that the shape required to fulfil a stiffness criterion is not very different from that required to fulfil a strength criterion. If this is so, even if a structure is designed for a minimum mass for a particular stiffness, it is unadaptive for it to have a very high value for *R/t*.

The second, and usual, limitation to the thinness of the walls is the marrow fat in lumen of the bone. We have just seen that, if buckling is not a problem, the greater the size of the internal cavity, the less the mass of bone necessary to make a cylinder of the required stiffness. However, the long bones of vertebrates, except some bones of some birds and pterodactyls, are (or were) filled with marrow. This is of two kinds: red and yellow. The red marrow is blood-forming and, though important in young bones, is restricted in adult mammalian long bones to the very ends, if present at all. In humans, virtually all long-bone marrow is hematopoietically inactive by late adulthood (Richardson and Patten 1994; Zawin and Jaramillo 1993) In all mammals investigated there is a strong tendency for the concentration of blood-forming marrow to decrease in the proximal–distal direction. In the adult, skull bones and the marrow of the spine do form blood cells, and the limb bones barely do so (Ascenzi 1976). The rest of the marrow, the yellow marrow, is just fat. This seems to have little physiological function, and has a very low rate of turnover. For instance, yellow marrow is not mobilized during starvation, in the short term anyhow (Tavassoli 1974), though it may be used in extremis (Thouzeau et al. 1997). It appears to be acting mainly as a packing material. The calculations made above assume that the bone material is the only part of the anatomical bone to have mass, and that all this mass is contributing to the stiffness. But fat has a minute effect on the stiffness and the strength, though it does contribute to the mass. The density of marrow fat is about 930 kg m^{-3},

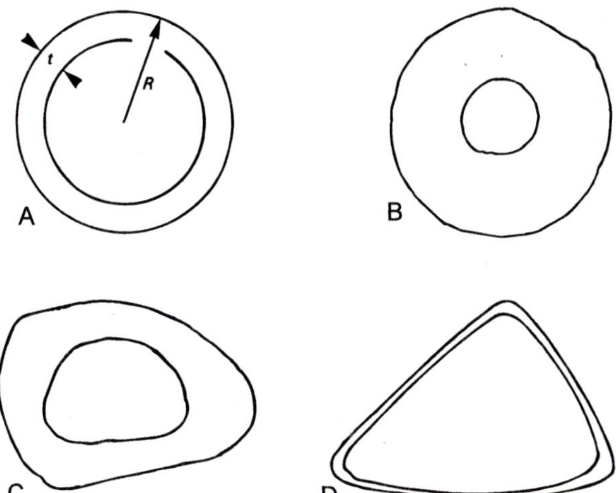

FIG. 7.6 (A) Diagram of the conventions used in figures 7.7 and 7.8. (B–D) Sketches of sections of bones of different values of R/t. (B) Alligator femur, $R/t = 1.5$. (C) Camel tibia, $R/t = 2.4$. (D) *Pteranodon* first phalanx, $R/t = 11$. (From Currey and Alexander [1985] by permission of the Zoological Society of London.)

that of bone about 2100 kg m^{-3}. What is the effect of taking marrow into account?

Neill Alexander and I investigated this, making theoretical calculations and also examining many bones (Currey and Alexander 1985). The conventions we used and the results of the calculations are shown in figures 7.6 and 7.7. We did the calculations for a number of design criteria for bending: minimum mass for stiffness, yield or fatigue strength, ultimate strength, or impact strength. The details of our assumptions are set out in the paper, and it would be tedious to reiterate them here. The critical points are (1) that each of the design criteria has an intermediate minimum; that is, the minimum value for mass is found not in a completely solid bone or in a balloonlike thin-walled one, but in some intermediate shape; (2) that the minima for different design criteria are different; and (3) that the saving in mass by adopting the best value for wall thickness is not very great, 18% at most, compared with solid bone. For bending strength the best possible saving is 5% at most. These results imply that for a hollow cylinder of bone filled with more or less useless fat, the minimum mass solution for stiffness or strength requires rather low values of R/t, with walls much too thick to be likely to fail by local buckling. The range of thicknesses producing mass values quite close to the minimum is large, and selection is unlikely to act strongly to produce great uniformity between different

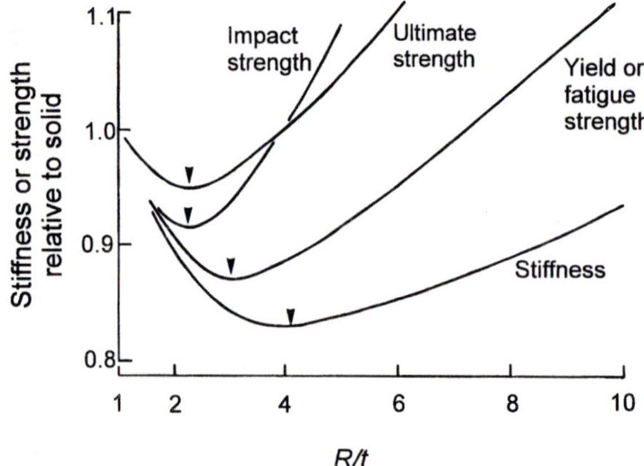

FIG. 7.7 Diagram of the effect of altering the value of *R/t* of a bone on the mass necessary to achieve a particular value for various mechanical properties in bending. The bone is assumed to be filled with marrow fat. When the value of *R/t* is 1 (the bone is solid) the mass is defined arbitrarily as 1.0. Note that all properties have a minimum at an *R/t* greater than one, and that the mass saving is not great. (From Currey and Alexander [1985] by permission of the Zoological Society of London.)

bones. An important feature of this analysis is that the actual bending moments, the length and deflection of the bone, and the bone material's stiffness and strength do not figure in the solution; the minimum mass solution gives the best *shape* of the cross section.

Theoretical calculations of the shape producing minimum mass should be matched with the real world. Figure 7.8 shows values for *R/t* of a rather haphazard collection of bones that Neill Alexander and I collected. The results are rather striking. The values for the nonflying mammals' bones were mostly taken from the literature, so we do not know what kind of marrow they contained. However, most adult mammals have yellow marrow in their long bones. If this is so, they have, in general, ratios of *R/t* that would be appropriate for minimizing mass while resisting impact loading or bending loading. Of the birds' bones, most had red marrow and so the marrow still had a hematopoietic function. These bones have a wide spread, the median value of *R/t* being 3.5. They are, therefore, roughly minimum mass solutions for fatigue or strength or stiffness, but, because the marrow is physiologically useful and therefore has to be put somewhere, one might have expected them to have had a somewhat higher value. The marrow of the gulls was yellower than that of the others, suggesting a less active physiological role, and the gulls' values for *R/t* are rather lower than those for the

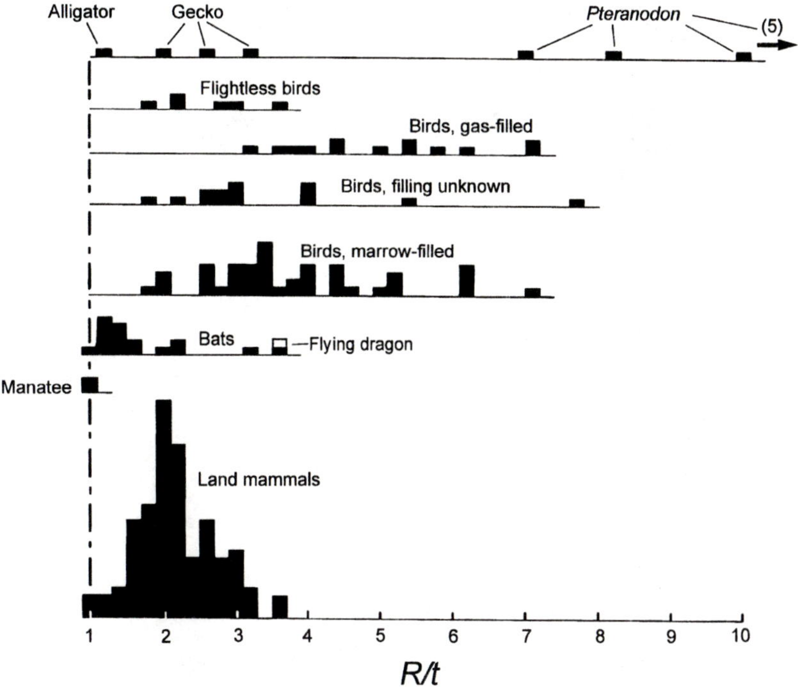

FIG. 7.8 Diagram showing the values of *R/t* for various bones. The smallest rectangles represent one bone. (From Currey and Alexander [1985] by permission of the Zoological Society of London.)

other birds, though not as low as for the main sample of mammals. The empty bird bones did not have startlingly high values of *R/t*, though they were among the highest in the birds, having a median value of 6. The flightless birds' bones were similar to those of land mammals. This general agreement between the contents of the lumen of these long bones and the relative thinness of the walls is pleasing, but it would be good if it could be tested in a much more thorough way. Cubo and Casinos (2000) provide data for a larger set of bird bones than we have shown. Their findings broadly agree with ours.

The bones of bats, *Pipistrellus pipistrellus* and *Plecotus auritus*, are particularly interesting here. Two, the humerus and the ulna of the pipistrelle, lie on the right-hand side of the general mammal distribution in figure 7.8. The second to fifth phalanges are extraordinarily long and thin, and the marrow cavity of some is a very small proportion of the total cross-sectional area. I suspect that the reason these bones have a meager marrow cavity (which is filled with yellow marrow) is that there

TABLE 7.1
Values of *R/t* for Various Long Bones of *Pteranodon*

Bone	R/t
Humerus	20
Radius	10
Ulna	20
Metacarpal	14
Phalanges	17, 14, 8, 7

is a limit in the outer diameter of the bones because they are part of the generally extremely thin wing. If this is so, then the advantage of a reduction in weight produced by a larger, thinner-walled bone would be overcome by aerodynamic disadvantages, and the bones are not, therefore, least-weight structures. On the other hand, the value for the rib of the flying dragon *Draco volans*, a lizard that glides using a winglike flap called a patagium, lies to the right of the main mammalian distribution, even though the ribs supporting it are, like the digits of the bats, extremely long and slender (Russell and Dalstra 2001). One would need to know more about the aerodynamics of *Draco* before being able to comment on this, but I suspect that aerodynamic constraints are less stringent for a glider that for aerobatic fliers like bats.

The bones of the pterodactyl are extraordinary. They belong to a specimen of the genus *Pteranodon*. It is almost certain that the bones were filled with gas. The values of *R/t* are shown in table 7.1 The values of *R/t* are so high that most of these bones would fail by local buckling before they failed by breaking in tension or compression. However, most of them are also very slender, so they would also be in great danger from Euler buckling if there were a significant axial compressive component in the way they were loaded. Such thin-walled bones would be very vulnerable to locally applied loads, and the general feeling one has about these pterodactyl bones is that they have been pushed to the limit in the pursuit of lightness. de Ricqlès et al. (2000) point out that there are rather large trabecular struts going from one side of the bone to the other, particularly near the ends of the bones. These would have had an important role in preventing local buckling. There has been a fair amount of debate about the lifestyle of pterodactyls (Pennycuick 1988; Wellnhofer 1991; Hazlehurst and Rayner 1992). However, most of what is known about pterodactyls suggests that these animals led sedate, well-ordered lives. The hurly-burly of the life of, say, a crow *Corvus* or blackbird *Turdus merulus* would have been too dangerous for them, especially the larger ones. The metacarpal and the three phalanges of the fourth digit of *Pteranodon*, as in other pterodactyls, formed the leading edge of the membranous wing. Although they are extremely thin-walled, it is interesting that, as is the case in the bat, the bones forming part of the outboard wing are less relatively thin-walled than the more proximally situated bones. It would good to know whether these pterodactyl bones were relatively highly mineralized. If so the bone would be stiffer, with a small cost in density. This might have been advantageous: these bones would fail because they were not stiff enough long before they failed by not being strong enough.

Analysis of a Permian reptile *Coelurosauravus jaekeli* shows it to have had wings, presumably used for gliding rather than flapping flight, which were stiffened by totally new skeletal elements, springing from near the vertebrae, which were not modifications of ribs or vertebral spines (Frey et al. 1997). These new elements, like the wing bones of pterosaurs, and rather unlike those of bats, and even the ribs of the flying dragon, were thin-walled. It is interesting to see bones, not themselves developed from hollow long bones, being hollow if the mechanical situation requires it.

Among the animals Neill Alexander and I examined, only the manatee *Trichechus manatus* and the elephant had solid bones, although the bones of the alligator and many bat metacarpals and phalanges were also virtually solid. This seems adaptive for the water-living animals, because they have lungs and so would tend to be positively buoyant in water. Having thick-walled bones is a way of achieving neutral buoyancy, while getting rid of useless fat. Domning and de Buffrénil (1991) have suggested a slightly more subtle reason. They point out that the total skeletal mass as a proportion of the total body mass is not greatly different from that of land mammals. Furthermore, not all the bones of the manatee and the dugong are thickened and solid; this phenomenon is restricted to the bones near the middle of the length of the body. They suggest that the main function of these ribs is to act as a balancing organ, concentrating the weight of the animal over the lungs, so that as the lungs expand and contract they do not produce a change in the trim of the animal, causing it to go nose up or nose down, which would happen if, say, the skull were heavy and dense. The lungs themselves are greatly elongated and have a spatial arrangement that that tends to confirm the suggestion of Domning and de Buffrénil.

Whales do not have very dense bones. Wall (1983) attributes this fact, reasonably enough, to the great depth of dive of most whales. At depth the lungs are collapsed and barely contribute to buoyancy. For instance, at 30 m depth the lungs have one-fourth the volume they have at the surface. As a result, in whales, selection pressure to reduce density and mass will be present, as it is land mammals. For some unknown reason, whales' bones do not divide neatly into compact bone and marrow, and many of these bones are nearly entirely cancellous, so it is not possible to calculate meaningful values of R/t.

The plesiosaurs, large extinct sea-living reptiles, had bones in the juveniles that were solid like those of the manatee, whereas the adults had bones with large central canals, which look almost osteoporotic (Wiffen et al. 1995). The implication, which is backed up by the places in which fossils are discovered, is that the juveniles were confined to shallow inshore waters and used the passive negative buoyancy of their bones. The adults ranged much more widely and, like whales, needed to develop less dense bones appropriate for their greater, lung-squeezing, depth of dive.

Apart from the elephant metapodials, which Alexander and I found to be solid, Oxnard (1993) has found other examples of land animals with solid long bones. The extinct ground sloth and some very large extinct marsupials, *Zygomaturus* and *Palorchestes*, shared two characteristics: they were very slow moving and they had solid bones. He suggests that these bones may have been optimally adapted for withstanding axial compression. If they were loaded only in compression, that would indeed be optimal, because there would be no useless marrow fat to produce. However, it is difficult to believe that these bones were not frequently loaded in bending, and that bending would not be the most dangerous mode of loading. Elephant metapodials are not aligned vertically in life, and, judging from the skeleton alone, one would guess that they would be loaded in bending. However, they are supported by a large elastic pad (Kingdon 1979) and it is possible that this prevents significant bending loading. Unfortunately, as Neill Alexander pointed out to me, we do not know the distribution of force on an elephant's foot, which might resolve the issue, because no one has yet thought it sensible to walk an elephant over a force plate.

Whatever the loading system that is most important, and is being selected for, the weight saved by having hollow bones with marrow is not impressive. Indeed, Pauwels (1974, translated 1980) is very unimpressed. "The saving in weight [for strength] is maximum if the diameter is about 65% of the outside diameter. But, even in these optimal conditions, the saving of weight in the diaphysis would attain only about 8%. It would thus be minimal." Pauwels produces a reason for the presence of bone marrow, based on the remodeling control system, which I find unconvincing, but which the interested reader can refer to. However, the question here is whether a saving of 10% or so is "minimal." Surely not. In the long bones, particularly, mass does not merely have to be carried around, but also has to be accelerated and decelerated during each stride. Consider the implications of this.

In a splendid set of articles, Taylor and his co-workers (Fedak et al. 1982; Heglund et al. 1982a, b; Taylor et al. 1982) analyzed the contribution of internal energy changes and energy changes of the center of mass to the overall power requirements of running terrestrial birds and mammals. These papers are full of good things, but for our present purposes their relevant equation is

$$\frac{\text{Power required}}{\text{Body mass}} = 0.478V^{1.53} + 0.685V + 0.072$$

where velocity (V) is in m s^{-1}, and power/mass is in W kg^{-1}. This equation is derived from observed changes in the kinetic energy of the various body segments, and does not consider the possibility of energy storage in tendons or inefficiencies in the system. The first term on the right-hand side refers to internal energy changes, that is, accelerations of the segments relative to the center of mass. The other terms refer to kinetic and potential energy changes of the center of mass. (Like many equations in the literature on scaling relations this equation is dimensionally incorrect. They are written like this to avoid great equational periphrasis, and can be taken on trust.)

Consider a horse running at 15 m s^{-1}. The equation implies that 30.1 W kg^{-1} will be required for internal energy changes, and 10.3 W kg^{-1} for center of mass changes. So, about three-fourths of the energy required is for accelerating the limb segments. Other data in these papers show that about 88% of the 30.1 W kg^{-1} is accounted for by the three distal segments of the limbs. In the horse roughly 80% of the mass of these distal segments is bone, and so about 50% of the power required for galloping at 15 m s^{-1} is used to accelerate and decelerate the bones of the distal limb. In this context, therefore, a saving of 10% in the mass of the bone of the distal segments will produce a 5% saving in power required for running, which, from the point of view of natural selection, will be a very important saving. In fact, the analysis of Taylor and his colleagues pays no attention to the energy storage in tendons, so the savings may be a bit less; even so, it is clear that quite small savings in mass in the distal bones will be selected for strongly.

Pauwels produces another argument in favor of the uselessness of hollow bones for weight saving. The argument is that many long bones are expanded at the ends and that "the weight of the bone must be much greater than that of a solid structure of the same resistance to bending and of the same length because of the filling and of the considerably denser cancellous bone." This, though true, is irrelevant because, as I shall show below, the problems of the very ends of bones are much less those of resisting bending, and much more those of cushioning impacts and allowing sufficient footing for the soft and weak synovial cartilage. Because of these requirements, the ends of bones must be expanded and, if they were of solid bone, they would be disastrously rigid.

In this discussion of hollow bones I have so far supposed that natural selection is acting to produce a minimum mass solution to various mechanical problems. However, other constraints may be important. For instance, the metabolic costs of laying down and maintaining bone and fat may be very different. If bone is more expensive than fat, and is significantly costly, then one would expect the balance to be shifted slightly in the direction of having a larger marrow cavity. However, it is so difficult to obtain reasonable values for the metabolic costs of tissues, particularly bone, that I shall not discuss this further, except fleetingly in chapter 10.

7.3.3 How Stiff Should Bones Be?

If different bones are made of bone material with the same properties, the thicker they are, the stiffer they will be. They will also, of course be heavier. Therefore, it might seem that there is a simple trade-off between stiffness and mass. However, the situation is more complex. So far in considering stiffness, we have assumed that the problem set to the bone is that it should deflect only by some fixed amount under the influence of some load. However, because bones are controlled by muscles, it is sensible to think of the mass of the muscle plus bone system together. A reasonable criterion, often met in real life, is that the *system* should not deflect at all under the influence of a load. Suppose that the muscle is capable of bearing the load. To do this it needs to have a certain cross-sectional area. The load is applied at the end of the bone, which will deflect to some extent, but the deflection can be taken up by contraction of the muscle (fig. 7.9). The question now is, What is the optimum stiffness of the bone so that the mass of the system of muscle plus bone is a minimum?

In a very clever, and to my mind unjustly ignored article, Alexander et al. (1990) have analyzed this problem. The analysis is too long to follow here, but the idea can be described in a qualitative way. Suppose the bone were thin, and therefore light but flexible. After the load was applied, the muscle would have to contract a long way to counteract the large deflection of the end of the flexible bone. Suppose, alternatively, the bone were stout, and therefore stiff but heavy. After the same load was applied the muscle would have to contract only a short way to counteract the slight flexibility of the bone. The muscle attached to the flexible bone would have to be longer, because there is more deflection in the bone to take up, than the muscle attached to the stiff bone, and

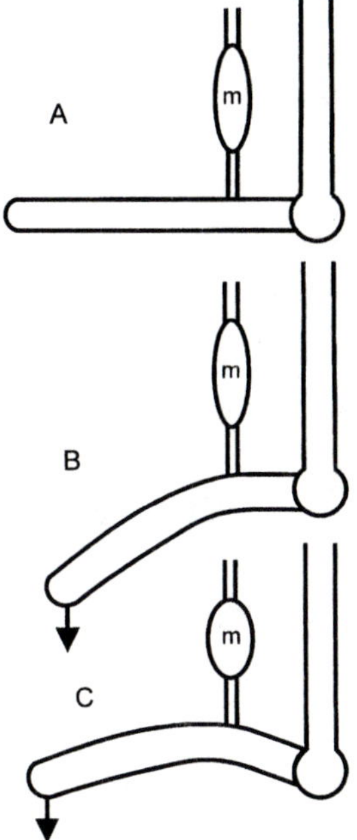

FIG. 7.9 Minimizing the mass of muscle plus bone. In these diagrams the tendons are assumed to be completely stiff. (A) The system is not loaded. The end of the bone is in the "correct" place. (B) The system is loaded. The muscle (m) does not stretch, but the bone deforms. (C) The muscle has to shorten to bring the end of the bone to the correct place.

since the muscles must have same cross-sectional area it would therefore be heavier. In other words, one is paying for lightness in the bone by heaviness in the muscles and vice versa.

Alexander et al. solved the problem for two systems, one in which the bone was considered to be a hollow cylinder, with the ratio of inner to outer diameter being a constant, and another in which the bone was considered to be a cone, again with the ratio of inner to outer diameter being a constant. To help fix ideas Alexander et al. worked the solution out in terms of what *stress* the most severely loaded part of the bone would experience, using sensible values for Young's modulus and other constants. The solutions are reasonably simple. For the cylinder it is

$$\sigma_{\text{bone,opt}} = \left\{ \frac{6\varepsilon\sigma_{\text{mus}} E[\rho_{\text{bone}}(1 - K^2) + K^2\rho_{\text{mar}}]}{\rho_{\text{mus}}(1 - K^4)} \right\}^{1/2}$$

where $\sigma_{\text{bone,opt}}$ is the stress in the worst affected part of the bone when the mass of the system is minimized, ε is the proportional shortening that the muscle undergoes, σ_{mus} is the stress in the muscle, E is Young's modulus of elasticity of bone, ρ_{bone} is the density of bone material, K is the ratio of the inner diameter of the bone to the outer diameter, ρ_{mar} is the density of marrow, and ρ_{mus} is the density of muscle.

It turns out that, whether considering the bone to be uniform or tapered, the peak stresses in the mid shaft are about 75 MPa when the system is lightest. There are many other factors that can be taken into account. For instance, it may be that the moment of inertia of the system is more important than its mass, which would tend to make the optimum stresses in the bone higher. On the other hand, it might be that the metabolic cost of keeping the bone–muscle system healthy would be important. In this case the muscles should be smaller since they are more expensive metabolically, and the bone should be larger, and therefore the optimum stresses would be lower. What is interesting about this result is that the optimum stress value is roughly the same as the maximum stresses found in the legs of many mammals during strenuous activities, such as running fast or jumping. In other words, if bones were designed to have a flexibility that would minimize the mass of the bone–muscle system, the stresses imposed by the muscles would be of the order of 75 MPa, and this is what we find that bones are exposed to.

These two analyses by Alexander and his co-workers have shown that after making fairly simple but reasonable assumptions about the way bones are required to function mechanically, one can show that their shape is well adapted to these functions.

7.4 Flat or Short Bones with Cancellous Bone

In chapter 5, I dealt with the arrangement of the trabeculae of cancellous bone and the properties of its material. However, cancellous bone sits inside cortical shells and we need to consider what the cancellous bone is doing there. This is most clearly seen in sandwich bone and in short bones like the ankle and wrist bones.

7.4.1 *Sandwich Bones*

Cancellous bone is found in sandwiches, that is, in bones that have two of their dimensions much greater than the third. Examples are the bones of the vault of the skull, scapulae, parts of the pelvis, the pectoral process of birds, and the carapace of turtles. They have a layer of cancellous bone sandwiched between two thin layers of compact bone. The cortical bone gives way quite abruptly to cancellous bone; there is no gradual increase in porosity as is seen in the cancellous–cortical transition in the ends of long bones (compare figs. 5.4 and 7.10). These flat bones are, in fact, almost classical sandwich constructions, such as are used by engineers in making lightweight panels. Unlike long bones, their

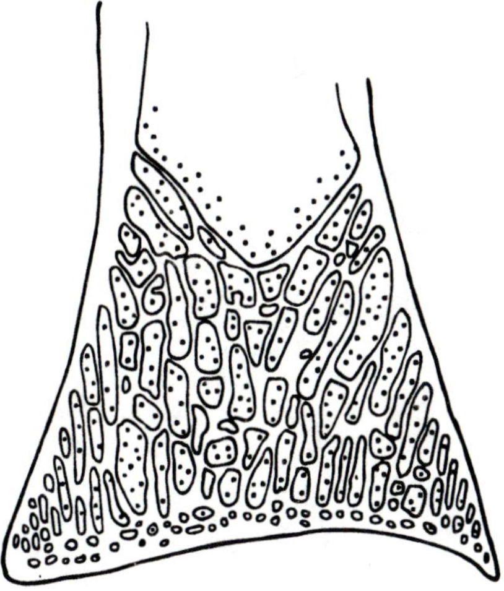

FIG. 7.10 Longitudinal section through the glenoid of scapula of a sheep. Marrow is shown dotted. Note that the cancellous bone does not start abruptly underneath the joint surface; there is a zone of transition. The cancellous struts lead from the joint surface to the cortex.

direction of loading and therefore the orientation of their neutral axis are in effect fixed.

The analysis of sandwich panels can be made in various ways. I shall use a rather simple one, which, nevertheless, brings out the main properties of sandwiches. A much more rigorous discussion than mine of the optimization of sandwich structures, which, in particular, takes into account the effect of shear stiffness, is given by Gibson and Ashby (1997, chapter 9).

Assume that the flat plate is a beam loaded in pure bending (which results in there being no shear stresses to worry about). For purposes of calculating the stiffness of the plate we can assume that, instead of being made of two materials of different Young's modulus, it is made of a single material, but that the width (not the depth) of the cancellous filling of the sandwich is proportional to its modulus (fig. 7.11). There are therefore two variables in a sandwich of bone that has a particular depth: the ratio of the depth of the cancellous bone to the overall depth, and the ratio of the Young's modulus of the cancellous bone to that of the cortical bone. Fortunately, it seems from measurements I have made that usually the thickness of the compact bone is very similar on the two sides of the sandwich. This means that the neutral plane is in the middle of the section, which simplifies the calculations.

One can perform calculations such as Neill Alexander did in Currey

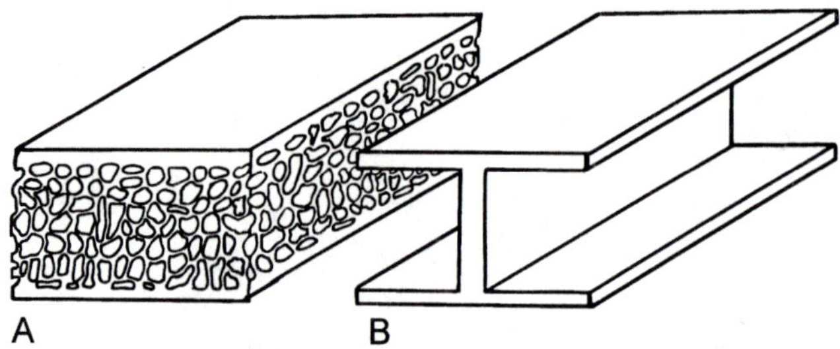

FIG. 7.11 For bending, a sheet of sandwich bone (A) can be considered as two flanges with a web connecting them (B). The ratio of the thickness of the web to the width of the flanges should be equal to the ratio of the modulus of the cancellous bone to the modulus of the cortical bone.

and Alexander (1985), and which were explained in section 7.3.2. As with the long bones, one varies the shape of the structure, and finds the minimum mass of the structure when the mass of both the bone *and* the marrow inside is taken into account. Assume that the marrow has a density half that of bone, which is roughly correct.

Figure 7.12 shows the results for three situations in which the structure of a beam of constant width is varied while the stiffness is held constant. In one it is assumed that there is only marrow between the

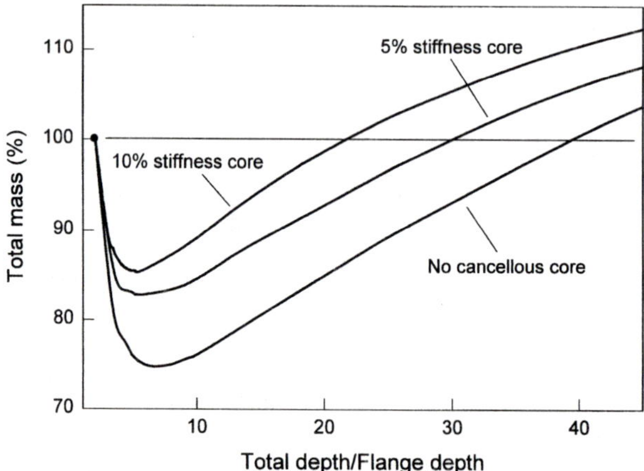

FIG. 7.12 Diagram of the mass of three types of sandwich bone as a function of the ratio of overall depth to the flange thickness. All bones have the same stiffness in bending. The mass of solid bone is taken as 100%. All types of bone show a minimum. The thin horizontal line is the 100% line. The wobble in each line just before the minimum is a glitch in the spline function, and not real.

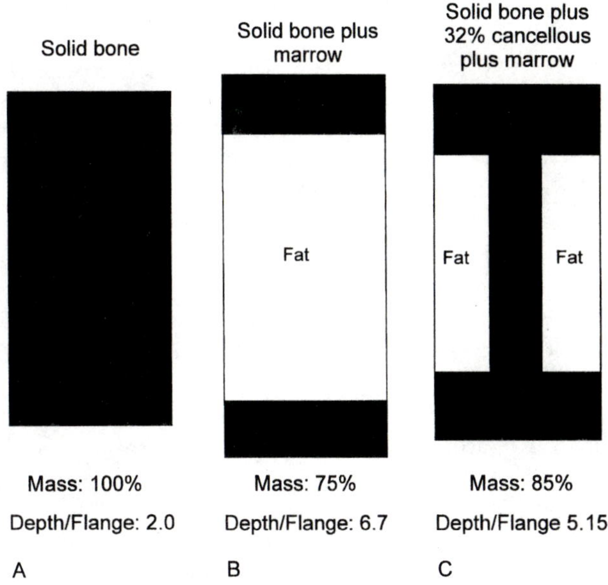

FIG. 7.13 Cross sections of beams that all have the same stiffness in bending, but different masses. (A) A beam made of solid bone. (B) The bony sheets (webs) are separated only by marrow. (C) The bony webs are joined by cancellous bone that has 10% of the stiffness of solid bone and is represented here as a central flange occupying the amount of space (32%) that would be taken up with cancellous bone of this stiffness. The mass of bone plus marrow is shown as a percentage of that of solid bone. In the cases of (B) and (C) the mass shown is the minimum mass that can be achieved relative to that of solid bone. A beam with cancellous bone of 5% stiffness would be intermediate between (B) and (C) in all respects. The depth-to-width ratio is irrelevant; the stiffness and mass both increase *pari passu* with breadth.

two sheets. This is not feasible in real life, because the sheets would have no stability, but it is instructive. In the others I have assumed that the modulus of the cancellous bone is either 10 or 5% of the modulus of bone. Because cancellous bone seems to obey a quadratic relationship between density and stiffness, 10 and 5% values for modulus imply densities of 32 and 22% compared to that of solid bone. The lines plotted are the mass of bone plus marrow that all have the same stiffness, as a function of total depth divided by the thickness of the sheet at the top and the bottom. Cross sections of three bones are shown in figure 7.13.

The maximum saving in mass is not great, somewhere between 15 and 20%, depending on the density of the cancellous material. Not surprisingly, despite my having taken a slightly different ratio of bone to marrow density, this is similar to the saving in mass found in tubular bone (fig. 7.7).

The pattern for a constant value for *I*/depth, which is a measure of bending strength, is similar to the curves shown in figure 7.12, though the curves are flatter, with their minima nearer to 100%, and the optimum ratio of cortical depth to cancellous shifted slightly to the left. Although the strength, as well as the stiffness, of cancellous bone is a function of density, as is its stiffness, we can ignore this. This is because the fracture should always occur in the compact bone. The strain at fracture of cancellous bone is slightly greater than that of compact bone, and the strains in the compact bone cortex are slightly greater than those borne by the most highly loaded cancellous material. The strain in bending is proportional to the distance from the neutral axis and so the cancellous bone, lying within the compact bone cortices, will inevitably undergo less strain than the compact bone and so the cancellous bone will not be loaded to its breaking strain before the compact bone to which it is attached is loaded to its breaking strain. I suspect, also, that the principal function of sandwich bones is stiffness, and they are probably overdesigned for strength.

What is perhaps surprising is the small extra depth of the whole beam required to allow for a considerable reduction in the amount of bone compared with the solid beam. This is brought about by the cubic relationship between the value of a volume of bone as a stiffener and its distance from the neutral axis. Bone near the axis has virtually no effect and can be dispensed with. (In a solid beam 60% of the resistance to bending is produced by the outer 20%; 6% is produced by the inner 50%.)

If the function of the cancellous bone in sandwich bones is to keep the shape of the bone constant and to resist such shear stresses as exist, another general feature of cancellous bone in this kind of position is explained: the rapid changeover from completely solid bone to completely cancellous bone whose porosity does not change much through the depth of the section. There would be no advantage, in the performance of either of the functions of the cancellous bone (keeping the cortices apart and resisting shear), in having the Young's modulus of the cancellous filling greater near the flanges. In fact, shear stresses are somewhat *higher* near the neutral axis (Young 1989, p. 97). Things are not nearly so straightforward in the cancellous bone of joints, where impact loading and complicated geometrical shapes are encountered. The porosity in these regions is much more variable.

In this discussion of sandwich bones I have always assumed that the bone has marrow in it. Indeed, it is the mass of the marrow that dictates there is a maximum in the weight saving that can be achieved. However, some bird bones have no marrow, so we should consider briefly what would happen in a sandwich without marrow. It will always be advantageous, from the point of view of decreasing the mass, to increase the second moment of area of the bone, by moving the sheets further apart and making them thinner. Probably the reason this is not taken to extremes is that having the sheet very thin will simply make it very weak and prone to local buckling. Even so, this external sheath can sometimes be very thin. Bühler (1992) figures a section of the skull of a nightjar *Caprimulgus ruficollis* in which the outer sheath is only about 25 μm thick.

Producing marrow-free sandwich bone is obviously a very difficult trick to bring off. Most bones of most birds are not hollow, and most that are hollow are long bones. It may be difficult to produce air sacs around the intricate structures of cancellous bone. However, this is quite often achieved in the brain case, and Bühler (1972, 1992) figures some extraordinary pictures of the skulls of birds in which there are a series of plates kept apart (and together) by short trabeculae. Overall the skull table is thick, and such an arrangement would be too heavy if the bone were filled with marrow. Presumably the requirement that a bone should not be too thin is also the reason, mentioned in section 7.1, why some small animals do not have sandwich bone in their skulls, but instead have a solid sheet. There must come a size at which the requirement for an adequately *robust* structure overcomes the need for a bone performing the main function of the skull—to be stiff—to be of minimum mass.

7.4.2 *Short Bones*

Long bones usually have cancellous bone only near their ends. Many shorter bones, however, are essentially thin shells of compact bone enclosing a core of cancellous bone. Examples of bones showing such a structure are wrist and ankle bones and the centra of vertebrae. The mechanical factors making it advantageous to have cancellous rather than compact bone in a stubby bone like the scaphoid are quite different from those making it advantageous in the middle of sandwich bone. In sandwich constructions the bone is subjected to bending. The outer stiff sheets bear most of the load. The cancellous bone acts mainly to keep the outer sheets of compact bone apart and to resist shear stresses. In the short bones the loads are mainly compressive and are taken through the bone by cancellous material, which therefore bears a major part of the load.

If we consider a series of bones, loaded in compression on their ends, starting as conventional long bones, and becoming progressively relatively wider and shorter, the kind of deformation they undergo will

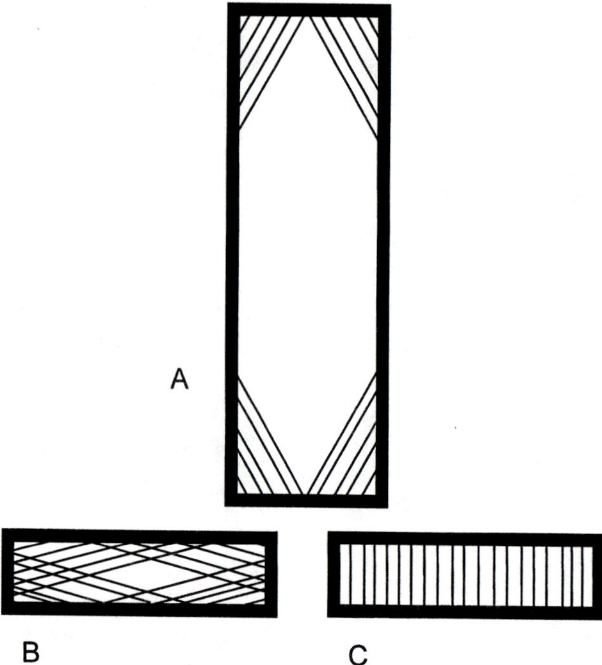

FIG. 7.14 The best arrangement of the trabeculae in relatively long and short bones loaded longitudinally. (A) In a long bone the trabeculae can transfer the load from the ends to the cortical walls. (B) In a short bone similarly arranged trabeculae would be acting at a very large angle to the longitudinal force, and would therefore be loaded severely in bending and be ineffective in preventing the endplates from deforming. (C) The trabeculae can transfer the load directly from one face to the other. In none of these diagrams are the laterally oriented trabeculae, which prevent Euler buckling, shown.

change. If the bone is long, the major deforming stresses will be longitudinal, tending to shorten the bone as a whole. For this the best arrangement of the trabeculae under the end will be to arc sideways into the sidewalls, leading the load from the end to compact bone, which is much better, weight for weight, than cancellous bone (fig. 7.14). In the very short, flat bones the end loads will tend to deform the ends by loading them as a plate, so that they bend. To prevent this bending it will not be useful for the cancellous struts to go sideways, because they would be at such a large angle to the load that they would themselves be bent. Rather, they should travel directly from one face of the bone to the other.

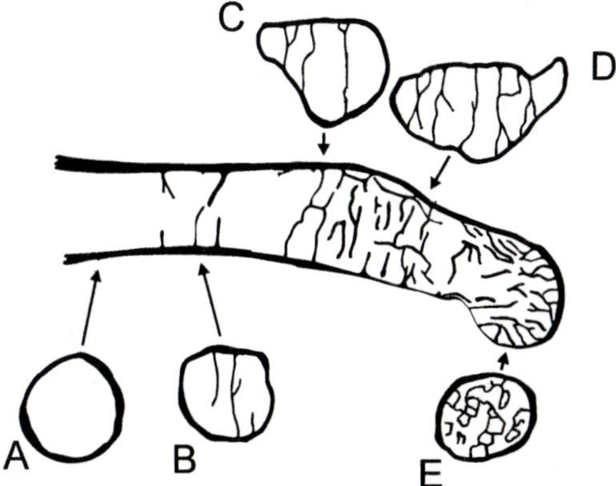

FIG. 7.15 Sagittal and transverse sections of a herring gull's humerus.

7.4.3 Synergy Between Cortical and Cancellous Bone

The way cancellous bone is nicely suited to the build of a bone is shown, for example, in the humerus of the herring gull *Larus argentatus* (fig. 7.15). In the middle of the length of the bone the shaft is a hollow cylinder with rather thin walls. Near the proximal end the bone becomes flattened on what we may call its dorsal side. The bone then dips down to the articular surface. Where the bone is circular in section A, with a fairly small radius of curvature in relation to the wall thickness, there is no cancellous bone. There is no need for it because the forces acting on the bone are carried in the shaft walls and there is no danger of buckling. In region B, however, the bone becomes flattened quite suddenly, and just where this happens, trabecular struts appear, running mainly from the flattened dorsal surface to the opposite side of the cavity. Notice, too, that these struts first appear in the middle of the bone, farthest away from the side walls. This again is what we should expect. When (D) is reached, the bone is fairly flat on both sides and is really just a conventional sandwich structure. Most of the trabeculae run dorsoventrally. At (E), just below the joint surface, the cancellous bone has a rather different function: to lead the loads from the articular surface to the cortex. Therefore, the trabeculae form tubes and are oriented in a different direction. This appearance of trabeculae underneath places where the cortex becomes flattened is very characteristic of bones in general.

It is difficult to test the contribution of cancellous bone to the mechanical properties of whole bones, because effectively the only way of doing this is to destroy the cancellous bone and to see how the bony shell behaves afterward. The destruction is a drastic treatment that might affect the cortical shell itself. Rogers and LaBarbara (1993) used this method on the humeri of wild pigeons, in which the trabeculae are very thin and could be destroyed with little force. Despite their delicacy the trabeculae had an significant bracing effect on the behavior of the humeri. Comparing treated and untreated bones, which were loaded as cantilevers, Rogers and LaBarbara measured the *discrepancy*. This is the percentage difference between the bones on the two sides normalized for the average value of the two sides. The discrepancies were 23% for stiffness, 21% for bending moment at failure, and 47% for the work required to break the specimens. In each case the braced humeri had the greater values. These large differences are especially striking since the actual mass of the trabeculae was very small compared with the mass of the bone as a whole.

One final feature of cancellous bone worth mentioning is that it very rarely occurs without an external sheath of compact bone. The exceptions are found occasionally in such peculiar bones as the long bones of whales, which may have neither compact cortex nor a medullary cavity, about whose adaptations we have little idea (de Buffrénil and Schoevaert 1988). We have seen that there are good reasons from considerations of least mass for this. There is also a more prosaic reason: naked cancellous bone would not be a good structure for interacting with the other tissues of the body. The muscles that run over the flat bones would be caught on the spiky ends, which would themselves be damaged. It will always be a good thing to have at least some solid bone on the outside of flat bones.

7.5 Paying for Strength with Mass

7.5.1 *Minimum Mass of Compact Bone Material*

So far in our discussion of whole bones we have taken values for strength and Young's modulus of the compact bony material and have worked out the consequences of differences in architecture. However, as we saw in chapter 4, bone material can have a whole range of mechanical properties, so we need to consider whether different kinds of loading imply different adaptive bone properties.

In chapter 4, I showed that the mechanical properties of bone change considerably with variations in the amount of mineralization. In particular, as bone becomes more mineralized it gets denser, stiffer, and less tough. If a bone is being selected for minimum mass that will deflect by not more than some given amount, the question arises, Will it be adaptive to increase the mineralization, thereby increasing the density, if the stiffness is thereby also increased? The obvious answer is that it depends on how much stiffer it gets for a particular increase in density. Less obvious, but also true, is that it depends on how the bone is being loaded. The answers for a bone being loaded as a short column or a beam are, for instance, different.

1. *Bone loaded as a short column.* Consider a short column (*short* meaning here that sideways buckling deflections can be ignored). It has some length L and must support a load W and undergo a compressive deformation of Z only. What value of the density ρ and the Young's modulus E will produce a bone of minimum mass? We assume that the bone is solid, square in section, and of side B. (In the cases here the results do not depend on the shape of the section as long as the shape is constant along the length.)

Deformation $Z = \text{Strain}(\varepsilon)L$
Strain $= \text{Stress}(\sigma)/E$
Stress $= W/B^2$
Therefore, $Z = (W/B^2E)L$
Now the mass $= \rho LB^2$
Therefore, the mass $= \rho WL^2/ZE$

But L, W, and Z are all given, so mass is proportional to ρ/E. In other words, the mass is proportional to the density, and inversely proportional to the Young's modulus.

2. *Bone loaded in bending.* Consider a beam of length L and second moment of area I, loaded as a cantilever with a load at its free end of W, that must deflect by an amount Z only. Again assume that the section is a solid square of side B. Then, from beam theory,

$Z = WL^3/3EI$
$I = B^4/12$, so $Z = 4WL^3/EB^4$
Mass $= LB^2\rho$
So mass$^2 = L^2B^4\rho^2$
Mass$^2 = 4WL^5\rho^2/EZ$

Again, L, W, and Z are given, so mass$^2 \propto \rho^2/E$ or, mass $\propto \rho/E^{1/2}$.

These results show that if a bone is to be loaded in tension or compression, the effect on the mass of reducing the density of the bone material is the same as increasing the Young's modulus by the same proportion. On the other hand, if the bone is loaded in bending, the Young's modulus appears as a square root term, and so changes in Young's modulus are less important than similar proportional changes

TABLE 7.2
Expressions That Must Be *Maximized* to Achieve the *Lowest* Mass for a Particular Function in Compact Bone

	Rupture	*Elastic energy absorption*	*Stiffness*	*Euler buckling*
Short column or tensile member	S/ρ	$S^2/E\rho$	E/ρ	—
Slender column	—	—	—	$E^{1/2}/\rho$
Beam of constant shape	$S^{2/3}/\rho$	$S^2/E\rho$	$E^{1/2}/\rho$	—

Note. ρ, Density; E, Young's modulus of elasticity; S, yield strength.

in density. Table 7.2 shows the expressions that must be maximized to achieve a minimum mass for bones loaded in various modes and with different imposed requirements. (The situation for rupture in a beam is slightly complicated by postyield deformation in bending [see section 3.10], but we can ignore that here.)

Although there are differences between *structures* in the function that must be minimized for minimum mass, the more obvious differences are between the *loading systems*. In particular, if deformation or buckling is likely to be deleterious, then Young's modulus should be high, but if energy absorption is important, then Young's modulus should be low. If a bone has merely to achieve a certain stiffness with minimum mass, it should be as fully mineralized as possible because, as we shall see soon, stiffness will increase more rapidly than the square of density (a condition that must be fulfilled if increasing density is to be an advantage in the case of a slender column or a beam). But, in life, it is rarely true that only one property is being selected for. The ear bones discussed in chapter 4 are exceptional in their single-minded pursuit of stiffness.

As shown in chapter 4, it is possible, by looking at the relationships between mineralization and other features and at mechanical properties, to predict what would be the effect of changing the amount of mineralization. For instance, in two papers (Currey 1987, 1990) I examined how the Young's modulus, the strain at fracture, and the work under the tensile stress–strain curve varied with the proportion of the bone that was occupied by mineral (the mineral volume fraction, MVF). The strain at fracture and the area under the curve are both quite good (though not independent) measures of the toughness of the bone. Recalculation of the calcium values in the 1987 paper allows them to be expressed in terms of MVF.

MVF can vary between 0 and 1. The density of the material when MVF is zero, that is, when it is just water and collagen, is about

1000 kg m^{-3}. When MVF is 1, that is, the material is just mineral, the density is about 2700 kg m^{-3}. As a result, MVF varies much more, proportionally, than does density. One would expect, therefore, that the power laws relating density to the mechanical properties (the exponents in the equations in table 7.2) would be larger than those for MVF. They are. Simple, though not very precise, calculations suggest that the power law for the effect of density on Young's modulus is about +4½. What this implies is that very small changes in density have quite disproportionately large effects on the mechanical properties.

Stiffness and Euler buckling will therefore both be maximized by increasing Young's modulus as much as is feasible, because the effect of increasing mineralization on density will be *much* less than its effect on Young's modulus. The optimum for elastic energy absorption and rupture of a short column is not clear because, as shown in section 4.2, there is no obvious relationship between the amount of mineral and strength. In the case of a beam, section 4.2 showed that Young's modulus and bending strength are proportional to each other. Therefore, increasing mineral until just before the bone becomes brittle (figs. 3.12, 4.7D) will be optimal for rupture and, probably, for elastic energy absorption.

This is all very well, but table 7.2 says nothing about toughness, and this cannot be forgotten. Calculations for work under the curve and ultimate strain, both good indicators for toughness, as a function of density, suggest power laws of −14 and −17, respectively! Although natural selection is probably tending toward a minimum mass solution to any particular mechanical problem, the analyses above suggest that it will be the balance between the need for stiffness and the need for toughness that will determine the value of MVF selected. Any concomitant effect on the density that changes in the mineral volume fraction produce will be unimportant.

Natural selection acting on the mechanical properties of bone cannot produce a material that is best in all circumstances. The compromise that is reached will depend on the relative importance of the different loading modes, and possible modes of failure, to the success of the animal. As we saw in chapter 4, the compromise may differ, both between bones and between different times of life.

The whole question of the design and selection of materials to perform particular functions, under various constraints, is very interestingly and clearly set out by Ashby (1999). Although that book is almost entirely concerned with human technology, I recommend it to anyone wishing to get a better understanding of minimum mass analysis.

7.5.2 Minimum Mass of Cancellous Bone

Using a minimum mass table we can also try to make sense of the location of cancellous bone. Here, density is certainly more important than it is in compact bone. Carter and Hayes (1977a) suggested that bone obeys the relationships $E \propto \rho^3$, where ρ is the density; compressive and tensile strength $S \propto \rho^2$. However, all the more recent work (for example, Rice et al. 1988; Hodgskinson and Currey 1992) suggests that the exponent for elasticity should be roughly quadratic, as it is for strength. Although the bony *material* itself can show some differences in density produced by differences in the degree of mineralization, the large differences in density shown by cancellous bone are caused almost entirely by the porosity of the bone. Overall density is almost inversely proportional to porosity and we can ignore differences in stiffness of cancellous bone produced by differences in mineralization (Hodgskinson and Currey 1990a, b). Substituting $E \propto \rho^2$ and $S \propto \rho^2$ into table 7.2, we get table 7.3.

The cells labeled Constant are those in which the efficiency is the same whatever the density of the bone. None of the cells has a function *decreasing* with density, so usually there will not be an advantage in having porous bone; in fact, just the opposite. Five of the seven cases considered here suggest that maximum density—that is, when the bone is solid—will minimize the mass needed to perform a particular function. Therefore, the statements, so frequently seen, to the effect that cancellous bone makes the bone structure lighter are strictly wrong. Moreover, this analysis, like the analysis of the hollowness of bones, ought to take account of the mass of marrow. The density in table 7.3 refers to the mass of the bone material in a unit volume of tissue. In all bones, except for some in birds and pterosaurs, there is marrow in the space left between the bony struts, and this will contribute to the mass but not to the mechanical properties. The effect of this marrow on the values in the table above depends on its density relative to that of bone. Marrow fat has roughly half the density of bone. Inserting this factor in the table above would make the table very nonintuitive. If there were no bone, and therefore the structure had no mechanical virtue, the density would nevertheless be half that of bone. The mechanical property would increase even more rapidly as a function of density than was the case when marrow was ignored. In every case, taking account of the marrow will make the cancellous bone *less* competitive with compact bone than table 7.3 suggests.

Even so cancellous bone is an important constituent of long bones, as well as wrist and ankle bones and vertebral centra. The reasons for this

TABLE 7.3
Expressions That Must Be *Maximized* to Achieve the *Lowest* M
Particular Function in Cancellous Bone

	Rupture	*Elastic energy absorption*	*Stiffness*	*Euler buckling*
Short column or tensile member	ρ	ρ	ρ	—
Slender column	—	—	—	Constant
Beam of constant shape	$\rho^{1/3}$	ρ	Constant	—

Note. This is table 7.2 modified, assuming $S \propto \rho^2$; $E \propto \rho^2$.

are various and have been dealt with above. Tables 7.2 and 7.3 refer to structures made of a single type of material. They do not refer to composite structures. Except in a few, obviously peculiar, cases, cancellous bone does not exist without at least a thin outer covering of compact bone. As shown in section 7.4, the combination of cancellous bone with a thin cortex of compact bone can have properties very different from either type on its own. It is interesting that the echinoderms (the starfish, sea urchins, and so on) have adopted a cancellous structure for effectively all their skeletal elements. The reasons for this are completely obscure.

7.6 THE SWOLLEN ENDS OF LONG BONES

Returning to long bones, we find that there are considerable geometric constraints imposed on the shapes of the ends. The shafts of long bones have particular sizes, governed by the minimum mass necessary to bear the imposed loads. For reasons discussed in the next chapter, synovial joint surfaces have to be large compared with the cross-sectional area midway along the bone of which they form the ends. The forces across the joints are broadly similar to the forces in the shaft. These forces are often *much* larger than the forces the bone seems to be exerting (Hughes et al. 1970). This is because most muscles work at great mechanical disadvantages, for reasons I shall discuss in chapter 9. Figure 7.16 shows the result of this. The top diagram shows a bone, jointed at *J*, which has to exert a force *W* at the distal end. The origin of the muscle acting on the bone is at O. The middle diagram shows the forces in the muscle and across the joint, when the muscle has a fairly large turning moment about the joint, as might be produced by a large flange on the bone. The forces in the muscle and across the joint are five times

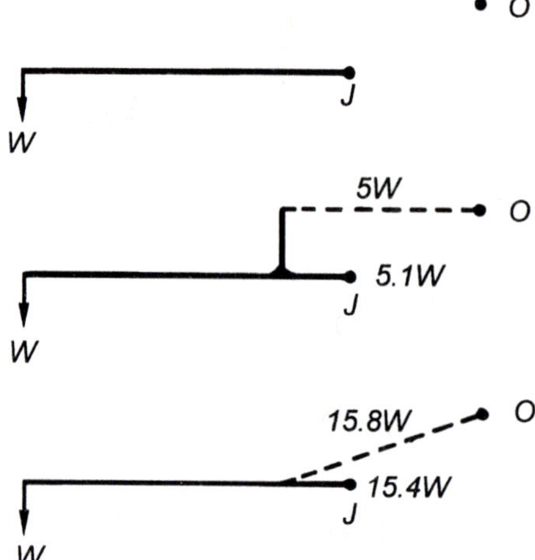

FIG. 7.16. Forces in a muscle and across a joint. A bone, jointed at *J*, exerts a force *W* at the distal end. The origin of the muscle acting on the bone is at O. Muscle is shown by the interrupted line. The diagrams show the forces in the muscle and across the joint, expressed in terms of the force at the distal end.

greater than *W*. (These forces are calculated according to the particular geometry diagrammed here and are not, of course, universally true.) The bottom diagram shows perhaps a more usual situation, in which the muscle passes quite close to the center of rotation of the joint. The forces in the muscle and across the joint are now over 15 times as great as the force being exerted at the far end of the bone.

Although I shall discuss the articulations between bones more extensively in chapter 8, there is one feature of joints that has such an important effect on the design of the shape of bones as a whole that I mention it now: bearing surfaces of most joints are lined with synovial cartilage. Synovial cartilage is not very strong. Its tensile strength (which is probably the relevant strength) is about 20 MPa, though there is an alarming decline with age (Kempson 1991), and its compressive strength is about 35 MPa (Kerin et al. 1998). Therefore, the loads being transferred from one bone to the next must be spread out over a rather large area of cartilage. The apposed surfaces in synovial joints have to slide past each other. Many vertebrate joints have large angles of excursion, and the requirement for low stress implies large radii of curvature (fig. 7.17). These together imply large ends to bones. It is informative to compare the situation in vertebrates with that in arthropods. Arthropods have rather hard bearing surfaces, capable of bearing high stresses. The load can be taken over small areas, the radii of curvature can be small, and, as a result, arthropod limb elements become pinched in at their ends, instead of being expanded.

The loads in vertebrates must, therefore, be transmitted from the cortex of one bone, with its relatively small cross-sectional area, to the

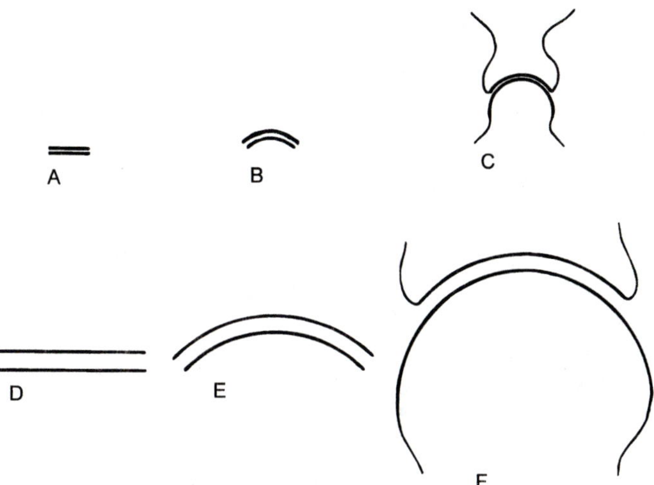

FIG. 7.17 The effect of articular surface strength on the shape of articulations. (A) A high-strength surface can bear loads over a small area, so (B) the radius of curvature can be small, allowing (C) the size of the articulation to be small relative to the size of the limb elements it joins. This state of affairs is characteristic of arthropods. (D–F) A weak surface, as is characteristic of the vertebrates, leads to large articulations.

cortex of the next via an interface of large surface area. It is universally found that the very thin subchondral bone lying underneath the cartilage is itself underlain by cancellous bone that leads the loads from the subchondral bone to the dense cortex. The adaptive reason for this is not quite as straightforward as might seem to be the case. Various possibilities for the design of the ends of bones, and their mechanical consequences, are shown in figure 7.18. If the bone underneath the cartilage were an unsupported plate, it would have to be thick if it were not to undergo quite large deformations. Large deformations would not be good for the functioning of the joint. Finite element analysis by J. Bryant shows that there is little advantage, in terms of weight and amount of material used, in having the loads borne by a plate in bending compared with having the bone end made of a solid lump of bone, in which most of the load can be taken as compressive stresses (fig. 7.18).

However, there is a grave disadvantage to this solution as compared with the solution of having cancellous bone underneath a very thin shell: solid bone would destroy the cartilage during loading in impact. Consider two blocks of bone of cross-sectional area A, one lying on top of the other, loaded statically by a load P. The stress in each block will

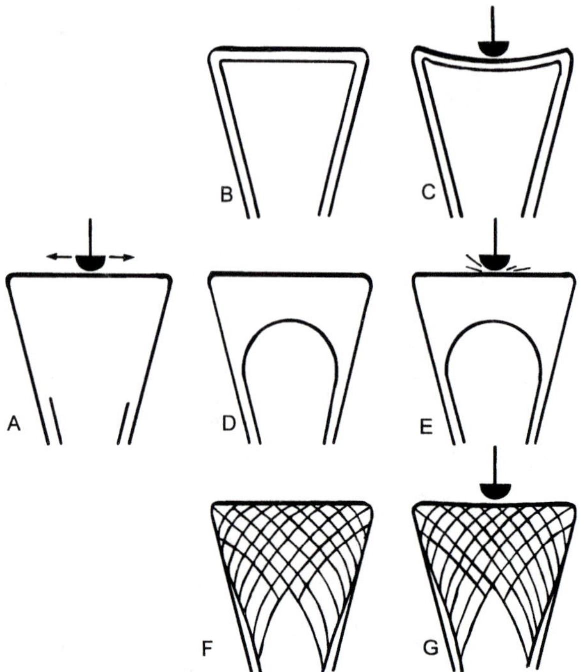

FIG. 7.18 Design for the ends of long bones. (A) The basic problem: a load, which may be variable in position, must be transmitted to the cortex away from the joint. (B) The thin cortex is continued right round the end of the bone. (C) Such a solution would allow large deformations on loading. (D) A thickness of bone that is stiff enough to prevent large deformations would (E) produce high local stresses on impact loading. (F) Cancellous bone when loaded (G) will produce adequate overall deformations to make impact loading innocuous, but will not allow large deformations of the joint surface.

be the same, *P*/*A*. The stress in one block of tissue will not be affected by the presence of the other. (I am ignoring the slight complication produced by the possible different sideways strain in the two tissues, which might lead to a complex stress system at their interface.) Probably the only important effect of a more compliant footing for the cartilage in joints would be that the bone's deformation would allow a rather larger area of contact between the two cartilage surfaces. In fact, calcified cartilage, which is a very thin layer between most cartilage–bone blocks, probably has a Young's modulus intermediate, about 0.32 GPa, between that of bone and that of cartilage (Mente and Lewis 1994). However, the layer is very thin, and I shall ignore it in what follows.

Further, consider how much energy has been absorbed by the tissue blocks when being loaded. Each block will have a stiffness directly proportional to its Young's modulus and inversely proportional to its thick-

TABLE 7.4
Amount of Energy Absorbed by a Two-Tissue Block, Loaded in Impact

		Relative thickness of bone						
		0	*1*	*2*	*5*	*10*	*20*	*50*
Young's modulus of bone (GPa)	0.1	1	2	3	6	11	21	51
	0.2	1	1.5	2	3.5	6	*11*	26
	0.5	1	1.2	1.4	2	3	5	11
	1	1	1.1	1.2	1.5	2	3	6
	2	1	1.05	1.1	1.25	1.5	2	3.5
	5	1	1.02	1.04	1.10	1.2	1.4	2
	10	1	1.01	1.02	1.05	1.1	1.2	1.5
	20	1	1	1.01	1.02	1.05	*1.1*	1.25

Note. Unity is the energy that cartilage can absorb on its own. Thickness of the bone is relative to that of cartilage, whose Young's modulus is assumed to be 100 MPa = 0.1 GPa. Young's modulus for compact bone assumed to be 20 GPa. Values in italics are mentioned in the text.

ness. The work done on the blocks will be stored between them in a ratio inversely proportional to their stiffnesses. Suppose that the material is loaded in impact and given mechanical energy that must be absorbed. The limit to the process will be when the weaker material ruptures. Because the system is supporting a particular load, and has the same cross-sectional area throughout, the *stress* in this system will be the same everywhere. The system should be so arranged so that the *energy absorbed* is maximized. In the case of joints, the strength, Young's modulus, and thickness of the cartilage are more or less fixed by the requirements of lubrication. Assume that the cartilage has a Young's modulus of 100 MPa, which is a reasonable value for high rates of loading (Unsworth 1981), though the modulus measured after two seconds or so is much lower, of the order of 4 MPa (Yao and Seedhom 1993). Suppose we have a layer of cartilage 1 unit thick and that this is able to absorb a unit amount of energy before rupturing. Table 7.4 shows the effect of adding a layer of bone underneath with various values of thickness and Young's modulus. The amount of energy that can be absorbed increases as the thickness of the bony layer increases, but this increase is much more marked when the underlying bone has a low Young's modulus. Table 7.4 shows a reasonable value for the thickness of the underlying bone: 20 times thicker than the cartilage. If the bone has a modulus twice that of cartilage, the system will absorb 11 times more energy than cartilage on its own. However, fully dense bone with a Young's modulus of 20 GPa would allow a total energy absorption only 10% greater than cartilage on its own. The presence of cancellous bone under joint surfaces, therefore, will reduce the total weight of bone needed, compared with a solid block, but will also have the probably much more important effect of allowing the ends of the bones to be loaded in impact without the cartilage being squished.

This discussion of cartilage and bone has introduced, by the back stairs, a topic that is of great importance in structures such as bones that are likely to be severely loaded in impact. These structures should if possible be built so that there is a uniform *stress* throughout the structure. If the stress is uniform, then the amount of energy that can be absorbed will be maximized.

Suppose we have a bar loaded in tension. Its length is L and it is composed of material of Young's modulus E. If the maximum stress the material can bear before it yields (which we can take as the critical point) is S and if the bar is of uniform cross section A, then the maximum energy it can absorb before failure is $(S^2/2E) \times L \times A = \frac{1}{2}(P^2L/AE)$, where P is the load. Suppose half the length of the bar were doubled in cross-sectional area. (We ignore the stress-concentrating effect of the sharp corner.) The energy that could be absorbed is now

$$\left(\frac{P}{A}\right)^2 \times \frac{L}{2} \times \frac{A}{2E} + \left(\frac{P}{2A}\right)^2 \times \frac{L}{2} \times \frac{2A}{2E}$$

$$= \frac{P^2L}{4AE} + \frac{P^2L}{8AE} = \frac{3}{8}\left(\frac{P^2L}{AE}\right)$$

Adding material to the structure has actually *reduced* the amount of energy it can absorb! Although this effect is undoubtedly important in bones, it is difficult to demonstrate that selection acts to bring it about, because selection is also probably acting to produce bones of minimum weight, which should be equally stressed everywhere, producing a similar result. At the ends of long bones, however, there is a conflict. The cancellous bone is compliant to shield the cartilage. This rather large compliance must not be carried right through the length of the bone, for it is the job of bones to be stiff. Therefore, energy from impact will be absorbed disproportionately at the ends of the bones.

An interesting case where bone should be designed *not* to absorb energy in impact is that of the head-butting dinosaurs, the Pachycephalosauria. It is obvious, from a number of lines of evidence, that a large subgroup of the adults (presumably the males) engaged in head-butting, presumably to establish dominance (Galton 1971). An analysis of the bony structure of the part of the skull that did the butting was carried out by Sues (1978). The structure of the skull consists of a rather dense array of trabeculae in the core, becoming even denser as the outside is reached. This dense bone was organized so that its grain was aligned in the direction of the forces that would be imposed on it by the impact loading. It would be very interesting to know whether the bone was very highly mineralized and therefore had a high Young's modulus. Calculating the actual stresses in stiff objects colliding with each other is extremely difficult (Young 1989, chapter 13). However, in a head-butting confrontation the best strategy might be to be as stiff as possible, so that the majority of the kinetic energy of the impact is transferred to the less stiff rival. (One of my formative moments as a student was when a technician in our geology department glued a thin glass sheet to the completely flat surface of a large rock specimen, and then hit it with a hammer. The glass, and the rock, were unharmed.) Unfortunately, because the rival is likely to have had an equally stiff head, it may be that some of the contests ended with a shattering of the heads. The fact that the bone was not completely solid does suggest that the skull was adapted to take some of the blow.

A somewhat different technique for receiving heavy blows is shown by the extinct armadillolike glyptodonts. They had very muscular tails terminating in a formidable bony club. They also had bony carapaces, and these often showed fractures, either fractures right across the carapace or depressed fractures. These were presumably caused by blows from conspecifics. There were very large spaces between the carapace and the vertebral column beneath, and Alexander et al. (1999) suppose, reasonably, that these were filled with fatty pads, which would have provided a squashy energy-absorbing base for the carapace.

7.7 Euler Buckling

If the bone becomes very thin-walled relative to its diameter, it is likely to fail by local buckling. It is thus likely to collapse before the value of load/cross-sectional area reaches the compressive or tensile strength of the bone. There is another situation in which buckling is likely to occur: this is when the bone is very slender.

If a reasonably cylindrical and stocky bone is loaded in compression with a force P, the stress at any level will be roughly P/A, where A is the cross-sectional area at that level. If, however, the bone is slender, the situation is more complicated. The bone is unlikely to be completely straight or, if it is, it must inevitably be subjected in real life to small sideways forces along its length. These sideways forces will deform the bone slightly. In either case, therefore, the straight line joining the two points at each end where the forces are acting will not pass down the middle of the bone (fig. 7.19). As a result, the axial force will have a

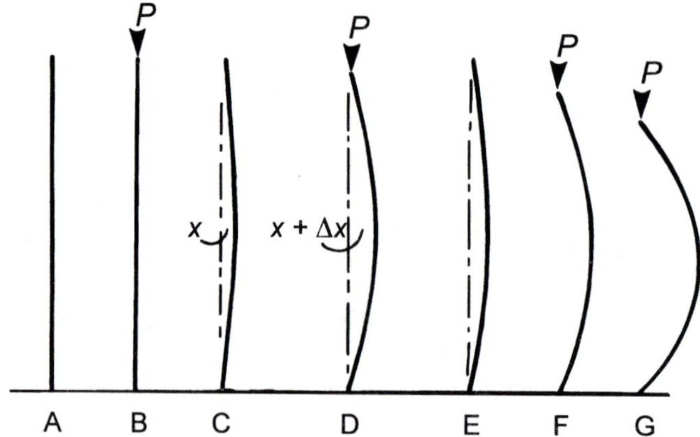

FIG. 7.19 Euler buckling. (A) A straight column is (B) loaded axially by a load P. (C) A slightly bent column, whose maximum deviation from the straight is x is (D) loaded axially. It deforms more than the straight column because it bows out sideways, but the stiffness of the column counteracts the bending moment $P(x + \Delta x)$, and the column remains stable. (E) A considerably bent column is (F) loaded axially. The stiffness of the column cannot prevent it from bending more (G), and it becomes unstable and collapses. In fact, all slender columns will buckle if the load is high enough, even if they are initially straight.

bending moment about the middle of the bone. This bending moment will tend to deflect the bone still more. This tendency will be resisted by the stiffness of the bone, which will be directly proportional to both the Young's modulus and I, the second moment of area about the appropriate axis.

However, if the bone is sufficiently slender, the axial force sufficiently large, and the Young's modulus and the second moment of area sufficiently small, an unstable situation arises. Now the deflection produced by the bending moment is great enough for the bending moment to be increased beyond the ability of the stiffness of the bone to prevent further deflection taking place. The bending moment is thereby increased still more, and so on. In this unstable situation the bone will collapse. The deformation becomes more and more extreme, resulting in large bending stresses near the middle of the length of the bone. The stress on the convex side of the bone, which started as compressive, becomes zero and then tensile. The stress on the concave side of the bone, which also started as compressive, becomes more and more compressive. Eventually, either the tensile or compressive strength of the bone material is reached and the bone ruptures.

This mode of collapse is called *Euler buckling*. It is characterized by a deformation of the whole structure rather than the deformation of a small part, which is what happens in local buckling. Euler buckling is easy to demonstrate with a long stick or a long piece of steel rod. When loaded in compression, a short piece of the material that feels completely rigid when its length is twice its diameter can be broken, or irreversibly deformed, if its length is, say, 50 times its diameter.

The real loading situations in life are more complex than the simple case outlined here, the bone of uniform cross section being subjected to bending moments by the action of muscles and by other adventitious forces. Take the simplest possible case, that of a straight, slender column whose ends cannot be *displaced*, but are not constrained against a *change in slope* (the round-ended condition). It is a standard theoretical result that the load F_e that will just cause the column to collapse by Euler buckling is $F_e = \pi^2 EI/L^2$. I is taken in the direction in which it is least. L is the length of the column.

We are interested in whether real bones are ever likely to fail by Euler buckling. Obviously, any bone can be loaded axially in compression until the load equals the critical Euler load F_e. However, if a load less than F_e ruptures the bone in compression, the value of F_e would be of academic interest only. Bone material has a compressive strength S, and so if a bone is loaded in compression it will fail by compressive rupture if force $= SA$, where A is the cross-sectional area. It will fail by Euler buckling if force $= \pi^2 EI/L^2$. From these formulae it might seem that the bone, loaded in compression, would fail first by Euler buckling if $SA > \pi^2 EI/L^2$. In fact, the force necessary to cause failure in a column of any particular slenderness tends always to be lower than the lower of these two values, unless the column is very slender or very short and stocky (fig. 7.20).

Instead of dealing with the factor I/L^2, it is often useful to think in terms of the *slenderness ratio* L/R_g, where R_g in a circular cross section is the *radius of gyration*. The radius of gyration is the distance that, were the whole area of a section concentrated in a ring that distance from an axis, would give a value of moment of inertia equal to the actual moment of inertia. It is somewhat analogous to the center of gravity of a body; for many purposes the whole mass of the body can be considered to be concentrated there:

$$R_g = \left(\frac{I}{A}\right)^{1/2} \quad \text{so} \ \frac{I}{A} = (R_g)^2 \qquad I = (R_g)^2 A$$

Therefore, $F_e = \pi^2 EA/(L/R_g)^2$. We can calculate the slenderness ratio at which bone should, theoretically, fail by both Euler buckling and by yielding in compression. Take $E = 20$ GPa $= 2 \times 10^{10}$ Pa, $S = 200$ MPa $= 2 \times 10^8$ Pa:

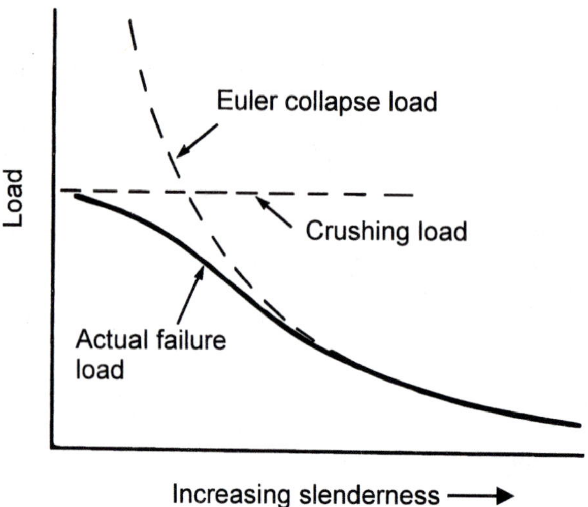

FIG. 7.20 The relationship between the theoretical Euler load, the theoretical crushing load, and the actual failure load of columns, all of the same cross-sectional shape and cross-sectional area, but of different slenderness. The actual strength is always less than the lower of the two theoretical strengths, and is considerably less in the region where the curves for the two strengths approach and cross each other.

$$2 \times 10^8 \times A = \pi^2 \times 2 \times 10^{10} \times \frac{A}{(L/R_g)^2}$$

So, $L/R_g = 32$. If the bone cross section is not circular, the least value of the second moment of area should be used, not the radius of gyration. (Parenthetically, interestingly, and somewhat counterintuitively, the least mass solution for resistance to buckling for a solid column is not a circle, but an equilateral triangle! [Keller 1960].)

The humerus of an immature gibbon *Hylobates lar* had a length of 173 mm and the midshaft had an external diameter of 7.35 mm and an internal diameter of 3.85 mm (Currey 1967). This produces a value of R_g of 2.07 mm and $L/R_g = 83.6$. Obviously, this gibbon's bone will be extremely prone to Euler buckling if loaded in compression. The Euler stress is about 30 MPa, well below 200 MPa, the compressive strength. Similarly, a stork *Ciconia* sp. tibiotarsus would theoretically undergo Euler buckling at a nominal stress of about 20 MPa, a flamingo *Phoenicopterus* sp. tibiotarsus at about 10 MPa, and a willow warbler *Phylloscopus trochilis* tibiotarsus at 20 MPa (Currey 1967). When we bear in mind that the actual collapse stresses will be somewhat lower than this, these are remarkably low values. The values for stiffness and compressive strength that I have assumed have been taken from specimens from other bones. It is likely that, because the bones are so slender and liable to buckling, the compressive and tensile strengths would never be the cause of failure in life. It would be adaptive, if this were the case, and would produce a bone of lower mass to perform the same function adequately, to increase the Young's modulus of the bone, even though it involved some small increase in density. We do not know whether this is so, but there is some evidence (fig. 4.2; Currey 1987) that slender bones have rather higher Young's moduli than shorter bones. For instance, the slender tarsometatarsus of the flamingo has bone with a Young's modulus of 28 GPa, compared with the 20 GPa I have assumed in the calculations above for the tibiotarsus.

Bones are usually loaded in such complex ways that it is difficult to know, after the event, why they failed. Borden (1974) made some observations on bones that must have started to undergo Euler buckling, yet did not fail. These are the arm bones of children who fell on the outstretched arm. The bones were bent into a bow and then, because they had been loaded into the postyield region of the load–deformation curve, remained bent when the shock was over. Adult bones do not show this behavior because they have much less ability to distort, without snapping, after yield (Höcker 1995; Simonian and Hanel 1996).

In general, however, limb bones that have a value of L/R_g sufficiently high for them to be likely to be severely affected by Euler buckling are found in animals that probably do not load them very severely. Most obvious among these are the long bones of brachiating apes and monkeys (fig. 7.21). In brachiators the large loads on the arm are tensile loads, produced when the animal is swinging on a bough while suspended by its hand. Although the arm is loaded in tension, it is almost certain (Oxnard 1971) that the bones themselves are loaded in compression. The muscles surrounding the bones exert a force slightly greater than the force exerted by the body, thereby putting the bones into compression.

Nevertheless the limb cannot do without the bones, however lightly they are loaded, because they are needed as a rigid strut to poke the hand onto the bough, where it can grasp and call into play the muscles, which do all the work. The adaptive response to these minimal requirements is a set of very slender bones, which are rarely, if ever, liable to fail through Euler buckling. This does not mean that such bones rarely break. In chapter 10 we shall see that such delicate bones are, in fact, quite often broken. The very slender tibiotarsal bones of birds are also probably never loaded very strongly except accidentally. In storks and flamingos, the legs are used in rather stately bipedal walking; such birds do not run more than a few paces.

In general, then, bones that would fail by Euler buckling if loaded in

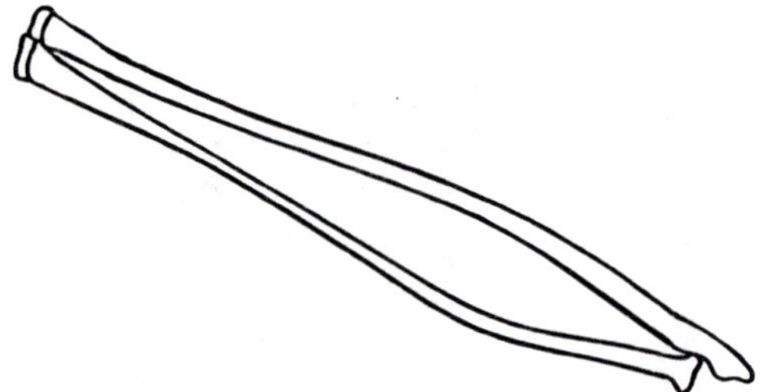

FIG. 7.21 Slender radius and ulna of an immature gibbon *Hylobates lar*. The ulna is 176 mm long.

compression are found in positions where they are rather lightly loaded or, as in the case of ribs, where the loading is not in compression.

7.8 Interactions Between Bone Architecture and Bone Material Properties

The mechanical properties of any bone are determined by two quite separate things, the mechanical properties of the bone material, and the size and shape of the whole bone—its architecture. We have seen in chapter 4 that the properties of the bone material can differ greatly between bones in ways that make sense in relation to the functions that it may have. Also, the architecture of different bones is obviously adapted to their functions. In this section I discuss cases in which we have some idea of differences in the architecture and differences in material properties, and so can see whether these are related to each other.

Brodt et al. (1999) and Keller et al. (1986) examined the mechanical properties of femora of growing mice and rats, respectively. Material properties were back-calculated from whole-bone properties and the geometry of the bones. Both groups found that there was a change in material properties and in whole bone properties, but it is not easy to determine how these changes related to each other. That is to say, it is not clear whether the bones were disproportionately thick or massive when the material properties were feeble.

Sometimes, however, it is possible to relate these changes to each other. Caroline Pond, a keen picker-up of unconsidered trifles, managed to obtain a set of 5 wild polar bear *Ursus maritimus* femurs of known weight and age (Brear et al. 1990a). The ages ranged from 3 months to 7 years (maturity occurs at about 2½ years) and the weight ranged from 9.5 to 400 kg. We found that the bone material was both weaker, having a lower yield stress, and less stiff, having a lower Young's modulus of elasticity, in the younger animals' bones, and that these differences correlated rather well with the lower degree of mineralization of the younger bones. Does this mean that the bones themselves were less strong and stiff? Before answering this, we have to have some idea about the loads on the bones. For instance, a rather feeble little bone might be adequate for supporting a little bear cub, while a much stronger bone would be necessary for a large adult male. We assumed that the characteristic loading on the bone was proportional to the weight of the bear times the length of the bone. This was not a totally arbitrary measure; a bear rearing up on its hind legs will be loading its femurs as a cantilever, and the bending moment round the ends of the femur will be proportional to weight times length, although there will also be large longitudinal stresses in the bone caused by the action of the muscles preventing the leg from collapsing.

We assumed that the strength of the whole bone would be proportional to the strength of the bone material times the second moment of area divided by the depth of the section. We assumed that the stiffness of the whole bone would be measured by the relative shape change, that is, the deflection of the end of the bone divided by the length of the bone. To calculate the deflection we measured the second moment of area at three places along the length of the femur and so obtained an estimate of the bone's architecture. Bones with the same relative shape change will be bent to the same shape, although, of course, the absolute deflection of the longer bones will be greater.

Table 7.5 shows how the strength and stiffness of the bone material and the relative whole bone strength and shape change when the size of the bear has been taken into account. Considering that the weights of the bears vary by a factor of 40 and the lengths of the bones by a factor of 3, producing bending moments that vary by a factor of 130, the range of resistances to yielding (a factor of 2.5) and in the relative shape change (a factor of 3.1) is rather small. The subadults and adults (the last three) are remarkably similar: the values for strength are virtually identical, and values for shape change, varying over a range of only 15%, despite the mass of the heaviest bear being 2.1 times greater than the mass of the lightest and its femur 23% longer, producing together a bending moment 2.5 times greater. The greater yield resistance of the cubs' bones might be an adaptation to their more rough-and-tumbly lifestyle.

The implication of these calculations, which, it must be remembered, are based on somewhat simplistic assumptions about loading, is that the architecture of the bones is rather well adapted to the loads placed on them *and* to the mechanical properties of the bone material. If the bone material in all the bones had identical mechanical properties, but the

TABLE 7.5
Properties of the Femora of Polar Bears

Mass (kg)	*Bone length (mm)*	*Yield stress (MPa)*	*Young's modulus (GPa)*	*Bone strength*	*Shape change*
9.5	160	63	6.7	100	100
58	240	88	11.2	51	182
197	396	107	16.5	39	273
251	435	123	18.6	40	292
407	490	129	22.2	41	314

Source: Derived from Brear et al. (1990a).

Note. Mass: mass of the bear; bone strength: relative strength of the bone, taking into account the mass of the animal, the length of the bone, the geometry of the bone, and the strength of the bone material; shape change: relative change in shape on loading, taking into account the mass of the animal, the length of the bone, the geometry of the bone, and the Young's modulus of the bone material. Bone strength and shape change are shown as percentages relative to the value for the 9.5 kg cub.

shapes were as we had measured them, then the resistance to yielding would have had a range of 5, rather than 2.5, and the shape change would have a range of 10.3, rather than 3.1.

A rather similar study on musk oxen, in which the authors unfortunately could not measure the mechanical properties directly, but had to estimate them for the mineral content of the bone, showed how safety factors were kept more or less under control by concomitant changes in the amount of mineralization and architecture (Heinrich et al. 1999). The situation is interestingly different from that of polar bears because the young calves have to move with the herd very soon after birth, and as a result have very slender femora compared with young polar bears.

A study in Californian gulls by Carrier and Leon (1990) showed them to have features similar to the polar bears and musk oxen, but here it was possible to study the different behavior of the leg bones (used almost from hatching) and the wing bones (which grew much in diameter only just before the juvenile started to fly). The bone *tissue* was initially weak in both arms and legs. In the legs this tissue weakness and compliance was compensated for by a relatively large cross-sectional shape. The wing bones grew in length quite steadily, though there was a spurt just before flight. However, they remained quite slender, and therefore very feeble and compliant, until just before flying started, when there was a very large growth spurt in diameter. Bones of both limbs were functional, therefore, when they were needed, but the extra diameter needed to compensate for the feeble tissue was needed only in the legs, not the wings.

7.9 The Mechanical Importance of Marrow Fat

Marrow fat at body temperature is a viscous fluid and so far I have assumed that, whatever mechanical properties it may have will not contribute to the strength or stiffness of the bone. I must discuss this assumption.

First, consider *static* bending of a long bone. One side is put into tension and the other into compression. Because a fluid is free to move and cannot resist shear forces, the marrow will merely move in the bone until it is unstressed. Therefore, marrow will be of no use in static bending. Because a fluid cannot resist shear, marrow will also be no good in resisting static torsion. Bones are probably not often loaded in tension overall (section 7.6). However, they are often loaded in compression.

Suppose bone material were isotropic and had the same bulk modulus as marrow. If a bone consisted of a hollow cylinder of bone tissue completely filled with marrow, then a longitudinal compressive load would be borne by the marrow and by the bone tissue in proportion to their cross-sectional areas. This would reduce the stress in the bone (fig. 7.22A–C). Another way the marrow might prevent bone failure would be in preventing local distortion of the thin cortex of bone near articulations. It could act by pressing hydrostatically on the walls, preventing them from collapsing (fig. 7.22D–F). For either of these strengthening mechanisms to work, however, fairly high hydrostatic pressures must be generated in the marrow. Swanson and Freeman (1966) applied cyclic loads to human femora at a frequency of 2 Hz, and showed that the hydrostatic pressures generated were very small. However, the rate of loading produced by cyclic loading at 2 Hz is much less than that imposed during falls.

Bryant (1983) has shown that in the sheep's tibia, at loading rates of 500 kN per second, which is roughly that occurring in falls, the greatest hydrostatic pressure in the marrow was only 50 kPa. This was in response to a load on the end of the bone of about 3 kN, which is about what is imposed in a severe fall, and is considerably greater than the loads imposed during locomotion. A pressure of 50 kPa is very low, and will be insignificant mechanically. Measurement of transient pressures in a viscous fluids is liable to "low" artifacts (Kafka 1993). However, Bryant (1995) calibrated for these. Furthermore, he was able to show that, if he arranged things so that the marrow was compressed by a mere 0.15% of its volume by a small cylinder pushed into the joint

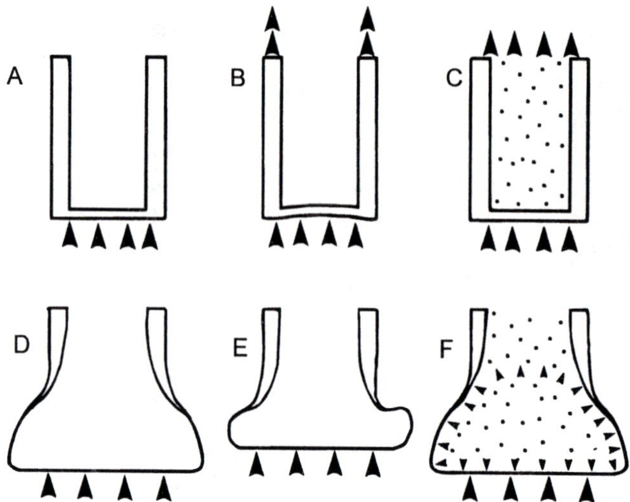

FIG. 7.22 Two possible mechanisms for a bone-protecting effect of marrow. (A) Load on the end of the bone is (B) transmitted to the cortex. In a marrow-filled bone (C) the load might be partially taken by the marrow. (D) Most long bones have thin walls near their ends. (E) Loads on the ends could cause the thin walls to buckle, or deform excessively. (F) Marrow might, by producing hydrostatic pressure, keep the thin walls stable.

surface by the load, the pressure in impact loading, measured with the same transducer, was as high as 500 kPa. The implication of this is that in impact loading of the normal bone the internal volume of the sheep's tibia changes very little. If we may extend these findings to other long bones, we can say that the marrow has little mechanical function to play in long bones, either in static or in impact loading. A theoretical analysis by Arramon and Cowin (1997) using reasonable parameters showed that the viscous interaction between the marrow and the trabeculae, another potential bone-stiffening effect of marrow, could have little effect except perhaps at very high loading rates. The question of whether it is important in bones such as the vertebral centra remains open at the moment.

An intriguing possibility, raised by Bryant (1988) as a consequence of the suggestion that there is effectively no volume change on impact loading, is that the cancellous structure is arranged so that its Poisson's ratio is about 0.5, instead of being about 0.3, characteristic of compact bone. An open foam structure can be designed to have almost any desired Poisson's ratio (Lakes 1987; Rothenburg et al. 1991). (A Poisson's ratio of 0.5 implies no volume change on loading. A negative Poisson's ratio implies that the material will increase in cross-sectional area when stretched, and decrease in cross-sectional area when compressed!) There is no evidence for or against cancellous bone being manipulated in this way.

7.10 Methods of Analyzing Stresses and Strains in Whole Bones

Determining the stresses and strains in simple specimens, such as were used to obtain the values described in chapter 3, is itself relatively simple (though not as simple as some investigators think). However, whole bones have complicated shapes and they are acted on by complex loads. It is usually very difficult to discover the stresses and deformations that are likely to occur in bone by using, for instance, the structural analysis applied to beams. Beam formulae will work well enough on long bones of fairly uniform cross section, though even here the analysis tends to break down near the ends (Huiskes et al. 1981). In complex structures, such as the ends of long bones, vertebrae, or the skull, analytical methods are almost useless. The situation is made worse by the presence of cancellous bone, so that the structures of complicated shape are also made of materials of different mechanical properties. I have dealt with some of these problems in section 3.3. Although An and Draughn (2000) deal mainly with analyzing strains in bone tissue rather than whole bones, they have several useful chapters on this subject.

Apart from beam theory there are other methods of analysis possible. One is to attach a number of strain gauges to the bone or to a model of it. This method is direct, allowing one to measure the actual strains on the surface, but it is expensive in time and allows only surface strains to be measured. The study by Huiskes et al., mentioned above, involved attaching more than 100 rosette strain gauges to a femur, which ended up looking like a Christmas tree. A short, clear account is by Biewener (1992a).

Another method is to use photoelasticity. Strained models of bones can be viewed or photographed after being placed between crossed polaroids. The strain affects the plane of polarization of light, and a colored pattern, indicative of the strain in the model, appears. Although this method produces visually convincing patterns, it is quite difficult to derive the actual strains in the model from the pattern. The method is readily applicable to plane models, but if it is extended to three-dimensional analyses, a feasible but complicated process must be carried out. The main disadvantages of this technique are the difficulties of making the models, and the fact that materials of different modulus cannot be modeled in the same structure. It is certainly capable of producing useful insights, but has not been used very often.

Two other methods can be briefly mentioned, as being in their infancy. *Holographic interferometry* makes use of a technique in which a hologram of the bone or whatever is made in the distorted and the undistorted state. The two images interfere with each other and interference fringes are superimposed on the image and can be used to determine strain (Shelton 1992). *SPATE* (stress pattern analysis by thermal emission) makes use of the fact that an object when stressed undergoes changes in temperature, the changes having different sign according to the state of stress. This is given by

$$\Delta T = \frac{-aTs}{rC_s}$$

where ΔT is the change in temperature, T is the absolute temperature, a is the coefficient of expansion, s is the sum of the principal stresses, r is the material density, and C_s is the specific heat. Tension usually leads to cooling and compression to heating. Infrared detectors scan the specimen as it undergoes cyclic loading and from the images produced the state of stress can be calculated (Duncan 1992). Both these techniques hold some promise for the future, but are tricky to set up and, in the case of SPATE, very expensive. A useful review of these and several other methods is given in Miles and Tanner (1992).

Another very powerful and extremely popular method is finite element analysis (FEA). This mathematical approach depends on the computer, because the calculations involved are unbelievably long-winded. I shall not discuss the theory of the method in any detail. Essentially, the structure is divided into a set of small elements of relatively simple shape, the mechanical behavior of each being known analytically. The program then has to calculate deflections in the structure as a whole in response to applied loads, making use of boundary requirements such as continuity of strain. The stresses are calculated from the strains.

The advantages of FEA are many. It can be used to investigate structures of any shape and complexity. There is no requirement that the Young's modulus of different elements should be the same. It is easy to alter the shape of the model, or of the mechanical properties of its constituent materials, by altering a few input variables. Graphical output is easily obtained, and stresses deep in the structure can be examined. The disadvantages of FEA are that it is an approximate method, it is *extremely* expensive in computer time, and it is not easy to check that the answers are even roughly correct, though considerable progress has been made in this respect in recent years.

Many people who use FEA use ready-made packages, which to a

large extent they have to treat as black boxes. There are methods for checking the accuracy of the results, but they are not completely satisfactory. A result of the large amount of expensive computer time required is a fatal tendency of people to do an insufficiently fine-grained analysis. In the early days of the use of FEA on bone, the late 1960s and early 1970s, there were some ludicrous cases of people applying two-dimensional analyses quite inappropriately to trabecular architecture. This is not always invalid, but in some cases it was like analyzing the stiffness of an I beam while ignoring the effect of the central web. There is still a great tendency to take the results as they come, without independent checks. FEA undoubtedly has a very important role to play in the analysis of bone, particularly as the programs are becoming able to cope with features such as anisotropy. Nevertheless, it is not rash to predict that a large number of bad and misleading papers using it will appear in the next ten years.

It is a good example of work following where the money is that by far the greatest amount of finite element analysis of bone has been done on cancellous bone, because this is of considerable clinical concern. There are few examples of really interesting objects being analyzed. One example where this has been done is the work by Rayfield et al. (2001) on the skull of a dinosaur. The analysis suggested that the safety factors in the skull were very high, leading the authors to suggest that the dinosaur used an unusual mode of subduing prey, by bashing it with its teeth, instead of the more sedate biting that *Tyrannosaurus* probably indulged in. This mode would have increased the loading on the bones considerably. Unfortunately, this paper is a preliminary one, and one must reserve judgment on its findings until one has seen all the methods described in detail.

Beaupré and Carter (1992) wrote a short, clear, account of the use of the method in biomechanics, emphasizing the need to run checks of various kinds on the results.

7.11 Conclusion

In this chapter I have shown that it is very often possible to show that the form of bones suits them well for their presumed mechanical function. With the growing power of FEA this will become easier and easier, though how often it will be *worthwhile* demonstrating once again that form follows function is another matter altogether.

I have assumed throughout that "all is for the best in this the best of all possible worlds" but, of course, things do not always work out to be exactly as theory would require, particularly if one takes only mechanical properties onto account. For example, in the semi-aquatic Nile monitors *Varanus niloticus* both males and males start off with about the same values for R/t in the femur, but in the females the endosteal cavity gets relatively larger as they undergo more egg-laying cycles (de Buffrénil and Francillon-Vieillot 2001). The females lose bone while producing the eggs, and do not fully regain it before the next eggs develop. As a result, the bones become relatively thin-walled. This is probably not mechanically the optimum state of affairs, but is probably the best that can be done from the point of view of the female monitors. So, it is always important to consider, as best one can, nonmechanical factors that may have an effect on the form of bones.

Chapter Eight

ARTICULATIONS

BECAUSE BONES are rigid, problems are inevitable when it is necessary to move one in space relative to another. The joints between bones have an important influence on the functioning and design of the bony skeleton.

The relative motions of two rigid bodies (elements) in space can be described by means of six independent modes of motion (fig. 8.1), three of translation and three of rotation. These independent modes are called degrees of freedom. If the two elements are independent, all six modes must be used to describe their relative positions and they are said to have six degrees of freedom. However, if the elements are connected in some way, then usually the number of degrees of freedom will be reduced. For instance, in the jaw joint of the badger, the condyle of the mandible fits very snugly into the mandibular fossa on the upper jaw. The lower jaw can be rotated from being tightly shut to being about 45° open. This is one rotational degree of freedom. There is also the possibility of the jaw moving as a whole laterally (fig. 8.2). The actual amount of movement in this mode is small. This is a translational degree of freedom. Apart from these two modes, effectively no other movement is possible. The position of the badger's jaw can, therefore, be uniquely defined by specifying two values only. The jaw of the cow, on the other hand, has six degrees of freedom (fig. 8.2).

The possession of a degree of freedom does not imply any particular value for the angular rotation or the translational movement possible (except that it is greater than zero). Furthermore, movement in respect of one mode may restrict or even completely abolish movement in respect of another. An example of this is the human knee. This has one degree of freedom in flexion–extension and another in rotation. When the knee is flexed at about 90° the lower leg can be rotated about its long axis. This second degree of freedom is abolished when the knee is completely straightened. The foot can still rotate, but this is achieved by rotation at the hip, not at the knee.

The number of degrees of freedom of a joint must be distinguished from the number of modes of movement possible. For instance, the acromion process on the scapula of the cat has three translational modes of movement relative to the manubrium of the breast bone, being attached to it only by a ligament whose length can be altered to some

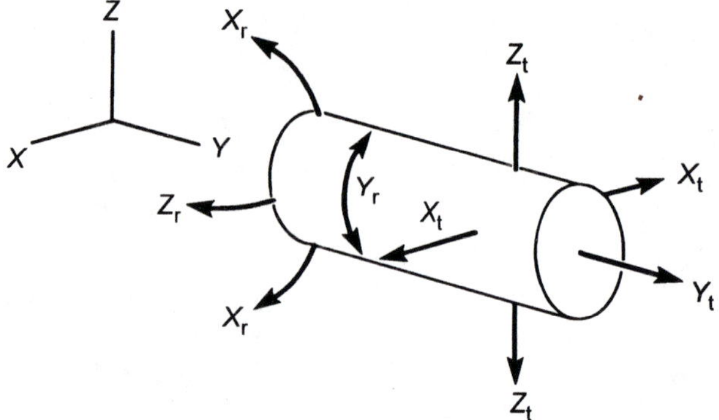

FIG. 8.1 The six possible degrees of freedom of a body with respect to fixed axes in space. Subscript r refers to rotation about an axis. Subscript t refers to translation along an axis.

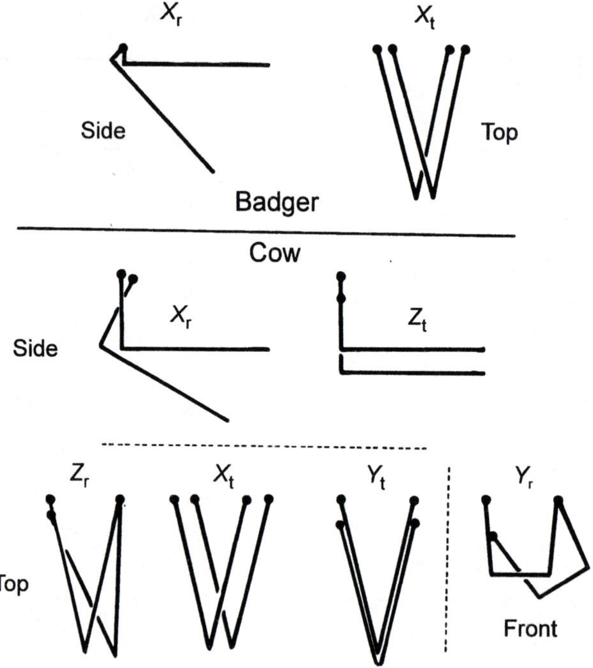

FIG. 8.2 Comparison of the degrees of freedom of the jaw in the badger and in the cow.

extent. This joint has, therefore, three translational degrees of freedom. (I am ignoring rotational freedom here.) In humans, however, the same two anatomical structures are connected by the bony clavicle, whose length is effectively constant and fixes the distance between the two structures. Therefore, although the acromion can move, relative to the manubrium, in three directions (anteroposteriorly, dorsoventrally, and lateromedially), it does not have three *independent* translational degrees of freedom. This is because if the distance between two points in space is fixed, a specification of their relative positions along two perpendicular axes immediately fixes it along the third.

8.1 The Synovial Joint

The synovial joint, or some slight modification of it, is found in all vertebrates with well-developed bones where two bones have more than a small amount of movement between them. In a synovial joint the ends of the apposed bones are capped with a smooth layer of articular (synovial) cartilage. These two surfaces are enclosed within a space, the synovial cavity, lined by a membrane, the synovial membrane, which secretes a fluid that fills the cavity, the synovial fluid. The membrane is itself enclosed in a strong collagenous capsule that may in places be so thickened as to form ligaments. There may also be ligaments (such as the cruciate ligaments of the knee) and meniscal pads, of which more later, that are functionally, though not strictly anatomically, inside the synovial cavity.

The three major constituents of articular cartilage are water (60–85%), collagen, mainly type II (15–22%), and proteoglycan (4–7%). The collagen in cartilage is different from that in bone, which is type I. The principal difference is that in type II collagen all three chains of the tropocollagen molecule triple helix are the same, whereas in type I collagen there are two chains of one type, one of another. The mechanical functional significance of this difference is not known. The proteoglycans consist of a protein core to which are attached a very large number of glycosaminoglycans, which stick out from the protein core like the bristles on a bottle brush. The proteoglycans themselves are attached, also like bristles on a bottle brush, but at a larger scale, to a hyaluronic acid molecule. The glycosaminoglycan's various groups become ionized, and these charges exert strong repulsive forces on each other, which tends to make the molecules swell. They are constrained from doing so completely because of the collagen network. Another cause of swelling is the osmotic pressure caused by the excess of ions interacting the with fixed charges on the glycosaminoglycans compared to the number in the synovial fluid. The collagen fibrils have some covalent cross-links with each other and also become entangled with the proteoglycans. The upshot of all this is that the collagen provides the tensile stiffness and strength, while the proteoglycans, which are pushing against the restraining collagen, provide the compressive stiffness.

The articular cartilage has an overwhelmingly important function: to lubricate the relative motion of the two bones it separates. This it does remarkably well. Measures of the coefficient of friction (the force required to slide the two surfaces along each other relative to the force pushing them together) are difficult to obtain, and the results in the literature are rather varied. Somewhere between 0.002 and 0.02 seems reasonable. Such values are not as good as for a ball race (0.001), but the lower values are rather better than those of a bearing hydrodynamically lubricated with oil. The methods of lubrication found between the two opposed cartilage surfaces are still under intense investigation. The main difficulty is that the properties of synovial cartilage are very difficult to investigate, because it comes in such thin layers and its properties are extremely time-dependent. Fluid is able to permeate and exude under pressure and, for instance, the apparent Young's modulus decreases by over an order of magnitude in a second or so. The system is physically, experimentally, and to some extent mathematically complex. The main point about synovial cartilage in the context of bones is that, to achieve its low coefficient of friction, it must have a rather watery consistency and as a result is very weak. This has implications for the design of skeletons, as has already been mentioned in chapter 7, and will be discussed further in section 8.3.

A description of many aspects of the synovial joint is given in Archer et al. (1999). Unfortunately, little is said about its biomechanics.

8.2 The Elbow

I shall describe one human joint, the elbow, to give some idea of design constraints that the necessity for movement impose on the shapes of the bones themselves. Three bones meet in the elbow: the humerus, the ulna, and the radius. The joint has two main functions: to allow the forearm to flex and extend relative to the upper arm, and to allow the wrist to pronate and supinate. The ulna has only one degree of freedom with respect to the humerus. The distal part of the humerus has an articulating surface, the trochlea, which is shaped like a pulley with a central groove. The trochlea articulates with the trochlear notch on the ulna (fig. 8.3). The trochlear notch has a ridge that corresponds to the groove on the trochlea. The most proximal part of the ulna is the

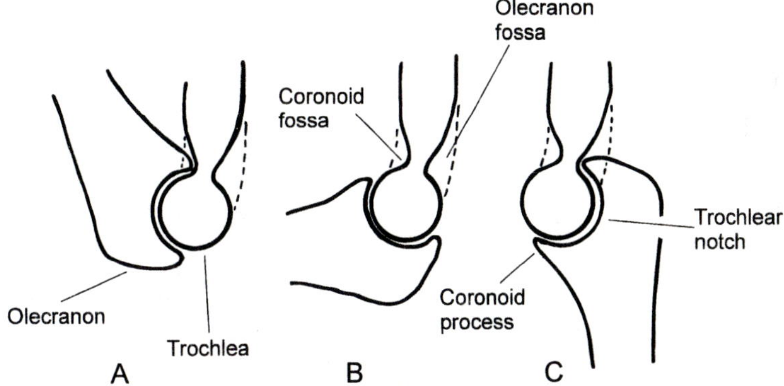

FIG. 8.3 The relationship between the distal end of the human humerus and the ulna; these bones are shown in sagittal section. The dotted lines show the thickness of the humerus out of the plane of the section. (A) Elbow fully flexed. The coronoid fossa of the humerus receives the coronoid process of the ulna. (B) Elbow at right angles. (C) Elbow fully extended. The olecranon fossa of the humerus receives the olecranon process of the ulna.

olecranon process to which various extensor muscles are attached. If the humerus had the same conventional, tubular shape at its distal end as it has at its proximal end, the olecranon process would bump against it during extension. To allow for full extension the humerus has a hollow, the olecranon fossa, into which the olecranon process can fit (fig. 8.3). The other end of the ulnar notch has the coronoid process, which in full flexion would similarly bump into the wall of a conventional tubular bone. So, there is a coronoid fossa to accommodate the process. These two fossae deeply indent the humerus; indeed, in some mammals, like the dog, there is a hole right through the humerus. The humerus is correspondingly thickened in two buttresses, which diverge sideways to the epicondyles (fig. 8.4). The articular surface of the trochlear notch occupies nearly a semicircle. As a result the humerus and ulna will be prevented from disarticulating, whatever the degree of flexion or extension, by bearing surfaces that are being pressed against each other (fig. 8.3). The penalty that must be paid for this stability—this lack of tendency to disarticulate—is that the lower end of the humerus has a near-hole in it. The sideways-flaring buttresses must be rather expensive in terms of mass compared with a simple tubular end to the bone. The ulna can rotate relative to the humerus, but about one axis only; it has only one degree of freedom.

The radius has two degrees of freedom with respect to the humerus, which allow it not only to rotate *with* the ulna during elbow flexion or extension, but also to twist *around* the ulna. This movement allows the

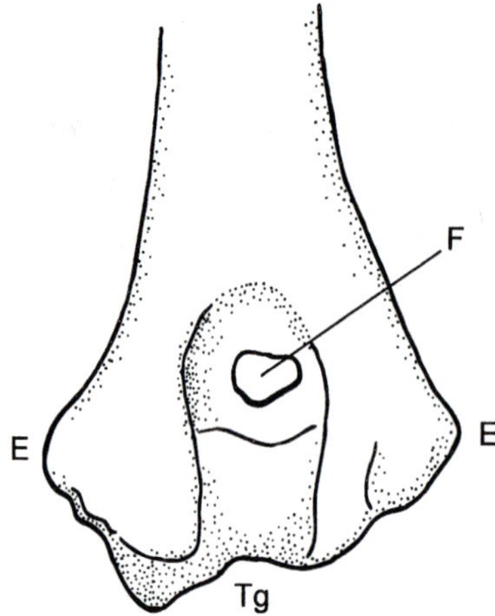

FIG. 8.4 Posterior view of the distal end of the humerus of a dog. The main shaft divides into two, and the bone is actually pierced by the coronoid fossa (F). The ulna rotates in the trochlear groove (Tg). The epicondyles (E) are the sites of the origins of ligaments that stabilize the elbow in the lateral direction.

wrist, which is attached to both the radius and the ulna, to pronate and supinate. To permit this to happen, the proximal end of the radius is hollowed out into part of the inner surface of a sphere, and this articulates with the capitulum on the humerus, a convex spherical surface. Although the joint looks like a ball and socket, it does not have the three degrees of freedom usually associated with a ball and socket. The ulna has one degree of freedom only, and the radius is bound to the ulna at top and bottom. It can rotate with the ulna, and if this were its only possible movement, it would be better to have some form of cylindrical, rather than spherical, bearing surfaces. However, because the radius must rotate around the ulna, it needs another degree of freedom at the elbow. This is allowed by the spherical bearing surfaces (fig. 8.5).

The bearing surfaces are kept in contact by a whole network of ligaments (fig. 8.6). An annular ligament wraps around the head of the radius and keeps it from separating from the ulna. Two groups of lateral collateral ligaments stretch from the epicondyles to the top of the ulna. Their function is to prevent the ulna from rotating laterally relative to the humerus and to keep the humeral, radial, and ulnar joint surfaces in contact. These potential lateral rotations of the ulna cannot readily be prevented by the action of muscles, and so the ligaments are very important. The joint can also potentially undergo hyperflexion or hyperextension. These movements can to a large extent be prevented by muscular action. There is also an anterior ligament, running from the top of the coronoid fossa to the tip of the coronoid process, which helps to prevent hyperextension.

The way in which ligaments function is not always obvious. They tend to be slack at some orientations of joint movement and taut at

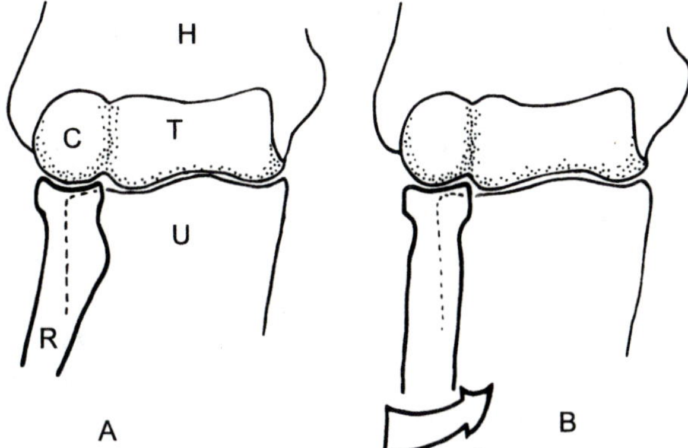

FIG. 8.5 Anterior view of the human elbow joint in the extended position. (A) The wrist is supinated. The ulna (U) engages with the trochlea (T) at the end of the humerus (H). The dished head of the radius (R) engages on the hemispherical capitulum (C). (B) The wrist is pronated. The head of the radius has rotated on the capitulum, producing the movement shown by the arrow. The ulna can move with one degree of freedom only, but the radius can rotate on the capitulum, and so can pronate or supinate the wrist at all angles of extension of the elbow.

others. Their precise functioning is a matter of geometry that is difficult to analyze in three dimensions. Alexander and Bennett (1987) have described the principles, and applied them to the sheep's tarsal joint. As they say, "This paper has used methods developed by engineers for studying the kinematics of machinery. In simple cases, little more than common sense is needed to understand ligament function, but the more formal approach is helpful when more complex systems are to be explained."

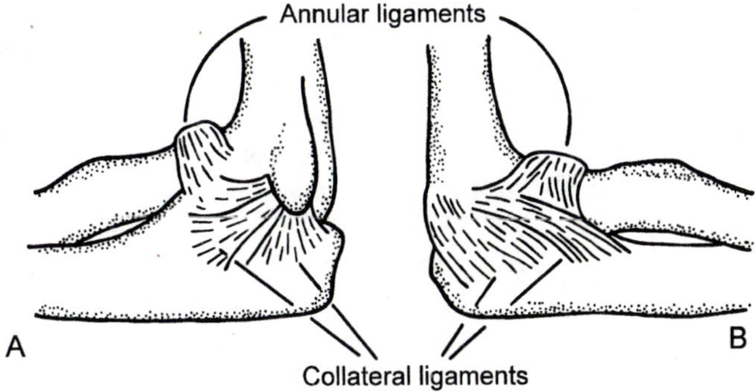

FIG. 8.6. The ligaments of the human elbow joint: (A) medial view; (B) lateral view.

Although the humero-ulnar joint looks like a simple one-degree-of-freedom joint, there is an adaptive subtlety in its anatomy. There is an incongruity between the two joint surfaces, so that the deepest part of the ulnar socket is not involved under light loading (Eckstein et al. 1995a). However, as the loading increases, more and more of the two joint surfaces come into contact, the humeral head prising the two lips of the ulnar socket apart. Finite element analysis of the situation shows that this arrangement reduces the maximum stresses in the joint because the load is more uniformly loaded in extremis than would happen if the joint were congruous when loaded lightly (Eckstein et al. 1995b). This kind of subtlety is no doubt true of many joints that have not been analyzed so fully.

The elbow is a simple joint, the main articulating bones having only one degree of freedom (more or less). Nevertheless, it does show that joints impose on the bone's anatomy in a profound way. The bones swell at their ends, there are notches, and in the case of the dog's elbow there is even a necessary hole in the bone, and there are numerous ligaments binding the bones together so they do not fall apart, and move only in particular directions.

A joint such as the knee, in which the expanded ends of the femur skate around on the relatively flat plateau of the tibia, is clearly more difficult to control. There are various devices that keep things in order. The cruciate ligaments determine how much the two bones rotate on each other and how much they slide on each other. The semilunar cartilages, or menisci, are two C-shaped pads of fibrocartilage lying between the femoral condyles and the tibia. In essence, the menisci function to fill the space left by the lack of congruity between the femur and tibia. The mensici are tethered to the tibia at the ends of the C but are able to squirm around to accommodate the movements of the condyles (fig. 8.7). They have a marked and beneficial effect on the distribution of load, making it much more even, and so lessening the stresses. When the knee is extended the menisci bear about 50% of the load crossing the joint, and when it is flexed to 90° they bear about 85% (Ahmed and Burke 1983). They also have an important joint-sparing effect when the knee is loaded in impact (Fukuda et al. 2000).

The patella is a sesamoid bone lying at the distal end of the quadriceps tendon just before it inserts onto the anterior crest of the tibia. The part of the tendon distal to the patella is called the tibial tendon or ligament. A sesamoid bone is one that develops within a tendon. The patella is probably the most important sesamoid in the human body. It articulates with the femoral condyles. Its function is to keep the line of action of the quadriceps tendon away from the axis of rotation of the knee joint. In this way the quadriceps is able to exert a large turning

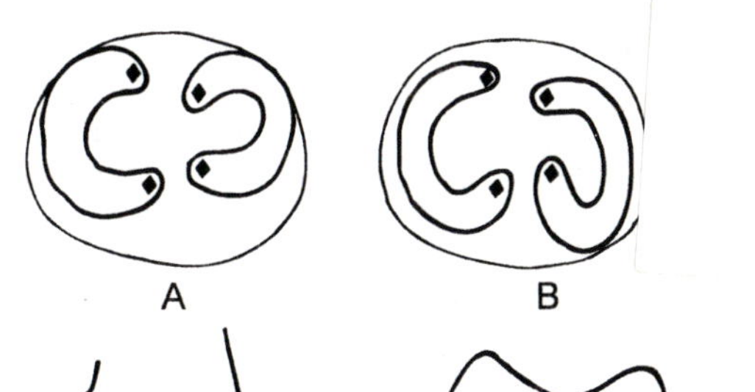

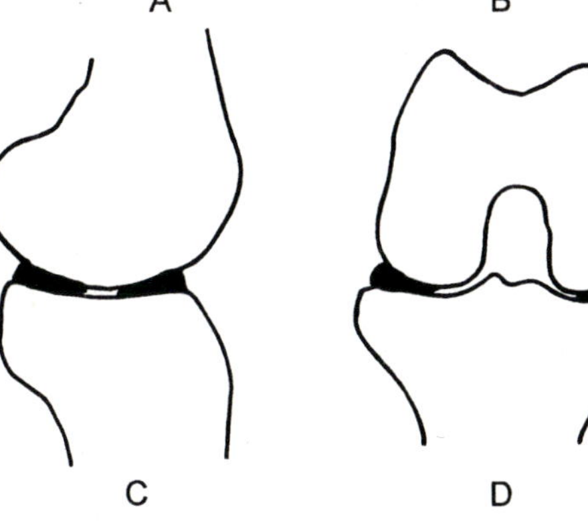

FIG. 8.7. The semilunar cartilages of the human knee. (A) Top view of the tibia with the cartilages (shown by thick lines) as they are when the knee is fully extended. The cartilages are free to move around except that they are anchored to the tibia at the points marked by solid lozenges. (B) Similar view, showing the shape adopted by the cartilages when the knee is fully flexed. (C) Side view of the fully extended knee. The cartilages, shown solid black, greatly increase the area over which the load is transmitted between femur and tibia. (D) Frontal view of fully flexed knee.

moment about the knee joint at all degrees of flexion (see also section 9.2). The geometry and mechanics of the human knee joint are well described by O'Connor et al. (1999).

A ball and socket joint, such as in the hip, has a different set of problems. The advantage of the ball and socket joint is that it has three degrees of freedom, all of rotation and not of translation, of course, and it is also stable, in that the structure of the joint itself keeps the bones in their correct relative positions. The main difficulty with this type of joint is that, if it is to have a reasonable range of excursion, there is a geometrically imposed conflict between stability of the joint and the strength of the neck of the male segment (fig. 8.8). Some of the worst effects of this can be overcome by making the greatest rotation to be a spinning of the neck about its own long axis. The main part of the bone can then be angled to the neck, and great rotatory freedom is possible. This is what happens in the human femur, which has about 160° of excursion in flexion–extension. Unfortunately, this solution has its own problem: the shaft is offset from the head, so that longitudinal forces on

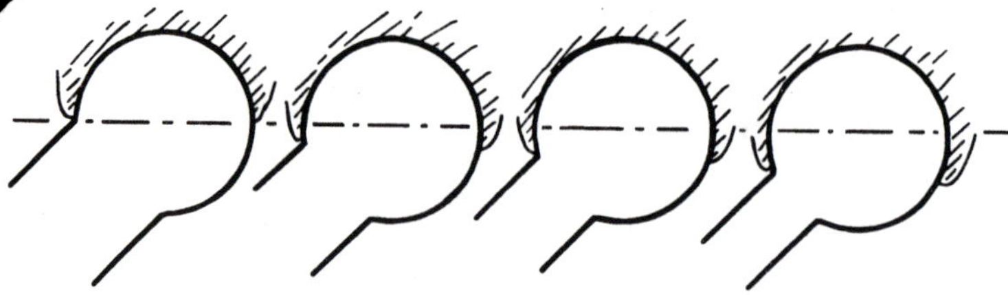

FIG. 8.8. The problem of stability in a ball and socket joint. Each of these joints has 90° of excursion in the plane of the paper. The interrupted line passes through the center of rotation of each. As the female portion (lightly hatched) surrounds the joint more and more completely, and so prevents disarticulation more and more successfully, so the neck of the male portion must get narrower relative to the head.

the shaft have to be led into the joint via bending stresses in the neck. The incidence of fractures of the femoral neck in elderly people shows that this is mechanically a very demanding design.

As a result of all these complications, the ends of bones change from being rather idealized tubular structures and can become very complexly shaped. It is difficult to draw any general conclusions about how function will influence form, but in the remaining pages of this chapter will try to demonstrate some particularly striking features of the articular ends of bones.

8.3 The Swelling of Bones Under Synovial Joints

It is very characteristic of long bones that they swell at their ends, this swelling being capped by synovial cartilage. Beneath the thin layer of cartilage is a thin layer of subchondral cortical bone. Beneath this, in turn, is cancellous bone, which extends for some distance from the cartilage before merging with the compact cortical bone of the shaft. This matter has been discussed, from a somewhat different point of view, in section 7.6.

Being watery, cartilage is very viscoelastic, but a reasonable value for the "instantaneous" Young's modulus in tension is 100 MPa (Unsworth 1981). This is two orders of magnitude less than compact bone. The tensile strength is in the region of 20 MPa, which is about one-eighth the value for compact bone. It is difficult to obtain a very secure value for the compressive strength of synovial cartilage. As Kerin et al. (1998) write, "Testing a structure such as articular cartilage requires some compromise to be made between the need to preserve structural integrity . . . and the need to isolate the properties of that structure from surrounding structures, especially bone." The values that these workers found for young bovine cartilage had a mean of about 35 MPa.

Suppose the cartilage were absent; what would be the likely compressive stresses in the bones at the joints? (This analysis ignores the hideous wear and pain that result when bony surfaces rub together.) If the joint surfaces were completely congruent, the stresses across the joint would be less than in the cortex of the bones, because the area taking the load would be greater. However, if, as is usually the case, the diameters of the joint surfaces are different, say D_1 and D_2, with the male surface having the smaller radius of curvature, then the stress is given by

$$\text{Stress} = 0.62\left[PE^2\left(\frac{D_1 - D_2}{D_1 \times D_2}\right)^2\right]^{1/3}$$

where P is the load across the joint and E is Young's modulus of the material of the apposed surfaces (Young 1989). The greater the difference in the radii of curvature, the more the load will be concentrated near one point and the greater will be the stress there. Furthermore, the stress increases by the ⅔ power of Young's modulus as this increases. If we assume Young's moduli of 100 MPa and 15 GPa for cartilage and bone respectively, the stress in naked bone would be about 30 times greater than in the cartilage-covered joint with the same radii of curvature. It is characteristic of bone in osteoarthritis that the bone beneath the joint surface thickens up, so assuming a value of 15 GPa is not out of order, though probably somewhat too high.

Day et al. (1975) investigated the human hip. On applying a load of three times body weight they obtained stresses in various regions of the acetabula of cadavers, the higher values being about 3 MPa. Such values are in line with stress values from other joints (Yao and Seedhom 1993). This would result in a stress at the bony apposed surfaces of about 80 MPa. The synovial cartilage also makes the radii of curvature of the male and female parts of the joint more similar. Synovial cartilage certainly acts, therefore, to reduce local stresses by increasing the area of contact, though the size of this effect is not clear.

An old article by Freeman and Kempson (1973) illustrates the stress-reducing properties of a soft sheath well. Plexiglas models of the two parts of the hip joint are pressed together and viewed through crossed polaroids. When they are unsheathed, there is a region of intense stress concentration, shown by tight-packed black and white fringes near the area of contact. When sheathed by rubber the stresses are clearly much lower.

(Playing with Plexiglas sheets is quite a valuable Friday afternoon-type activity to get some insight into the effects of forces acting on bones. In particular, one quickly gains a feeling for St. Venant's principle. This states that if all the forces acting on the surface of a small part of a body are replaced with an equivalent force, with the same resultant force and couple, the stress state is negligibly changed at large distances compared with the dimensions of the part. So, for instance, if a piece of bone on a testing rig is able to withstand the local forces, it does not matter if the forces are exerted by knife edges, rounded probes, or even rubber-covered probes, the general state of strain in the specimen will be the same. In the Plexiglas model of the femur described above, the states of strain in the sheathed and unsheathed model are virtually the same a little distance from the joint surface.)

The cartilage might also act as a cushion for damping dynamic loads. This has been discussed theoretically in section 7.6. It was investigated experimentally by Radin and Paul (1971). If the greatest stress in the cartilage is to be small, the cartilage should be compliant, loaded over a large area, and thick. In fact, it is compliant, but it is also very thin, particularly compared with the thickness of the subchondral cancellous bone. In their experiments Radin and Paul (1971) found that the subchondral bone was only about ten times stiffer than the cartilage capping it. Therefore, as the area loaded was effectively the same in the two tissues, we should expect that 10 mm of bone would have the same impact stress-attenuating effect as 1 mm of cartilage. In fact, it turned out that the cartilage was rather ineffective. Radin and Paul dropped weights onto the phalanges of cows and measured the peak load. Removing the cartilage increased the peak force at the far end of the phalanx by about 9%, whereas removing about half the length of the phalanx (this would include compact as well as cancellous bone) increased the peak force by a further 31%. Therefore, the cartilage, though significant, clearly does not have a very important function in the attenuation of impact forces.

Burgin and Aspden (2000) have shown that the boot is really on the other foot, and that the bone protects the cartilage. Loading a bone + cartilage sample in impact produced little damage to the cartilage, but if the bone is removed and the cartilage sits on a steel plate, it is severely damaged by the same loading. Fukuda et al. (2000) investigated impact loading of the knees of pigs in various orientations. They found that the menisci were very important in reducing the load just underneath the joint surface, the maximum load increasing two- to fivefold, but that removing the layer of cartilage had considerably less additional effect, the maximum load increasing by about 20%. The relationship between loading rate and any protective effect is unclear.

As we have seen, the swelling of bones at their ends to a large extent allows a sufficient area for cartilage. However, there is another function that this swelling will serve in many joints. This is to increase their

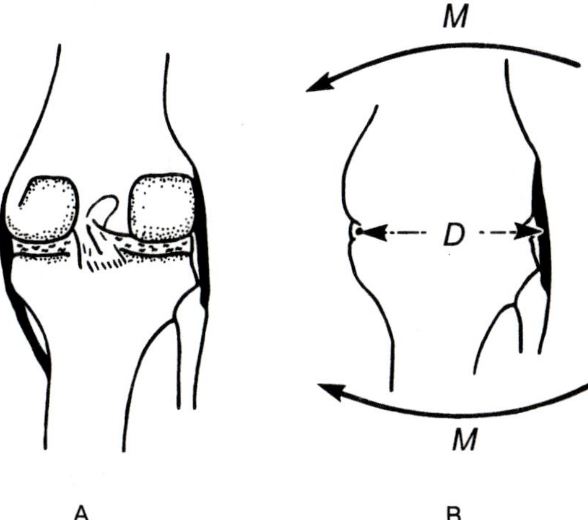

FIG. 8.9. (A) Posterior view of the human knee joint, showing the collateral ligaments (solid black) and semilunar cartilages (squiggles). The femur and tibia are both expanded laterally near the joint. (B) Diagram showing how the expansion reduces the tensile load in the collateral ligament when the joint is subjected to a bending moment *M*, because the force in the ligament is *M*/*D*. The larger *D*, the smaller the force in the ligament.

lateral stability (Pauwels 1950). Synovial joints are almost without exception stabilized by ligaments, which prevent movement in inappropriate directions. For instance, the tibial and fibular collateral ligaments run down the sides of the human knee joint and prevent the leg from bowing inward or outward. If the knee is hit, or otherwise subjected to a bending moment in the lateral direction, it is the ligament on the tensile side that is primarily responsible for preventing disarticulation. (If the adventitious loading is in the direction of usual movement, then the muscles will also be concerned with resisting it, as well as ligaments such as the cruciates.)

For a bending moment *M*, which will not depend on the morphology of the joint, the tensile force in the ligament will be *M*/*D*, where *D* is the distance from the ligament to the furthest point in the joint that is bearing load, in effect the hinge point. So, the greater the value of *D*, the more resistant the joint will be to movement in the wrong direction (fig. 8.9). A survey I made of a good number of mammalian hinge joints showed that the widest point of the bone, which usually lies near the axis of rotation of the bone with respect to its neighbor, is about twice as wide as the bone in its midshaft region, measured in the same direction. If the bones did not expand at the ends, the collateral ligaments

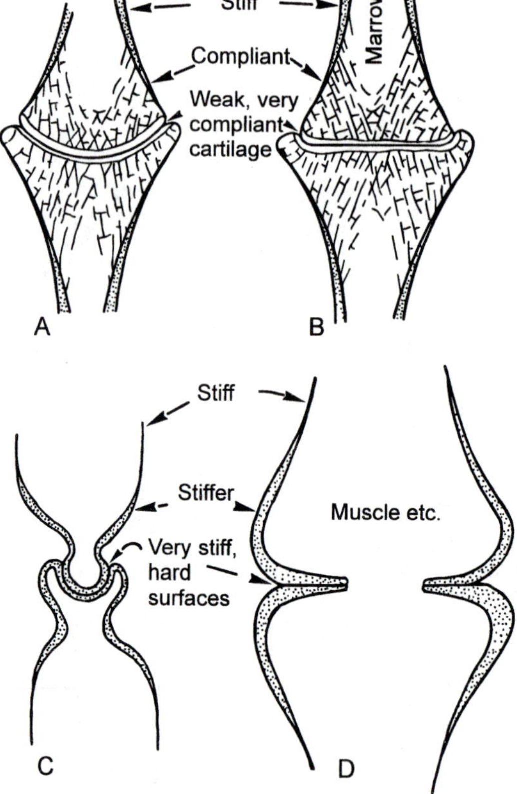

FIG. 8.10. Comparison of a typical synovial joint (A, B) with an arthropod sliding joint (C, D). In (A) and (C) the view is along the axis of rotation. In (B) and (D) the view is normal to the axis of rotation. (See also fig. 7.17.)

would be exposed to about twice the force for any given bending moment.

An instructive comparison can be made with the arthropods, to show that joints do not all have to be like vertebrate synovial joints. The arthropods have exoskeletons, with the living tissues lying inside the practically nonliving cuticle. The sliding joints, which are usually simple hinges, have an almost unmodified cuticle at the bearing surface, and this is strong and stiff. So, the joint surfaces can deal with high contact stresses, and there is no need to spread the load out. As a result, the joint surfaces have a relatively *small* cross-sectional area compared with that of the middle of the element. These differences are shown in figure

8.10. Indeed, as the length of the hinge joints is restricted because the living tissues have to pass through from one element to the next, the loads on the structure near the joint are higher than elsewhere, and the cuticle is stiffened. This is quite the reverse of the situation found in synovial joints. One point of similarity in the two types of joints is that in both there is a tendency for the elements to broaden out in the line of the hinge, reducing the disarticulating moment.

Unfortunately, we are not out of the wood yet in our discussion of the consequences of the weakness of articular cartilage. The expanded ends of the bone are themselves weaker than the main part of the shaft, and this has bad effects on the ability of some bones to absorb kinetic energy without breaking, such as during falls. In my laboratory we studied the load/strain behavior of the human radius at two levels when it is loaded in compression (Horsman and Currey 1983). At the distal level there was much cancellous bone, with a thin sheath of compact bone around the outside. At the more proximal level the bone was completely compact. The amount of bone material at the two levels (determined by photon absorptiometry) hardly differed. However, the stiffness, as measured by strain gauges, was about 1.7 times greater at the more proximal level than nearer to the joint surface. The bones were loaded until they broke; this always happened distally. Even though the strength varied greatly according to the amount of bone material, which itself varied greatly according to the age of the subject, the *strain* at fracture was almost constant. The bone structure fails when the strain in the cortex exceeds some particular value.

The significance of these results is this. Imagine a radius that in a fall has to absorb a certain amount of energy. What is the best disposition of the bone material so that it nowhere reaches a strain at which it will break? We can consider the bone as consisting of a number of segments of length L. If a segment has a cross-sectional area of bone material A, behaves linearly, has a Young's modulus E, and is loaded elastically to a load P, then the energy stored is $P/2 \times \Delta L$, where ΔL is the change in length ($= \varepsilon L$, where ε is the strain). The energy stored

$$= \frac{P}{2}\varepsilon L$$

$$= \left(\frac{P}{2}\right)\left(\frac{P}{AE}\right)L,$$ where E is Young's modulus (strain = stress/Young's modulus)

$$= \frac{P^2 L}{2AE}$$

For equilibrium, the force all along the bone must be the same. There will be some value of strain, P/AE, which will be greater than the material can bear. Since P and E are, by definition, the same at all levels, so too should A be, because then the strain will be the same everywhere and the whole bone will, in the best situation, be loaded to just under the failure strain. (If one part of the bone has twice the cross-sectional area of another, the strain and, therefore, the total energy stored will be half that stored in the more slender part.) However, things are more complicated if the Young's modulus varies as well as the area. If the value of failure strain is fixed, but the value of E in the fatter segment is half that in the slender segment, then the energy stored in each segment will be the same.

Now the radius is more compliant near its end than in the midshaft. The adaptive reason for this—mainly to reduce the peak stresses in the cartilage—has been discussed in section 7.6. I also discussed there the advantage of having uniform stiffness in a bone loaded longitudinally in impact. The situation in the radius is not good, therefore, because the bone will have to absorb a disproportionate amount of the energy. Because the end of the radius consists of a very thin sheath of compact bone with cancellous bone inside, the stiffness will, to a large extent, be determined by the stiffness of the compliant cancellous bone. At the end of the radius the compact sheath and the cancellous core are acting almost in parallel; that is, the strains in each will be the same. Stress = Strain × Young's modulus. Because the Young's modulus of the cancellous core is less than that of the sheath, the core will be at a lower stress than the sheath. This is not in itself significant, because the core is also weaker. What is important is that the cancellous bone will not yield first, because it yields at a higher strain than does compact bone. The compact sheath has to bear a disproportionate share of the load and yields and breaks at a load that is determined largely by its own meager cross-sectional area; this results in a breaking strain being reached at a fairly low load. The radius, when loaded longitudinally, tends to break at its end, producing, in humans, a Colles' fracture. Although the end of the bone has to be compliant in order that the cartilage be spared, the whole bone cannot be compliant, because it is the function of bones to be stiff. As a result, bones like the radius are not designed particularly well from the point of view of energy absorption.

There is a final twist to the story in these particular experiments. The impact and static strength of the radius declines with age, concomitantly with a reduction in the amount of bone mineral. However, the stiffness at the lower level declines more rapidly with age than the stiffness at the higher level. Because the deformation at fracture is effectively constant, and because the bone fails when any part of it is loaded to a strain greater than it can bear, the disparity in the amount of energy absorbed by the end, and by the shaft, increases with age.

8.4 Intervertebral Disks

So far, I have described synovial joints only. The joints between vertebral centra work on a different principle, because the relative motion of the bones is allowed by a flexible connection rather than by the sliding of two cartilage surfaces. The flexibility is produced by the intervertebral disks. A disk consists of a central nucleus pulposus surrounded by an annulus fibrosus (fig. 8.11). The nucleus is a ball of very hydrophilic jelly, bounded peripherally by collagen fibers. The annulus consists of layers of lamellae. Each lamella has collagen fibers, oriented almost in the same direction, and its fibers are at an angle to those of its neighbors, usually at about the same angle from the long axis of the spine, but in opposite directions. Near the outside of the disk the fibers are at an angle of about 65° to the long axis of the vertebra; this angle decreases smoothly through the thickness and approaches about 45° close to the nucleus. The collagen fibrils are tied to the cartilaginous end plates of the disk. Near the outside the collagen is mainly type I, the proportion declining smoothly, being progressively replaced by collagen type II as the nucleus is approached (Cassidy et al. 1989).

The intervertebral disk is a device that makes use of the tensile strength and tensile stiffness of collagen, even though the disk is loaded as a whole in compression. This state of affairs is brought about by the nucleus pulposus. The loads applied to the disks are sufficiently small for the watery nucleus to be considered incompressible; that is, its volume cannot change, although its shape may, and indeed does, alter.

The hygroscopic gel of the nucleus draws water in, and does so until the tendency to increase in volume is matched by the resisting tension in the fibers of the annulus and the tone of the muscles of the vertebral

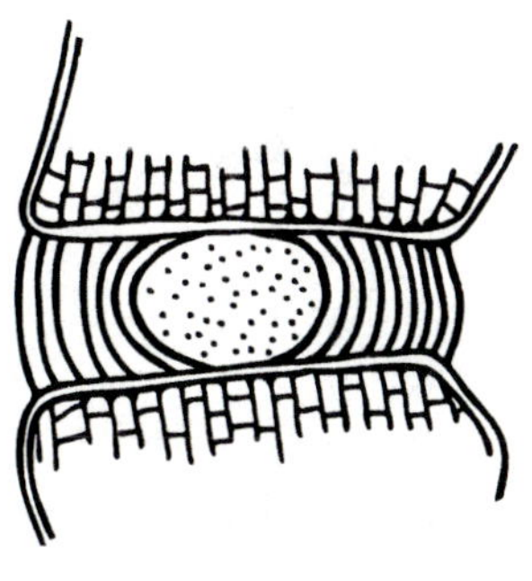

FIG. 8.11 Highly diagrammatic cross section of the intervertebral disk between a sheep's lumbar vertebrae. The nucleus pulposus is shown dotted and is surrounded by the lamellae of the annulus fibrosus. Beneath the thin cortical bone at the ends of the centra lies the cancellous bone.

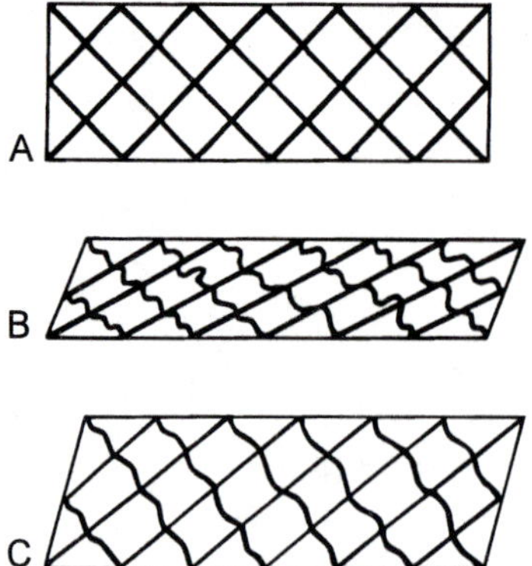

FIG. 8.12 How the intervertebral disk resists torsion and shear. (A) Undistorted shape. (B) The disk has sheared but the southwest–northeast fibrils are shown as having the same length, and therefore not being stretched. This would require a reduction in volume, which would be resisted by the nucleus pulposus. Alternatively, (C) there is no change in volume, and this would require the southwest–northeast fibrils to be stretched. Their stiffness would limit the amount of shear possible. In practice, both these effects come into play.

column. A hydrostatic pressure of about 2 MPa develops in the nucleus and there will be a tensile stress in the collagen fibers of the annulus to produce equilibrium. The fibers of the annulus are, therefore, slightly prestressed in tension before any compressive load is applied to the disk. When load is applied, the virtually incompressible nucleus is squashed into an ellipsoid that protrudes laterally. This increases the tension in the fibers of the annulus. Furthermore, when there is bending between two vertebral centra, the nucleus distorts so that extra tension is placed on the fibers on the tensile side. If there is shear between adjacent centra, the obliquity of one set of the fibers in the annulus will tend to make the distance between the centra decrease, and this will be resisted by the nucleus (fig. 8.12).

It is clear, therefore, that, for all normal loading situations, the two-component structure of the disk puts the collagen fibers of the annulus under tension. If the spine is loaded in tension, the nucleus plays no part, but in compression it is important in allowing the collagen fibers to be stretched. This description of the functioning of the intervertebral disk is highly simplified and, indeed, the complicated mechanics of the disk are seriously difficult, and not yet well worked out (McNally and Arridge 1995; Klisch and Lotz 1999).

Unlike the situation in bones with synovial joints, the presence of a disk imposes no particular shape on the ends of the centra.

8.5 Sutures

The joints discussed so far have obviously been designed in some way to allow movement, often a considerable amount of it, between the bones on either side. However, particularly in the skull, there are sutural joints, in which the adjacent bones are separated by a very thin layer of collagen. The structure of sutures does not immediately suggest their function. Are they designed to allow relative movement or to prevent it?

Sutures sometimes show a very precise adaptation to their function; at other times this is less obvious. Herring and Mucci (1991) described the sutural anatomy in the zygomatic arch of pigs. The suture has a vertical section, which strain-gauging showed to be loaded in compression, and a horizontal section in which masticatory forces tended to pull the bones apart. The vertical part had quite complex chevron-shaped interdigitating ridges, and the fibers joining the two bones were arranged so that compression of the joint would result in their being stretched. The horizontal part had a simple tongue and groove structure, but, again, the connecting fibers were arranged so that any tendency for the two bones to move apart would stretch the fibers. This seems, in general, good design, particularly if two parts of the bones in the horizontal section had to move past each other to some extent during growth, and therefore could not have complex interdigitations that would require constant remodeling if relative movement were to take place.

Skull sutures often, particularly early in life, have an intricate interdigitated suture line. This prevents the adjoining plates from moving much under shearing or distracting forces. However, for the skull vault to grow in size overall requires an enormous amount of bony remodeling. In later life, in most mammals, the skull sutures fuse, but in some animals they remain into adult life and may have mechanical function. Jaslow (1990) tested the static and impact strengths in bending of the frontal–parietal sutures of goats *Capra hircus*. The strength increased with the amount of interdigitation, as might be expected, but even the strongest was only about half as strong as specimens taken from nearby in the skull. On the other hand, the energy absorbed by specimens with a suture in them, when loaded in impact, was considerably greater than the energy absorbed by nearby skull specimens. Jaslow's specimens were cut from the skull. Jaslow and Biewener (1995) strain-gauged the horncores, the cranial bones, and the sutures of female goats' skull, but left the skulls intact. They loaded the skulls in impact in a more or less lifelike way. They found that the strains across the sutures were about ten times as great as those in the surrounding bone, and also that in going from one bone to another the strains about halved. This implies a considerable shock-absorbing effect by the collagenous tissues of the suture.

8.6 Epiphyseal Plates

There is another type of joint that, although it has an important function in the growth of bones, is a source of weakness. This is the epi-

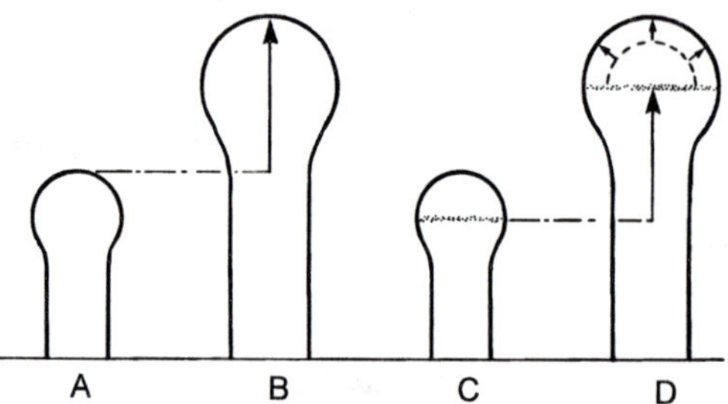

FIG. 8.13 The function of epiphyses. (A) The short bone has to become longer. (B) The amount of growth in length is shown by the arrow. (C) A bone with an epiphyseal plate (shown dotted) can divide its growth into two parts. (D) The main growth is produced by the epiphyseal plate. The growth of bone underneath the lubricating synovial cartilage can take place much more slowly.

physeal plate. Many bones of birds and mammals, and some reptiles, grow in length by means of epiphyseal plates. A problem with growing long bones is that the ends have to articulate with other bones, and it would clearly be awkward to have rapid growth taking place actually at a sliding surface, exposed to quite large loads. The main growth takes place well away from the articulating surfaces. In many anamniote vertebrates, such as fish and amphibia, there is a large cartilage epiphysis between the line of bone growth and the articulation. This, however, means that if large forces are applied across the joint, the ends of the bones, being cartilaginous, will deform considerably. The epiphyseal plate is a device to get around this difficulty. The epiphyseal cartilage develops one or more secondary centers of ossification, and the end result is that the cartilaginous epiphyseal plate is sandwiched between two bones and the whole system is fairly rigid. The epiphyseal bone grows, too, but not at the frenetic pace often necessary for the bone as a whole (fig. 8.13).

Epiphyseal plates are an excellent solution to the problems of growth, but they are mechanically weak, and this produces design problems. There inevitably remains, until skeletal maturity, the cartilage of the growth plate, and this tends to be weaker than the bone around it. Fracture across the epiphysis is a common form of injury in skeletally immature humans (Aronsson and Loder 1996; Ogawa and Ui 1996). Bright et al. (1974) showed that if the tibia of the rat is subjected to shear, the cartilaginous epiphyseal plate breaks at a rather low force, and even if the epiphysis does not break off, the cartilage plate may show tears in planes where the highest shear strains occur.

Smith (1962a,b), in two undervalued papers, discussed ways in which

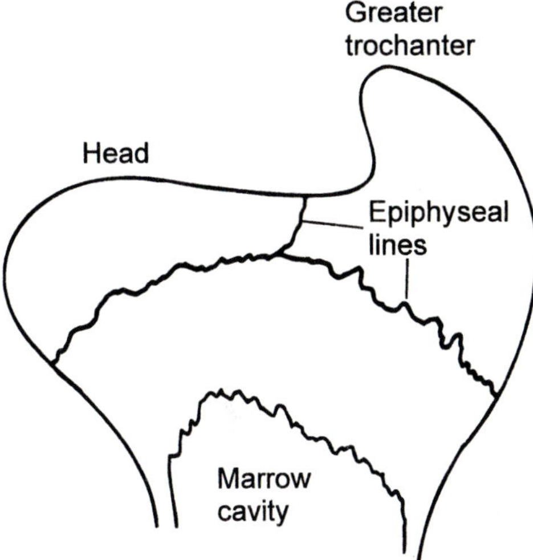

FIG. 8.14 Frontal section of the proximal end of a calf femur. Compact bone is not distinguished from cancellous bone. The epiphyseal cartilage is shown as a thick line.

mammals overcome the weakness of the epiphyseal plate of developing bones. Usually, if the general direction of the plate is straight across the bone, the plate will actually form an extremely crumpled sheet, so much so that the bony ends of the metaphysis and the epiphysis interdigitate (fig. 8.14). (This interdigitation is not usually apparent on radiographs, because the X-rays travel the whole breadth of the plate and so the peaks and craters of the interdigitations average out over the image.) As a result of the interdigitation, the shear between the epiphysis and the metaphysis is to a great extent borne by the bone. The cartilage is not subjected to much strain, and therefore not much stress (fig. 8.15).

Apart from these rather fine-grained irregularities in the line of the epiphyseal plate, there are, as Smith shows, larger-scale undulations that also have a mechanical function. Smith demonstrated the stress in various bones, mainly those of humans and the cow, by making Plexiglas models of them, loading them in appropriate ways, and observing the pattern of strain so produced through crossed polaroids. There are many objections to representing the three-dimensional structure of bone by a Plexiglas sheet. Even so, the bones Smith modeled were probably uniform enough for this technique to be reasonably valid.

The results of the analysis for the human calcaneus are shown in figure 8.16. When the heel is raised, it is compressed by the tibia and put into tension by the plantar ligament and the Achilles tendon. The directions of the principal tensile strains (that is, the directions in which there are no shear strains) are shown by interrupted lines, principal compressive strains by continuous lines. The line of the epiphyseal plate

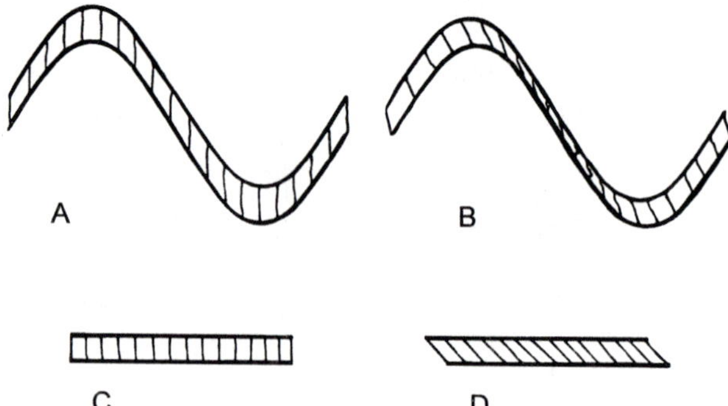

FIG. 8.15 The effect of interdigitation on shearing in epiphyseal cartilage. The subchondral bone is shown by thick lines. (A, B) When a wrinkled cartilage plate is sheared, the cartilage is not deformed much before the subchondral bony sheets butt against each other. (C, D) A straight plate can resist shear only by the strength of the cartilage.

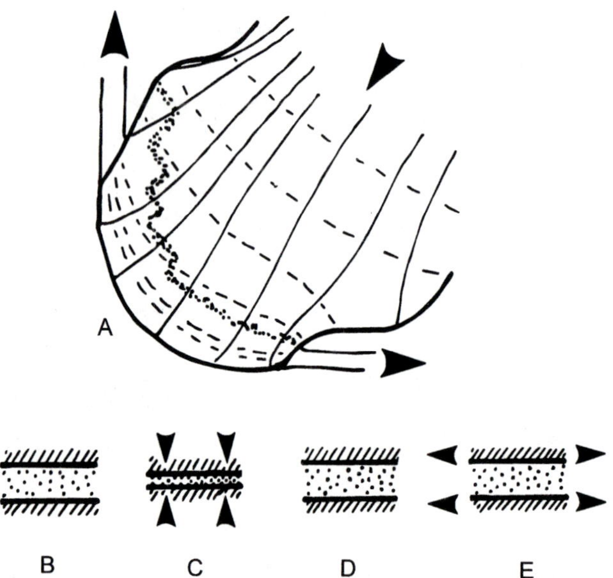

FIG. 8.16 (A) Diagram of a Plexiglas sheet modeling the human calcaneus. The arrowheads show the direction of the important forces. Principal compressive strains, thin solid lines; principal tensile strains, thin interrupted lines. The epiphyseal plate is shown dotted. (B, C) The plate may be loaded safely in compression. (D, E) If loaded in tension, the load is borne almost entirely by the bone acting in parallel with the cartilage.

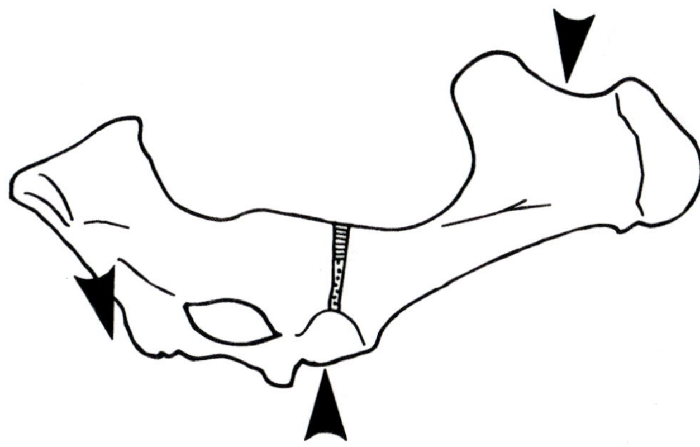

FIG. 8.17 Side view of the pelvis of a calf. The three major forces acting on it are shown. The epiphyseal plate passes dorsally from the acetabulum. The lower part is hyaline cartilage (dots), the upper part is fibrocartilage (lines).

is shown stippled. The greatest shear strains are at 45° to the lines of the principal strains. Along the bottom of the calcaneus the plate is normal to the line of the compressive strains, so it will be compressed. The plate is also in line with the tensile strain. However, the compliant cartilage will be strained in parallel with the much stiffer bone, which will prevent it from being greatly strained in tension.

In the upper part of the calcaneus near the insertion of the Achilles tendon, things are not so simple. The epiphyseal plate passes at about 45° to the direction of both the principal strains. This is the worst possible direction for shear. However, the danger of shear failure is mitigated by the plate being stepped: it lies alternately parallel to the principal compressive strains and to the tensile strains. Similar analysis of the distal end of the femur and the proximal end of the tibia with load across the joint and tension in the cruciate ligaments (a situation occurring often during locomotion) shows that the plates are arranged everywhere nearly normal to the principal compressive strains, a most adaptive arrangement.

In general, then, epiphyseal plates seem to be arranged so that the cartilage is in compression. If it is in tension, the epiphyseal and metaphyseal bone take most of the load, and so the cartilage undergoes little strain. The geometry of the plate is such that it is rarely subjected to much shear strain. Occasionally, however, it is not possible to prevent at least part of the plate being loaded in tension. Here are two examples. When a calf is stationary, each femur bears about one-fifth of the animal's weight. The reaction to the weight is upward through the acetabulum (fig. 8.17). Equilibrium is achieved by the downward

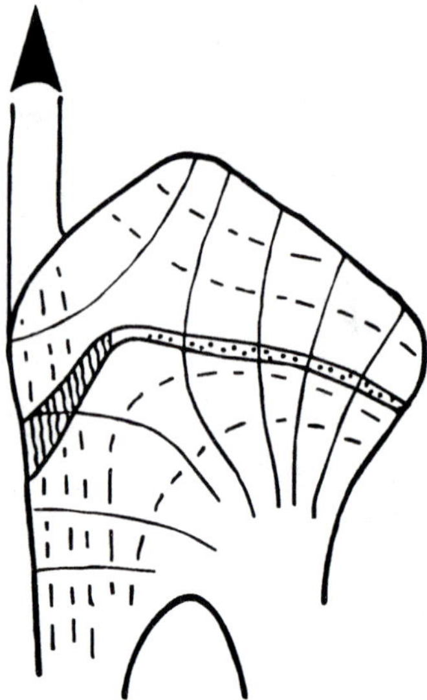

FIG. 8.18 Plexiglas model of the tibia of a calf, conventions as in figure 8.17. The action of the patellar tendon inevitably loads the anterior part of the epiphyseal plate in tension.

weight of the body and the force produced by the hamstring muscles on the ischium (Smith 1962b). The forces load the pelvis in what is effectively three-point bending. The ilium and the ischium are attached to each other by an epiphyseal plate, which runs dorsally from the acetabulum. The ventral part of the plate will be in compression and the dorsal part in tension. No artful deviation of the path of the epiphyseal plate from the direct one can really prevent the cartilage from being loaded in tension. Similarly, in the tibia of the cow the action of the patellar ligament, which is inserted onto the epiphysis, makes it inevitable that the plate be loaded in tension at its anterior end, although the reaction of the femur will load the posterior region in compression (fig. 8.18). In these places, where tensile loading of the epiphyseal plate is inevitable, the plate is quite different in structure from where it is loaded in compression: It is composed almost entirely of collagen fibers, which bridge the gap between the epiphysis and the metaphysis and are orientated in the direction of the tensile forces. Smith describes several plates in which the cartilage changes from being of typical, hyaline form with much watery matrix to being fibrous, as the loading changes from being compressive to being tensile.

8.7 Joints in General

Because the type of joint at the end of a bone has such an effect on the shape of the bone, we need to consider what kinds of movement between adjacent bones impose what kinds of joint geometries.

First, in the vertebrates, at least, flexible joints, in which there is no relative sliding of joint surfaces, seem confined to three situations: (1) The bones may show little relative movement; (2) they may show a great deal of relative motion, with all six degrees of freedom; or (3) the loads across the joint may be small. The hyoid bones show the second state, with adjacent bones often being connected by quite long strips of fibrocartilage. In such joints the form of the end of the bone is not constrained by any requirements of the joint material. Flexible joints that bear little load are, for instance, seen in some rib–sternum joints. Again, the general shapes of the bones are unaffected by the fact of being part of a joint. In the vertebrates, flexible joints that carry large loads are restricted almost entirely to the vertebral centra, to some of the joints of the pelvis, and to the joint between the two rami of the lower jaw. The amount of relative movement allowed in the lower jaw, for instance, may be small but it is essential. Scapino (1981) shows how various degrees of flexibility of the carnivore's mandibular symphysis are correlated with different masticatory actions. Ride (1959) showed how the flexible symphysis in kangaroos is crucial in allowing the two procumbent first lower incisors to act as scissors against the upper incisors, cropping grass and then freeing it so it can be manipulated by the tongue.

There is a great advantage in having a joint made of flexible material rather than of synovial cartilage—There is no need for the extremely delicate synovial lubrication mechanism. Against this is the difficulty that large joint excursions are not possible if large forces act across the joint. This is shown in figure 8.19. Large excursions of flexible hinges inevitably mean either very large strains in flexible links or a hinge so compliant that the movement of one bone on the other is very imprecise. In the rib cage this limitation is unimportant.

Flexible joints, like intervertebral disks, are good for damping out impulsive loads. If the spine were a rigid bone, then rapidly applied loads on the hip joint would be transferred with little attenuation to the base of the skull. (In a hungover state this seems indeed to be the case.) In fact, the *elasticity* of disk material greatly prolongs the time over which the force reaches a maximum; as a result, the maximum force in the bone is lower. Furthermore, the *viscosity* of the disk material results in energy being lost as heat, and so the energy transferred up the spine becomes progressively less. (Energy is also lost in giving kinetic energy to the parts of the body to which the spine is attached.) The other side of the coin is that the viscosity of the disk material results in all movements of the spine requiring work that is lost as heat. Synovial joints are so nearly frictionless that effectively the only work done in moving them is that required to deform the soft tissues surrounding the joint.

I have until now considered bones to be entirely stiff. Of course, they

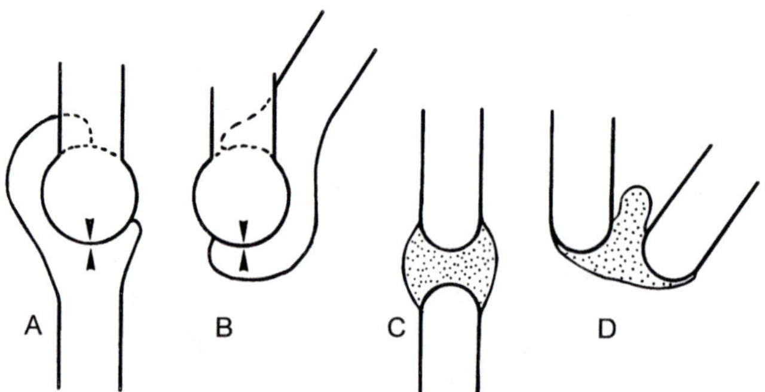

FIG. 8.19 Comparison of a synovial and a flexible joint. (A) Synovial joint fully extended. (B) Synovial joint fully flexed. In both positions there are compressive forces (arrows) between the two surfaces. These prevent disarticulation. (C) Flexible joint extended. (D) Flexible joint flexed to the same amount as in (B). Such a large excursion requires a considerable strain in the hinge material. If the hinge were larger in volume, the strain would be less, but the compliance of the hinge material would result in a geometrically very vague joint that would be difficult to control.

are not, and sometimes they are sufficiently flexible as to behave as if they were jointed. An example is the wishbone of birds (Jenkins et al. 1988). During the wing beat cycle of the starling, and no doubt many other small birds, the wishbone, which is the fused clavicles, bends and recoils. The probable function of these movements is to aid in the filling and emptying of one of the bird's respiratory air sacs. The wishbone is essentially two cantilevers joined at the base. It is possible to produce large movements at the end of a cantilever with small forces and without inducing dangerous strains if, as here, its length is sufficiently large compared with its second moment of area (section 2.2.4).

Joints with flexible hinges inevitably have six degrees of freedom, though the amount of excursion possible may be very small in some or in all directions. Synovial joints, on the other hand, have no such necessary freedom. In fact, the great majority of synovial joints have one, two, or three degrees of freedom. The most important effect of synovial joints on the bones is, as I have suggested, the necessity for the expansion of the bone under the weak synovial cartilage. Unless joints take small forces, which would require little expansion, the amount of the expansion needed will vary according to the geometry of the joint. Some joints carry low loads, and impose low stresses on the cartilage as a result. This is true, for instance, of the joint between the malleus and the incus, and between the incus and the stapes, in the middle ear.

Where forces in the bones are large, certain types of joint may necessarily impose large loads on the cartilage. In general, if there is a considerable lack of congruity between the two halves, the synovial cartilage surfaces will meet, notionally at a line or even a point. This situation is found in the knee. The stresses would be very high here, were it not for the compliance of the cartilage, which increases the contact area somewhat, and the presence of the semilunar cartilages, which spread the load out a great deal. Sometimes the lack of congruity in synovial joints is remarkable. In the joint between the mandible and the skull of the horse the opposing surfaces are two convex, coaxial half-cylinders, about as unstable an arrangement as it is possible to imagine. However, the kinematics of this joint are determined mainly by the big bags of muscle on each side, in which the mandible is slung, and the synovial surfaces engage only at small parts of the chewing cycle.

On the other hand, where the joint surfaces are nearly flat, the joint pressures can be much lower, and little expansion of the ends of the bone may be needed. The small bones of ankles and wrists, which tend to have rather flat facets, show no external morphological features associated with the synovial cartilages. However, they do require cancellous struts going from one end of the bone to the other.

8.8 Conclusion

In this chapter I have tried to show that the necessity for joints imposes very strong constraints on the design of bones. In general, the higher the loads and the greater the number of degrees of freedom, the more complex and specialized must the joint become.

Chapter Nine

BONES, TENDONS, AND MUSCLES

THE BONY SKELETONS in a museum give such a good feeling for what the animals must have been like that we tend to forget what a skeleton is. We are looking only at the compression and shear-resisting members of a structure whose tension members have rotted away. In some ways, the whole of this book can be thought of as an extended Hamlet without the Prince. Bones have no functional meaning without their muscles, tendons, and ligaments, which move them and hold them together. Contrarily, muscles and tendons must always act against something comparatively unyielding, such as bone, cuticle, or water. This fact is frequently ignored by people working on muscle. A colleague who knew much about the anterior byssus retractor muscle of the mussel admitted to me that he had only the haziest idea of what the byssus was. In fact, the mechanical properties of the byssus (collagenous threads that anchor the mussel to the substrate) are peculiar and impose awkward constraints on the muscle that tightens them, but also have mechanically useful effects, since by having a long postyield strain, many threads can be recruited and loaded to high stresses (Bell and Gosline 1996).

Tendons, ligaments, and muscles have mechanical properties quite different from those of bone. We shall see in this chapter how all these tissues have to act in concert, and how they often make uneasy bedfellows and need complex adaptations to allow them to work in harmony. We shall also see how the skeleton makes use of the different mechanical properties, so that the good features of each tissue are allowed to show themselves, while the bad features are, as much as possible, shielded from selection.

Table 9.1 shows some important physical properties of bone, tendon, and muscle. Bone has a compressive strength of about 250 MPa and a tensile strength of about 150 MPa. Tendon has a tensile strength of about 100 MPa (Elliott 1965; Schechtman and Bader 1997), and it is not meaningful to talk about its compressive strength, because it is so flexible that it cannot really be loaded in compression unless it is confined laterally, a most unlifelike situation. The densities of bone and tendon are about 2000 and 1200 kg m^{-3}, respectively, so the values of the tensile strength/density at 7.5×10^4 and 8.3×10^4 Pa kg^{-1} m^3 are effectively the same. Therefore, the weight penalty for replacing tendon by bone of the same ability to resist tensile loads would be very small.

TABLE 9.1
Some Physical Properties of Bone, Tendon, and Muscle

	Bone	*Tendon*	*Muscle*
Compressive strength (MPa)	250	–	–
Tensile strength (MPa)	150	100	0.35[a]
Young's modulus E, (GPa)	20	1.5	–
Shear modulus (GPa)	5	Negligible	Negligible
Density (kg m^{-3})	2000	1200	1200
Tensile strength/Density (Pa kg^{-1} m^3)	7.5×10^4	8.3×10^4	2.9×10^2
Tensile strength^2/2E (Pa)	5.6×10^5	3.3×10^6	–
Strain energy storage (J kg^{-1})	2.8×10^2	2.8×10^3	4.7[a]

[a]Taken from Alexander and Bennet-Clark (1977).

Table 9.1 also shows that muscle has an extremely low value for strength/density (and, incidentally, for energy storage per unit mass). It would be extraordinarily expensive to resist tensile loads by using muscle. However, this is not what muscles are for, and muscles can perform work of about 250 J kg^{-1} in a single twitch, whereas bone and tendon can do no active work at all; at best they can pay back some of the work done on them. It is the flexibility but high tensile modulus of tendons that makes them so suitable for transferring the force exerted by the muscles to the bones themselves.

(Tensile strength)^2/2E is a measure of how much strain energy can be stored in a volume of tissue, assuming that it is a linearly elastic material with no postyield deformation. (The area under the curve for a linearly elastic material is (Stress × Strain)/2. Strain is Stress/Young's modulus, hence the expression. It is informative to put it this way because it shows how a large Young's modulus reduces the amount of energy that can be stored.) The expression is a little unrealistic to apply to tendon, because it does not have a linear stress–strain curve, and for bone the expression refers only to the region before yield occurs. Nevertheless, the table does suggest that tendon is considerably superior to bone. The bottom line of the table shows the effects of taking density into account and shows how much energy can be absorbed before damage occurs, per unit mass.

9.1 TENDONS

Tendons attaching to bone have three main functions. One is to allow the muscle to be reasonably far away from the joint or joints it activates. Another is to make forces go around corners. We shall see how this is brought about by the use of retinacula and pulleys. The third function is to act as an energy store. The first two functions, which usually go together, require tendon to have a high stiffness and high strength-to-mass ratio in tensile loading, but to have a very low shear stiffness—to be very flexible. The energy storage function is best carried out by tendon with a high specific energy-absorbing capacity. This is given by a material with a high strength and *low* modulus. This last requirement runs counter to that needed for the first two functions. In some tendons energy storage is a very important function.

What happens is that when an animal is running in each step it has to raise and lower its center of gravity and has to accelerate and decelerate the center of gravity in the direction of motion. If, as is usually the case in running (though not in walking), the body has to be raised at the same time as it has to be accelerated, work must be done. This can be done by the muscles, but it saves energy if work can be done by the shortening of the tendon, which has been passively stretched, and has stored strain energy during other stages of the locomotory cycle. This is an important energy-saving device, much used by mammals (Alexander 1988).

Interestingly, there are no important differences in most of the material mechanical properties between the tendons of the two types, differences in the tendons' behavior being produced by differences in the cross-sectional area (Pollock and Shadwick 1994). However, the fatigue behavior of the types is different. Those tendons that are subject to high stress, which are the tendons that save energy repetitively in running, are more resistant to fatigue than more lightly stressed tendons (Ker 1999; Ker et al. 2000). The molecular or histological bases of these differences are unknown.

Figure 9.1 is a diagram of the human foot sketching how the extensor muscles of the toes originate on the shin and their tendons insert on the toes after turning through a large angle. The tendons are forced to keep close to the bones by the two bands of the inferior extensor retinaculum. A few words about retinacula. Tendons run freely in collagenous sheaths, which separate them from the surrounding tissues. Retinacula are highly specialized sheaths; they are bands of collagenous tissue that function to control the position of tendons as they pass from their muscle of origin to their insertion. Usually, like the inferior extensor, they prevent the tendon from straightening out while allowing the tendons complete freedom to slide. Where the tendon passes through the retinaculum it is surrounded by a synovial sheath, which ensures that the relative movement of the two is virtually frictionless. Figure 9.2 shows the directions of the forces in the neighborhood of retinacula situated at each end of a bone. The greater the change in direction

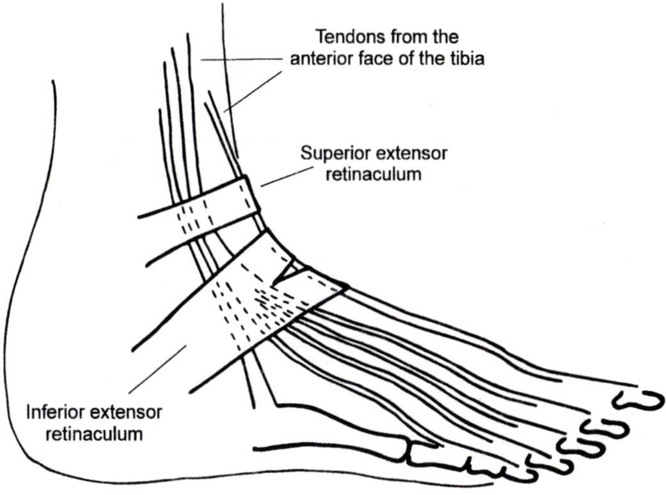

FIG. 9.1 Extensor tendons of the human toes, passing under retinacula.

caused by the retinaculum, the greater the force on it. The force acts in the direction of the bisector of the angle made by the tendon before and after passing under the retinaculum.

The tendons of complicated structures such as the foot adopt a whole variety of shapes when the foot and the toes are in various positions. It would be impossible to accommodate such excursions with bonelike tendons. It is interesting that many arthropods do have tendons, made of cuticle, that are rigid. However, these tendons generally span only one joint, and even so they always have one or two short regions of flexible cuticle intercalated in their length to allow flexibility at critical points (fig. 9.3). Some birds, such as turkeys, have ossified tendons, which work in much the same way. They are discussed below (section 9.2).

As well as being strong, tendons are flexible, yet have quite a high Young's modulus. That is, tendons have a very low shear modulus but are quite stiff in tension. Tendons have a sigmoid stress–strain curve that is nowhere straight; that is to say, it does not have a linear region from which one can calculate a true Young's modulus. However, if we take the steepest region of the curve, the modulus is about 1.5 GPa. Using beam theory we can roughly calculate the *minimum* radius of curvature of a tendon made of material of tensile strength 100 MPa and modulus 1.5 GPa. The maximum tensile stress in a beam of depth $2X$, which was originally straight, but which has been bent into a radius of curvature R, is given by Stress = EX/R (fig. 2.10). Therefore, Tensile

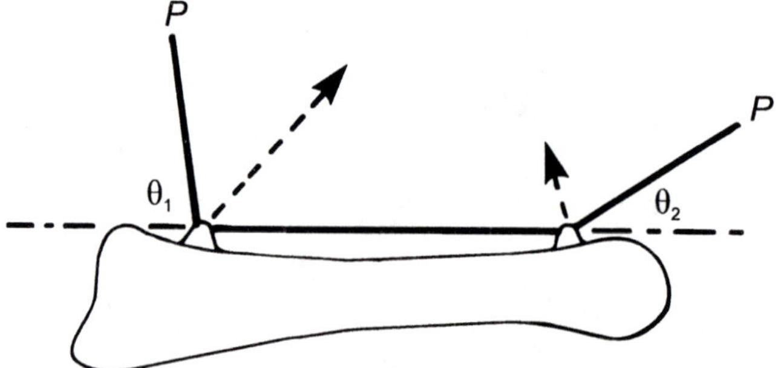

FIG. 9.2 The action of retinacula. The tendon (thick line) is under a tension P. The two retinacula near the ends of the bone allow a total angle change of $\theta_1 + \theta_2$. The resulting force on the retinacula is greater, the greater the change in angle they are responsible for. The vectors (with arrows) show the relative directions and magnitudes of the forces acting on the retinacula.

strength/E = X/R = (100 MPa)/(1.5 GPa) = 0.067. This suggests that tendon cannot be bent into curves of radius of curvature less than about seven times their thickness without rupturing. This is clearly not really the case: tendons can be tied into quite tight knots without damage. Something must be wrong. What is wrong is that tendon cannot be

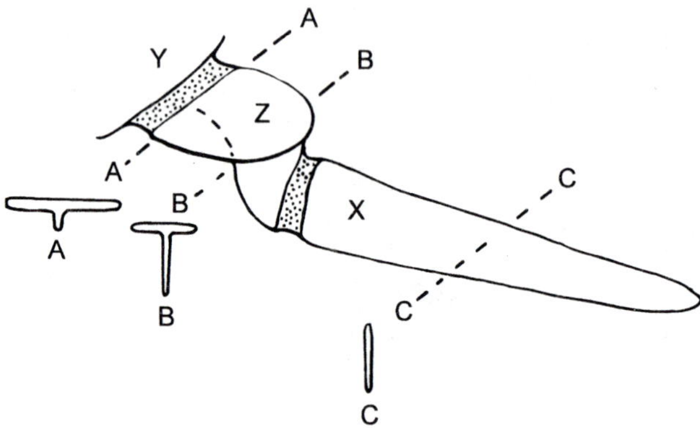

FIG. 9.3 Tendon of a joint of the American lobster *Homarus americanus*. The muscle inserts onto the blade-shaped part on the right X. This is made of rigid cuticle. The muscle has to move the rigid piece of cuticle Y, which is oriented at right angles to X. This is made possible by the intercalation of two short pieces of flexible cuticle (dotted) and the complexly shaped piece of rigid cuticle Z. A, B, and C show the section of rigid cuticle at three levels.

considered to be a beam. It consists of many fibrils, which can shear past each other. All these fibrils cooperate to produce a fairly high modulus in tension, but in bending they do not cooperate, and the tendon can be thought of as being made of innumerable beams, each with a very small cross section and therefore a small second moment of area. The result is that tendon has a very low shear stiffness. The same principle is made use of in wire ropes. Bone, on the other hand, is not made of independent fibrils; it has considerable shear stiffness. It is not flexible, and is stiff in both compression and bending.

Usually, tendons need to move without friction through their sheaths and under retinacula. However, occasionally tendons have a ribbed surface and are designed to lock onto the tendon sheath, which has complementary ribbing. This arrangement is found in some perching birds and some bats, and allows them to perch, or to hang, with very little expenditure of muscular effort, because the tendon is held in one place by its sheath (Quinn and Baumel 1993). Of course, there have to be clever mechanisms to allow the tendon to be freed from its constraints when the time comes to move.

9.2 Sesamoids and Ossified Tendons

Sesamoid bones are found quite frequently in mammals and reptiles (Haines 1969). They develop in tendons in the neighborhood of joints. Usually their function, as was first clearly stated by Gillette (1872), is to ensure that the moment arm of a tendon about the center of rotation of a joint remains large at all joint angles (fig. 9.4). Sesamoids occur particularly often on the plantar surfaces of feet, where it would be awkward to have large protuberances, like the olecranon of the ulna, sticking into the ground when the foot was dorsiflexed. Where there are large protuberances, there tend not to be sesamoids, and vice versa.

When a tendon passes over a bony prominence, it will be subject to both tension along its length and compression normal to its length. An example of such a tendon is the flexor digitorum profundus where it passes over the calcaneus. This has been studied by Merrilees and Flint (1980). They found that on the side of the tendon subject to tension only, the tissue was ordinary tendon, but that on the compressive side the tissue was much more like fibrocartilage. Vogel and Koob (1989) review the subject of parts of tendons loaded in compression, particularly their biochemistry. They show how, in the bovine deep flexor tendon the outer, tensile, side of the tendon is quite normal in its appearance, but the histology changes progressively toward the surface that is loaded in compression. Here the "collagen fibers. . . . are ar-

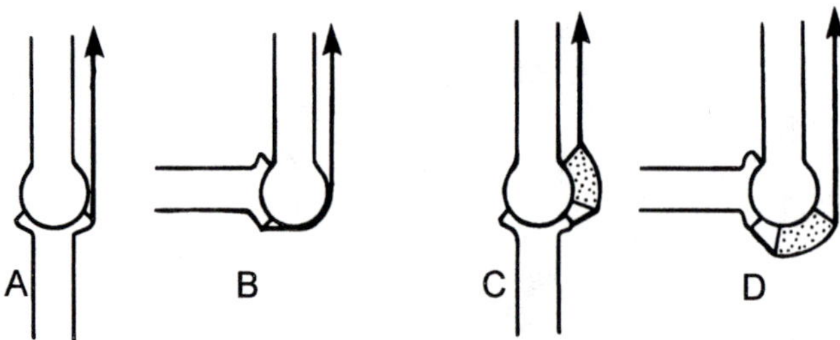

FIG. 9.4 The action of a sesamoid. (A) and (B) show a joint with a wrap-round tendon, and no sesamoid. Although in one position (A) the tendon is just clear of the joint surface, in another (B) the moment arm is smaller. (C, D) With a sesamoid (dotted) the tendon's moment arm is always larger than that of the wrap-round tendon. In (C) the sesamoid would seem likely to swing clear of the joint. Often this does happen but often, also, ligaments keep the sesamoid closely apposed to the joint surface.

ranged in a network or basketweave pattern enclosing regions of cartilage-like matrix." Such changes in the matrix will assist the tendon in acting like a sesamoid. Where tendons pass under retinacula there are forces on the retinacula equal to those on the tendon, and it is therefore not surprising that the retinacula often have developed fibrocartilage as well (Benjamin et al. 1995).

The mechanics of sesamoids are not totally straightforward. Alexander and Dimery (1985) have analyzed three ways in which joints can be extended: wrap-round tendons (as in a metacarpophalangeal joint), sesamoids (as in the knee), and retroarticular processes (as in the elbow). In a wrap-round tendon the forces in the tendon on each side of the joint are necessarily almost equal. This is not necessarily true of the tendons each side of a sesamoid. For instance, in rising from a seat, when the knee is nearly at a right angle, the ratio of the force in the quadriceps tendon (and hence the force exerted by the quadriceps) is about twice that in the patellar ligament, which attaches the patella to the tibia (Ellis et al. 1980). The difference between the forces becomes less as the knee is straightened. It is difficult to generalize about the comparative advantages of the three systems, except that, in general, the wrap-round tendon passes closest to the center of rotation of the joint, and so has the least mechanical advantage and the greatest velocity advantage. The other two systems have a greater mechanical advantage and also have a much more variable velocity advantage; the variation is probably greater in the retroarticular process than in the sesamoid.

Sesamoids do not, however, always function to keep tendons away from the center of rotation of joints. An example of this is a sesamoid embedded in the tendon at the origin of the gastrocnemius, just behind the knee, in many mammals. This sesamoid never comes in contact with the bones near it. The function of this tendon ossification is not known. Almost as obscure is the function of the ossification of many tendons in the legs of birds. Many of the birds that have ossified tendons spend much of their time walking, rather than flying; pheasants, turkeys, and chickens are examples. In these birds the tendon is replaced by bone over much of its length. In turkeys, ossified leg tendons do not span two joints, and there is always a small length of ordinary tendon intercalated between ossified tendon and the bone. The movements of the joints controlled by the tendon are simple, and geometrically there are no constraints requiring the ossified tendons to bend. Therefore, the tendon does not need to turn corners. As regards energy storage, such birds as turkeys and pheasants spend most of their time walking sedately. If they need to move fast or far they take to flight, and then the legs are not used in locomotion. It may be, therefore, that ossified tendons arise when neither energy storage nor bending are required. Bledsoe et al. (1993) have studied the state of tendons in the hind-limbs of woodcreepers (Dendrocalaptinae) in detail. Woodcreepers are birds that spend much of their time scampering over tree trunks, and their hind-limb tendons are generally all ossified. However, there are so many cases where birds do much walking, and yet the tendons remain unossified, and cases where the birds are not notable walkers, in some hawks, for instance, that this cannot be the whole story. Vanden Berge and Storer (1995) give a comprehensive review of the occurrence of ossified tendons in birds. These authors are clearly baffled by the spotty occurrence of ossified tendons.

Many dinosaurs had very well-developed ossified tendons, perhaps better called ligaments, in their spines. Particularly good examples are to be seen in the dinosaur hall of the American Museum of Natural History in New York. These form a lattice work alongside the dorsal processes of the vertebrae, crossing each other at about 30°. They are clearly not pathological. They must have made the spine extremely rigid.

As well as being a favorite subject for studies of ossification (section 1.1), ossified tendons are interesting because they are one of the few examples I know about of a bone that has a mechanical property that is probably almost irrelevant to its function in life. The mechanical properties of the Sarus crane's ossified tendon are given in table 3.3. A study of turkey leg tendons by Bennett and Stafford (1988) found a similar Young's modulus and strain to failure as we found in the Sarus crane, but a lower tensile strength. The crane tendon has a high Young's modulus; indeed, it is particularly stiff in relation to its mineralization. This stiffness is probably adaptive. Indeed, as I implied above, it is puzzling that many tendons are *not* mineralized. Many tendons, as shown so clearly by Ker et al. (1988), function best if they are very stiff. Bone is much stiffer than tendon, so that there would always be a weight advantage in turning tendon into bone, although there must always be flexible bits for the attachments and for turning corners. (On the other hand, Ker et al. also show that tendons such as the Achilles tendon in humans, which are used to store significant amounts of elastic strain energy, need to be extensible, and not stiff.) Sarus crane ossified tendon is very *tough*—It is almost impossible to make a crack travel across it; instead, any crack will travel up and down the lengths of the tendon. However, toughness is unlikely to be important to its function. The tendon is almost certainly overdesigned in relation to strength, since its important feature is its stiffness, and it is most unlikely to break in life. The toughness of ossified tendons comes from their developmental history: Turning tendon into bone produces a highly anisotropic material, with weak side-to-side connections of the fibers. This anisotropy is maintained during the intense secondary remodeling that occurs in Sarus crane. The Haversian systems have their fibers almost entirely in the longitudinal direction, and the osteocyte lacunae are spindle-shaped and elongated in the same direction—an unusual morphology for the lacunae of amniote osteocytes. This anatomy makes the bone tough, but also very weak from side to side. But that does not matter, because the tendons are never loaded significantly in this direction.

9.3 Attachment of Tendons to Bone

For engineers, the problem of attaching two materials of very different Young's modulus may be considerable. Usually the materials cannot be bonded to each other, because the strains each material undergoes near the interface are very different and lead to large stress discontinuities at the interface. These discontinuities in turn make adhesion difficult. Often, rather complex knots or other fastenings must be produced. In fact, there are few examples in engineering technology where a low-modulus material that is bearing a high tensile stress is attached to a high-modulus material. One example is the attachment of ropes to boats.

The junction of a tendon or a ligament to bone is called an *enthesis* (Benjamin and Ralphs 2000). Bone and tendon differ in Young's modulus by an order of magnitude, yet tendons rarely pull out of bones; avulsion of tendons seems to take place either in the bone or in the tendon itself (Clark and Stechschulte 1998). Furthermore, at the naked-eye level, there is no complex knotting. A tendon appears to run straight into the bone, with only a small increase in cross-sectional area.

Cooper and Misol (1970) describe the enthesis of the dog's patellar tendon and the patella. The patellar tendon inserts on both sides of the patella, the tibial and the femoral. This enthesis is probably typical. Cooper and Misol found, as had others, four regions. In order, from tendon to bone, these were

1. Ordinary tendon.
2. A fibrocartilage region, about 300 μm wide. In this region cartilage cells appear, lying in rows in the extracellular matrix of the tendon. The cross-sectional area of the tendon increases slightly to accommodate the cells.
3. A mineralized fibrocartilage region, about 200 μm wide. There is a sharp boundary between this region and the last; the mineralization does not start gradually, but appears as a clear tideline. The cartilage cells here are usually somewhat degenerate.
4. The mineralized fibrocartilage, containing mineralized tendinous fibers, merges imperceptibly into the rest of the bone, with no clear point where the fibers stop and the bone begins.

The orientation of the fibers from the tendon may be very similar to the fibers of the bone into which it penetrates, as in the anterior part of the patella itself (Clark and Stechschulte 1998) or it may be different from that of the generality of the bone. The collagen fibers on the surface of the bone belonging to the bone itself, the *intrinsic* fibers, are usually roughly parallel to the surface, whereas the penetrating tendon fibers, the *extrinsic* fibers, are often at a large angle to the surface. Where this happens the mineralized fibers are known as *Sharpey's fibers*. Boyde (1972, 1980) shows that if bone with Sharpey's fibers is deorganified, that is, it has the organic material removed chemically, and is then examined in the scanning electron microscope, the appearance is very characteristic. Sharpey's fibers appear as round depressions or bumps on the surface of the bone, while the intrinsic fibers, belonging to the bone proper, weave between them. If the surface is growing, the mineralization of Sharpey's fibers lags behind and they appear as depressions (fig. 9.5). If the surface is stationary, the mineralization more than catches up and Sharpey's fibers grow out along the tendon, appearing as bumps. This was presumably the state of the bone that Cooper and Misol viewed. Sharpey's fibers are mineralized from the outside in, and so they appear to have hollow middles in specimens from which the organic material has been removed.

This mode of inserting tendons is certainly mechanically effective. The difference between the arrangement here and that in man-made structures is that in the latter a low-modulus material must, in some way, be attached to a high-modulus material. In the tendon insertion the

FIG. 9.5 Sharpey's fiber bone. (A) Sketch of bone surface, with organic material removed. Sharpey's fibers are indicated by dots, with solid black representing the unmineralized core. The fibrils of mineral, which once had collagen inside them, run between the Sharpey's fibers. (B) Cross section of the insertion of three collagen fibrils into bone. The fibrocartilage cells are shown as ellipses between the fibrils. Incipient mineralization is shown as fine dots (this is removed along with the organic material). The fully mineralized bone and the Sharpey's fibers lie beneath the transverse line.

low modulus material penetrates the high-modulus one and becomes of high modulus itself within a millimeter or so. There are no real problems, therefore, in bonding the tendon to the bone, the only difficulty being that the grain of Sharpey's fibers and that of the rest of the bone is different. The critically important point is the continuity in the collagen fibrils right from true tendon into the heart of the bone; there is no line of weakness.

Cooper and Misol were puzzled by the need for the fibrocartilage. The reason for its presence is probably this: the collagen fibrils must separate from each other before they enter the bone, so that as Sharpey's fibers they can be gripped by the rest of the bone. Cartilage cells are presumably a handy way of producing packing material. Evans et al. (1990) show, in comparing the quadriceps tendon and the patellar tendon in humans, that fibrocartilage is more fully developed where there is a large change of angle between the tendon and the bone during joint movement. In these places there is a need for a greater length of material of intermediate shear stiffness to transmit the loads smoothly. Benjamin and Ralphs (1999) show a similar situation in comparing the human supraspinatus and deltoid. Where the inserting tendon is, because of its anatomical arrangement, loaded to some extent in compression (usually roughly normal to the collagen fiber direction) it will develop fibrocartilage (Matyas et al. 1995). Sometimes muscles seem to attach directly to bone, but, in fact, there is always a small intervening layer of collagen.

A difficulty with the bone–tendon system is that the insertions of the entheses must migrate as the bone grows. A bone grows at its ends; it does not expand, so if an enthesis is, say, close to the end of the bone, in remaining so it must move to a part of the bone that is newly formed. This involves erosion of the bone in front of the moving enthesis and deposition behind it (Wei and Messner 1996).

9.4 Muscles Produce Bending Stresses in Bones

The ability to bear bending loads without noticeably deflecting can be thought of as bone's raison d'être. Only because of this property can the muscles act effectively (section 2.1). But bending is dangerous, because quite small loads can impose large stresses.

Consider an idealized situation as in figure 9.6A, which is nevertheless quite like what one often sees in nature. A bone of length L, with a freely moveable joint, bears a weight P at its distal end and is prevented from rotating by the action of a tendon, exerting a force T, which inserts a distance J from the center of rotation of the joint. The value of T is PL/J. The bending moment in the bone is at its greatest at the level of the tendon insertion, and is $P(L - J)$. For a bone of symmetrical cross section the greatest stress at this position will be $P(L - J)c/I$, where c is half the depth of the bone and I is the second moment of area of the cross section. Suppose the bone has a length L of 100 mm, that the tendon inserts 20 mm from the center of rotation of the joint, and that it is a hollow cylinder of external and internal diameters of 10 and 6 mm, respectively. The force that would have to be exerted by the tendon is $5P$. The value of c/I is 0.0117 mm^{-3}, the bending moment is $80P$ mm. Therefore, the greatest stresses are ($80P \times 0.0117$) MPa = $0.94P$ MPa. If, instead of being bent, the bone were supporting the same mass as a column, in compression, the stress would be P/A. The cross-sectional area is 50.3 mm^2 so the compressive stress is = $P/50.3$ MPa = $0.02P$ MPa, which is far less.

The design features one might expect to find in skeletons would be that the tendons (and muscles) should be loaded in tension and the bones in compression. If we consider the bone bearing a load at its free end, as in figure 9.6, things could be arranged that the bones are loaded in bending as little as possible, even though, because of the jobs skeletons have to perform, the whole structure of bones plus muscles plus tendons must be loaded in bending. Figure 9.6B shows how this could be achieved. The tendon, instead of inserting close to the joint, inserts far away, where the load P is acting. The tendon still has to exert a force T^* sufficient to balance the force P tending to rotate the bone counterclockwise. The value of T^* is $P/\sin\theta$. There will be a compressive force $P/\tan\theta$ along the bone but, most importantly, there will be no

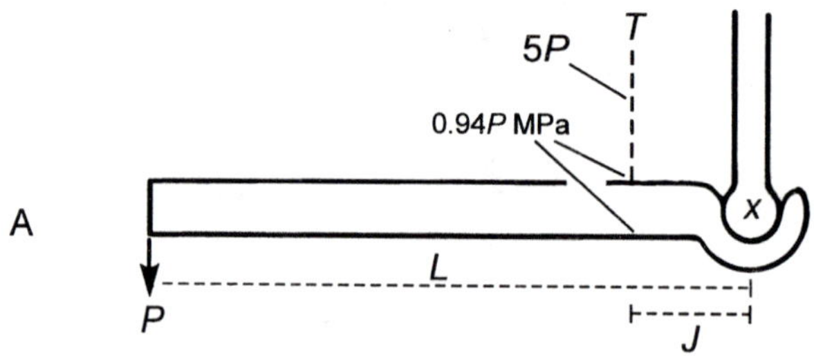

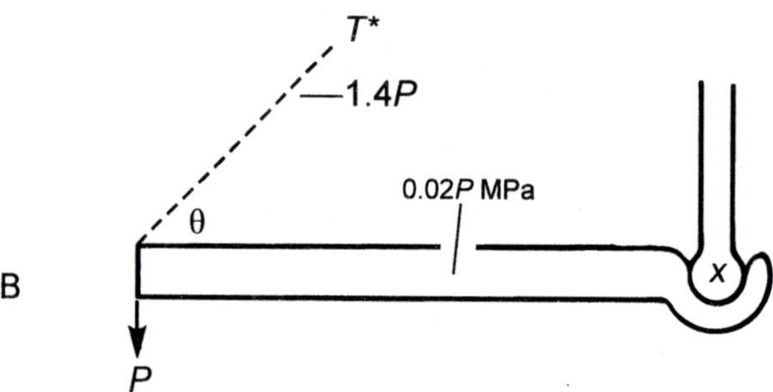

FIG. 9.6 The effect of position of the tendon insertion relative to the joint, whose center of rotation is marked x, on the bending moment in a bone. Explanation in text. Given the values suggested in the text, in (A) the load in the tendon is $5P$ and the maximum stress in the bone is $0.94P$ MPa; in (B) the load in the tendon is $1.4P$ and the (uniform) stress in the bone is $0.02P$ MPa.

bending moment. What will be the force in the tendon, and its associated muscle, and the stresses in the bone, if θ is, say, 45°? The force in the tendon is $P/0.707 = 1.4P$. In the situation shown in figure 9.6A it was $5P$. The bone undergoes a compressive force $P/\tan \theta = P$. The stress induced by this will be P/Area, and so the stress $P/50.3$ MPa $= 0.02P$ MPa. The arrangement as shown in figure 9.6B has, compared with that in figure 9.6A, reduced the stress in the tendon to 28%, and the maximum stress in the bone to 2.1%. This is a huge reduction, showing that the arrangement is figure 9.6B is much superior in reducing the stresses in the tendon and the bone. At first sight, therefore, it might seem strange that the kind of arrangement seen in figure 9.6B is very rare and that seen in figure 9.6A is very common.

The closer the muscles and tendons lie to the bones, the greater the bending moments in the bones, and the greater the tensile stresses in the muscles and tendons. This is, in general, always true, and is one of the most important features of the design of skeletons.

9.5 Why Do Tendons Run Close to Joints?

The skeletons of most vertebrates are arranged so that the muscles moving the joints usually have rather large mechanical disadvantages. For example, consider the human elbow joint when it is flexed at a right angle (fig. 9.7), a weight being held in the hand. The weight has a moment arm of roughly 300 mm about the center of rotation of the elbow, while the biceps brachii, brachialis, and brachioradialis have moment arms of about 55, 25, and 60 mm, respectively. Similar mechanical disadvantages can be seen in most of the muscles moving the long bones of

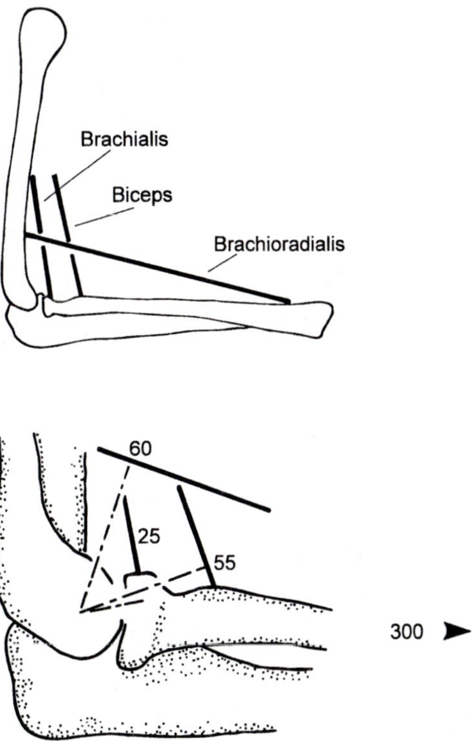

FIG. 9.7 (*Above*) The human elbow joint. The thick lines show the lines of action of the three main muscles resisting a weight on the hand. (*Below*) Enlarged view of the elbow. The numbers refer to the moment arms of the muscles, and of the weight, about the center of rotation, in millimeters.

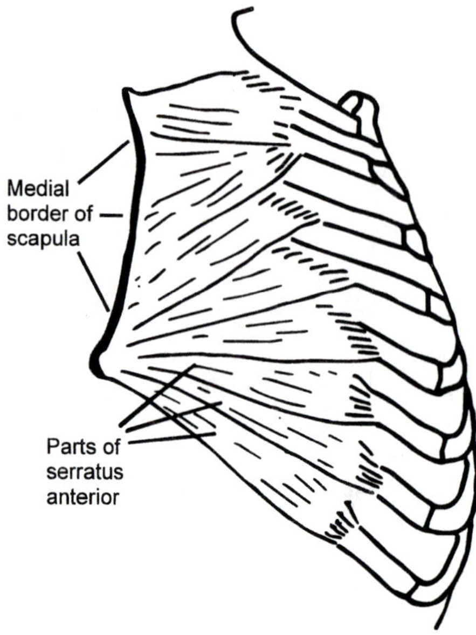

FIG. 9.8 Relationship, in humans, between the serratus anterior and the scapula.

reasonably agile animals. Usually there are retinacula, which keep the tendons close to joints (figs. 9.1, 9.2). These act to prevent the moment arm of the muscles increasing as the joint becomes flexed.

Of course, not all muscles have large mechanical disadvantages. In humans the serratus anterior draws the medial border of the scapula forward. This is the movement you make in punching. The movement of the whole bone is about equal to the shortening of the muscle, and so, in shortening, the muscle has a mechanical advantage of about one (fig. 9.8). The long bones of a few vertebrates have large flanges on them; these greatly increase the muscles' mechanical advantage. The forelimb of the mole is a good example of this (fig. 9.9).

The selective reasons for these bony shapes have nothing to do with bone itself, but concern the nature of muscle, and we need a digression into the physiology of muscle. Active muscle does two separate things: it shortens and it exerts forces. The way that these two properties interact affects the shapes of bones. If a muscle is made to contract in a situation with no external load, it will shorten at its maximum possible rate. (In reality it will be producing a small force, because it is accelerating its own mass.) On the other hand, if the external force it has to exert is made greater and greater, there will be some load that it is unable to move, but that conversely does not actually make the muscle any longer. This load is equal to the maximum force the muscle can exert. For any muscle it is possible to construct a *force–velocity* curve

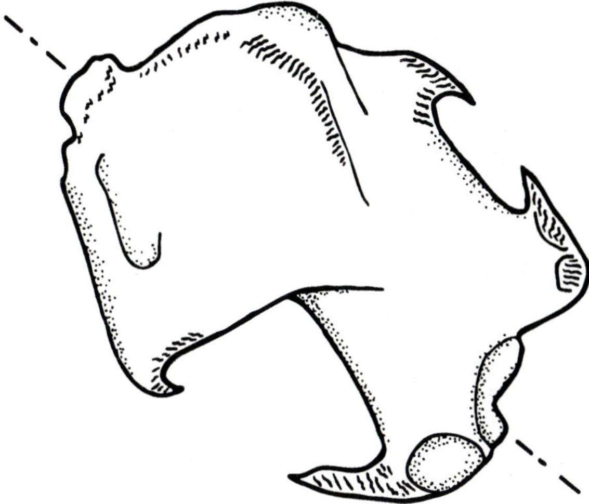

FIG. 9.9 The humerus of the mole *Talpa*. Squiggly lines show the major muscle insertions. The long axis of the bone is shown by the interrupted lines.

showing the relationship between the force the muscle is exerting (the load it is acting against) and the velocity with which it can contract (fig. 9.10A).

A. V. Hill, who did much of the pioneering work on muscle mechanics, showed that the force–velocity curve could be described by the equation $V(P + a) = b(P_0 - P)$, where P is the force the muscle is exerting, P_0 is the maximum force it can exert, V is the velocity of contraction, and a and b are parameters with units of force and velocity respectively (Hill 1938). Hill's equation applies only to muscle at its resting length (the length it naturally adopts if left unconstrained). The maximum force a muscle can exert changes with its length (Gordon et al. 1966); therefore, the value of P_0 will change as the muscle shortens (fig. 9.10b). The reasons for this relate to the way in which the actin and myosin in the muscle overlap. However, it turns out, fortunately for analytical purposes, that the parameters a and b hardly vary with muscle length (Abbott and Wilkie 1953), and so the equation above can be modified to $V(P + a) = b(P'_0 - P_0)$, where P'_0 is the maximum force at any particular length. Hill produced theoretical explanations for his equation having the particular form it does, but they need not concern us.

The external work done by a muscle is the product of the force it exerts times the distance it shortens. Some muscles function without doing much external work. These are the postural muscles, and muscles

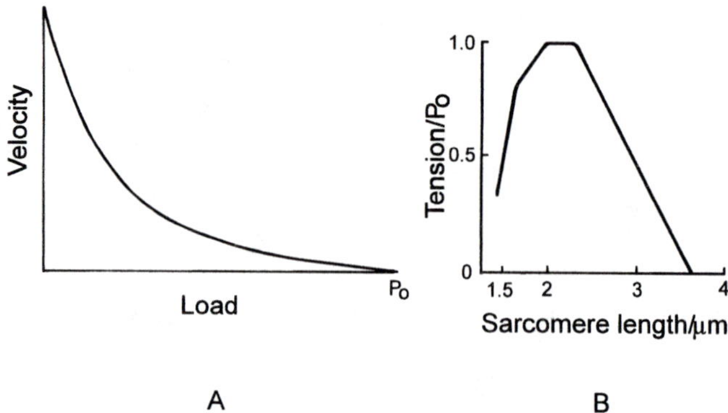

FIG. 9.10 Factors affecting the performance of muscles. (A) Relationship between the velocity of shortening and the external load that is being acted on. When the load is maximum (P_0) the muscle cannot shorten. The curve is a parabola, given by Hill's equation. (B) Relationship (idealized) between the maximum tension a muscle, frog semitendinosus, can exert, expressed as a fraction of P_0, and the sarcomere length. (Modified from Gordon et al. [1966].)

whose function it is to keep skeletal elements in constant spatial relationship against the action of external forces. When a chimpanzee is hanging with its hand wrapped around a branch, the digital flexor muscles are not causing any movement, and so are not doing any external work. However, they are preventing undesirable movement—the fingers do not unwrap—and so the chimpanzee remains suspended. Nevertheless, if an animal is to move around, its muscles must do external work.

Because Power = Force × Velocity, the force–velocity curve can easily be transformed into a power–force curve. It turns out that for many muscles the parameter $a \approx 0.2P_0$. If so, then the power is at a maximum when the force the muscle is exerting is $0.29P_0$. Now for most muscles, particularly those involved in locomotion, the property that is likely to be selected for is their ability to move their point of application as fast as possible. The point of application will be resisting this movement with some force, and so the power output of the muscle is what, in effect, will be selected for (fig. 9.11). The figure shows a muscle that is required to move the distal part of the bone at the maximum velocity. The resistance to movement is a constant force F. The force has a moment arm R, and the muscle is to be inserted where it will be most effective. Call the distance from the insertion to the joint kR. The muscle has a mechanical advantage k. Now $P = F/k$. The optimum value for P is when it equals $0.29P_0$. Therefore, $0.29\ P = F/k$, and k should be $F/0.29P_0$.

This analysis shows that the greater the force the muscle is able to

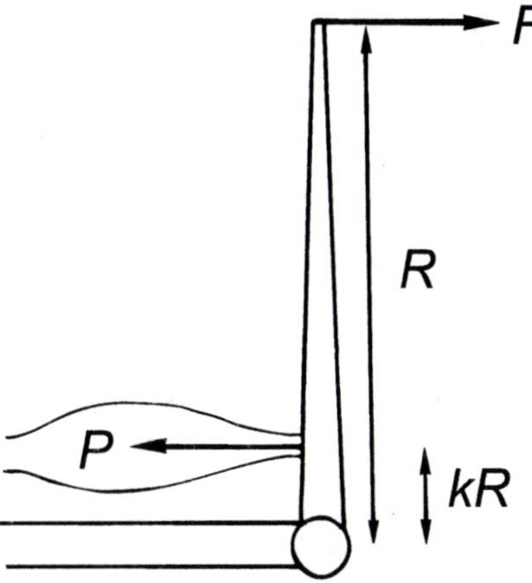

FIG. 9.11 The mechanical advantage of a muscle. See text.

exert relative to the force resisting it, the nearer the point of insertion should be to the joint. Thus, if the force resisting movement is 10 N and the muscle is capable of exerting a force of 100 N, then $k = 10/(0.29 \times 100) = 0.34$; that is, the insertion should be about one-third of the way along the bone. If $P_0 = 1000$ N, then k should be 0.034, and so on. (Note that the power exerted at the end of the bone is the same as the power at the muscle insertion. Call the velocity of the insertion V. The power of insertion $= PV$, the velocity of end of bone $= V/k$, and $F = Pk$. Therefore, the power of the end of the bone $= (V/k) \times Pk = PV$.)

This analysis was made by Calow and Alexander (1973). They showed, in the same paper, that if instead of considering the *greatest mean velocity* that a given resisting force can be moved at over the contraction of the muscle, one considers the best arrangement for accelerating a mass to the *highest final velocity* (the requirement in jumping, for instance), the optimum position of the insertion is slightly different. It is important to remember, when looking at figures such as 9.10, that they refer to muscles that are, artificially, maximally stimulated. In real life more and more muscle fibers are recruited to contract, as more and more force is needed, and unless the muscle *is* maximally stimulated, the situation is in reality more complicated than I have outlined here.

In 1974 Jack Stern carried the analysis further. Calow and Alexander's analysis was, for computational reasons, necessarily simplified. In particular, they could not allow fully for the fact that as the length of the muscle changes the value of P changes also. So, if the length of the muscle changes considerably during the course of its contraction, it will pass from some low value of P_0, through a maximum, and then de-

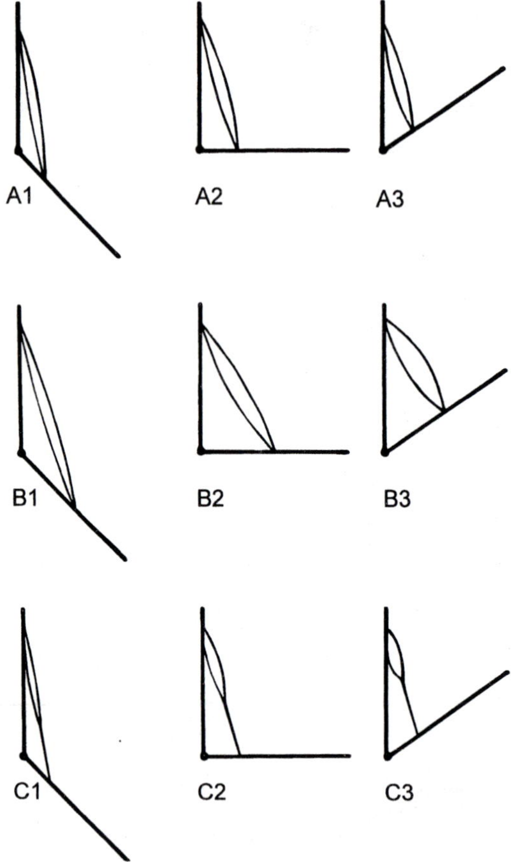

FIG. 9.12 Circumstances making a muscle, moving a joint through a given angle, contract more than a "standard" muscle. (A) The standard arrangement. The ratio of the distance of the joint to the origin and the joint to its insertion is called the attachment ratio. (B) The muscle is inserted further from the joint; the attachment ratio is greater. (C) There is a noncontractile element, tendon, in series with the muscle.

crease again (fig. 9.10). Figure 9.12 shows two circumstances that will make this effect of greater or lesser importance.

The first is that the positions of the origins and insertions relative to the hinge will make the muscle change in length to a greater or lesser extent as the joint angle changes. A1, A2, and A3 show a fairly long muscle with its insertion close to the hinge. As the joint angle decreases, the muscle does get shorter, but not greatly so. In B1 the muscle is inserted further from the hinge. As the joint angle decreases the muscle has to shorten considerably, much more, proportionally, than does the muscle in A1–A3 to bring about the same angular rotation.

The second circumstance producing great change in muscle length is that in which more or less of the distance between origin and insertion is occupied by noncontractile tendon. In C1 the insertion is again close to the hinge, but at this joint angle two-thirds of the length of "muscle" is, in fact, tendon. As the joint angle decreases, the muscle must shorten by a large proportion of its extended length. This shortening will decrease the sarcomere length and, as shown in figure 9.10B, considerably reduce the force the muscle can exert.

This inability of muscles to exert large forces when they are greatly shortened can be easily felt in the hand. With one hand, grasp the index finger of the other and force the metacarpophalangeal and first interphalangeal joints into right angles in flexion. Now try to flex the finger tip. Little, and very feeble, movement is possible. This is because the flexor digitorum profundus, which originates on the ulna and has a long tendon running to the fingertip, cannot shorten any further, once two of the three joints it affects are bent into right angles.

In determining the action of a muscle around a hinge joint, a number of variables must be taken into account: the distance of both the origin and insertion from the hinge; the proportion of the length of the muscle that is occupied by tendon; the amount of flexion the joint undergoes; the force resisting the movement; whether power over the whole range of flexion is important or whether it is final velocity that is critical, as in jumpers; and so on. Obviously, the number of variables can become daunting. Stern's computer analysis (1974) attempts to take a reasonable number into account.

Stern performed many analyses, and I shall merely mention a few points here. The first, agreeing with Calow and Alexander's, is that the results are different if the time taken to complete a movement is to be minimized or if the final velocity attained is to be maximized. In running, the total time to complete a movement is important, whereas it is the final velocity attained that is important in a jump, because this will determine the length of the jump. A sport fencer, for whom a mere touch is sufficient to gain a point, would wish to minimize the time elapsed during a lunge, to give his opponent less time in which to parry. A boxer must, notionally, settle for a compromise, a high velocity increasing the force of a blow, but a small elapsed time giving less time for blocking the punch. For any particular length of muscle, the muscle insertion should be closer to the joint (have a smaller moment arm and a greater "attachment ratio") if the aim is to achieve a high velocity at a particular point, but a greater moment arm, and therefore a lower attachment ratio, if the aim is to achieve the movement to that point in the minimum time. The reason for this is that if a muscle has a large moment about its joint, it can produce a high initial acceleration and velocity, and this will lead to high average velocities. However, the muscle will be shortening, and the force it can apply will fall off, both because the sarcomeres are shorter and because the muscle is shortening rapidly and so sliding down the curve of the Hill equation (fig. 9.10A). A smaller moment arm results in a lower early velocity and lower acceleration, but the acceleration is of longer duration.

If the contractile tissue of a muscle can be as long as desired, then it is always possible to improve the system's performance at the place where the work is done by increasing the moment arm. When the muscle is between the joint and the load, the larger the moment arm, the less the bending moments in the bone, which is advantageous. In fact, the muscles can rarely be made very long, because of anatomical constraints. The usual constraint is that if the origin of the muscle is a long way from the joint it moves, it is likely to be longer than the bone to which it should be attached. A very long biceps would have its origin on the base of the skull. There can also be difficulties if the insertion is far from the joint, particularly if the joint is concerned with grasping. Another point that comes from Stern's analysis is that smaller moment arms are increasingly favored as the excursion of the joint required becomes greater. This is because, for a given excursion, the shortening of the muscle is proportional to the moment arm. The more a muscle shortens, the farther it has to move from a sarcomere length where it can exert its greatest force.

If the total length of contractile tissue is held constant and only the length of tendon allowed to alter, it is no longer true that increasing moment arms will increase the effectiveness of muscles. As more and more of the length of the muscle plus tendon becomes tendon, the optimum attachment ratio increases, but the best position for the insertion remains about the same.

So, with muscles with a limited amount of contractile tissue there is an optimum position for the insertion. The precise position will depend on whether maximum or average velocity is being maximized, but will not be very different in these two cases. Most important, it turns out that if the muscle is reasonably long, the point of insertion should be close to the joint.

Of course, the simplifications in this discussion mean that reality is more complicated. For instance, often at the extremes of the excursion of a limb, the turning moment of a muscle may be so small as to make it very difficult for it to work properly. Other muscles, with larger turning moments, must initiate the movement, the more closely inserted muscle taking over when the movement has got under way and the turning moment has increased because of the relative movements of the

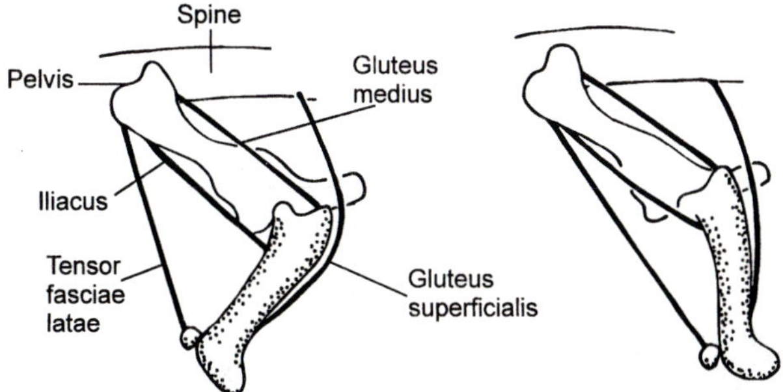

FIG. 9.13 The muscles protracting and retracting the femur of the horse. Right, the femur is protracted. The tensor fasciae latae is much shortened compared with its state on the left, before protraction.

bones. In the hip joint of the horse (fig. 9.13), the tensor fasciae latae always has a greater moment arm than the iliacus, and it is particularly well placed to initiate protraction. As the movement of protraction gets under way, the tensor fasciae latae will shorten considerably and exert less force even though its moment arm increases. However, the iliacus, which had a very small moment at the beginning of protraction, will become effective. Similar remarks apply to the gluteus superficialis and medius.

There are some situations in which neither power nor velocity, but force, is important. In such places the insertions are far from the joints, and often the muscles are very long and straplike. A good example is the forelimb of the mole (fig. 9.9). There are strong muscles, with large moment arms, acting against a large load, the earth through which the mole burrows.

The possibility of reducing bending moments in bone by altering points of insertion of the muscles that move the bone is limited in two ways. Probably less important is anatomical. As an example, I have mentioned already the hands of monkeys and apes, in which the tendon must lie close to the bone to allow room for the bough. It is a general feature of muscles that if they have long tendons these will lie close to the joints they move. This must be so, otherwise the total shortening necessary for the muscle and the tendon together would be greater than the muscle could manage. Therefore, the retinacula keeping the flexor tendons close to the phalanges of apes are what we should expect even if apes touched things with their fingertips only. In general, if muscles insert a long way from the center of rotation of the joint, the limbs will be bulky and undergo large changes of shape. The more generally important reason why bending moments tend to have to be large is because, as Stern, and Calow and Alexander, have shown, insertions need to be close to joints if the muscles are to work at anything like their maximum power output.

9.6 Muscles as Stabilizing Devices

This discussion has focused on the ability of muscle to produce the maximum power output at its point of action. This is not all muscles do, however. As a graduate student I went to a slaughterhouse to get some still warm cow's bones, and saw a cow being slaughtered with a captive-bolt gun. The standing, ruminating, cow was reduced to a lifeless body on the floor in such a short time that it was almost as if the acceleration due to gravity had been momentarily doubled. This brought home to me the fact that most muscles, most of the time, are not moving their points of attachment, but are in fact keeping them stationary. If they stop doing this, then the limb joints collapse under the weight of the animal. Muscles not only keep the joints in the right relative positions, but also *prevent* them from moving too far or too fast.

When a muscle is being extended while still exerting a contractile force it is said to be undergoing eccentric loading. This happens, for example, when we run. When our heel strikes the ground the weight of the body tends to make the knee flex. The knee is prevented from flexing too quickly or too much by the action of the quadriceps muscle, which extends in a controlled way, so that the body is in the right position when the quadriceps starts to contract and extend the knee. Eccentric loading happens fiercely in fast downhill skiing—hence the burning feeling in the quadriceps. The movement of a muscle when it is shortening against a load is called concentric loading. It happens that eccentric loading is more damaging to muscles than concentric loading. That is why climbing down a mountain is more painful than going up. It is not clear whether it makes any difference to a bone whether it is being loaded concentrically or eccentrically. However, it appears that muscles in general exert larger forces when working eccentrically. Possibly this larger force results in eccentrically loaded muscles having a greater osteogenic effect than when working concentrically (Hawkins et al. 1999).

9.7 Curvature of Long Bones and Pauwels' Analyses

Bending magnifies the effects of forces compared with pure compression or tension, and so it might be adaptive to have things arranged to minimize bending forces. A device that might do so is this. A bone is sub-

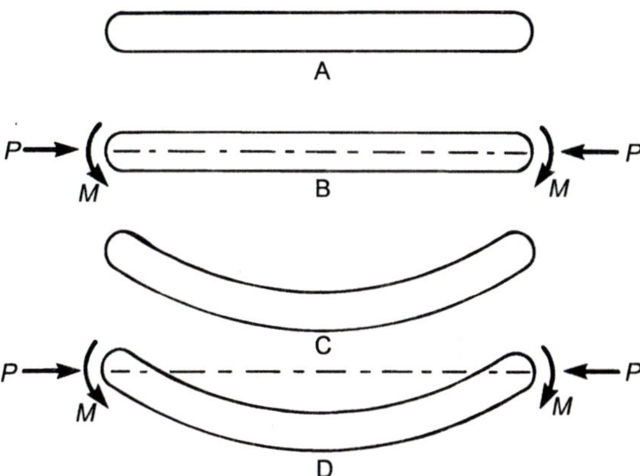

FIG. 9.14 (A) The unloaded bone is straight. (B) When an axial load *P* and a bending moment *M* are applied, their effects are additive at any point. (C) The unloaded bone is curved. (D) The force *P* is no longer axial because the midline of the bone deviates from the line joining the points of application of the load. This will result in the production of a bending moment, greatest in the midlength of the bone, in this case of the *opposite* sense to *M*.

jected to a moment of value *M* at each end, and also a compressive load *P*. If we assume that the alteration in the shape of the bone caused by the bending is negligible, we can calculate the stresses by adding the stress caused by the compressive load and by the (constant) bending moment (fig. 9.14). The effect of the moments can be reduced by curving the bone, because the force *P* will now itself have a bending moment about the bone, least at the ends and greatest in the middle. This bending moment is subtracted, if the curve is in the right direction, from the moments at the ends. It is possible to arrange matters so that in some parts of the bone the bending moment is completely abolished.

Long bones are quite frequently curved, so it might appear that this possibility is being realized. However, experimental studies (Lanyon and Baggott 1976; Lanyon and Bourne 1979) by no means support this hypothesis. Lanyon and his co-workers looked at surface strains on the radii and tibiae of sheep. Both these bones are longitudinally curved in the anteroposterior plane. However, they show great differences in the greatest principal strains on thé anterior and posterior surfaces, indicating considerable bending. So much was this so that one cortex was in tension, even though the bone as a whole was loaded in compression. In the tibia the tension side was the longitudinally concave side, so it is possible to suppose that the tibia is at least bent in the correct direction to minimize bending. However, in the radius the tension surface is the longitudinally convex surface. In this bone, therefore, the curvature is *enhancing* the effects of bending imposed during locomotion. Similar findings have been made in the horse (Biewener et al. 1983a,b).

The hand of the chimpanzee acts as a hook for grasping boughs. The bones of the hand, the phalanges and the metacarpals, are curved, and this allows room for the bough. A considerable bending moment is produced by the load (the bough in this case). The action of the flexor muscles that keep the hook hook-shaped put the bones into compression (Preuschoft 1973). However, *pace* Preuschoft, this does not, in general, put significant counteracting bending moments on the curved bones, because the retinacula keep the center line of the bones and the tendon parallel. But there are interosseous muscles that run from one end to the other of the metacarpals. One of their functions is to prevent the metacarpals from splaying out sideways. They also, by their contraction, put the metacarpals into compression, and this force will cause a counteracting bending moment, rather like the situation shown in figure 9.14. The importance of this effect is unknown, because the force exerted by the interosseous muscles is unknown.

This counterbending function of muscles has often been claimed by Pauwels and his school (Nachtigall 1971; Pauwels 1980) to be of general importance. Pauwels is very fond of the action of the iliotibial tract (IT) in counteracting the tendency of the human femur to bow out laterally. Although the tract must act in this way to some extent, its magnitude is uncertain, and there is no doubt that many of the proposals that have been put forward to explain the action of muscles in preventing bending should be regarded as interesting suggestions only. We do not yet have the experimental evidence to be able to assert their correctness.

Pauwels has written many papers on the subject of the design of skeletons, nearly all of them on the human skeleton, though many papers are of general applicability. Fortunately, many were collected and translated into English (Pauwels 1980). One particular concern of his is to show how the whole skeleton is designed to minimize the bending moments imposed on the bones. His procedure is to take some case of static loading of a structure, which is roughly like a limb, and to show how, as the structure is made more and more lifelike by the addition ties, lumps, angulations, and so on, the bending moments on the bones are progressively decreased.

Figure 9.15 shows, for example, his analysis of the human lower limb when the heel is lifted off the ground. As reality is approached, the maximum stress, produced by a combination of compression and bending, decreases from 28 to 7 MPa. Of particular interest here is the change between (C) and (D). The only anatomical differences between these two are in the shape of the femur and a concomitant change in the shape of the femorotibial joint. The femur in (C) is straight, but in (D)

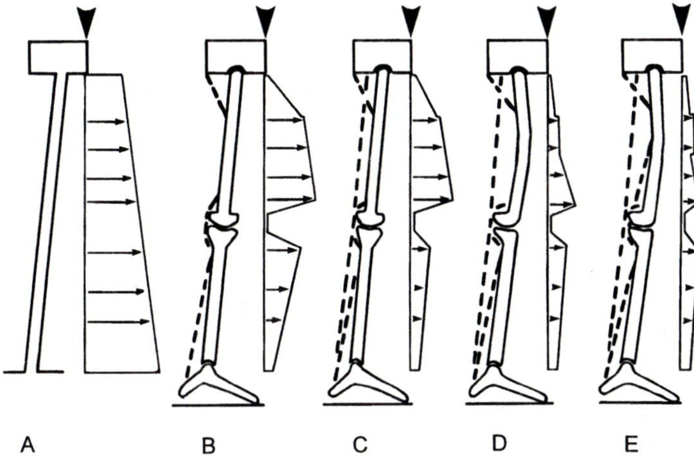

FIG. 9.15 These diagrams show Pauwels' method for demonstrating various devices in the skeleton that reduce stress in the bone. In this case it is the human leg, in walking, just at the point where the heel leaves the ground. The stresses, shown by the polygons with the small arrowheads, are the greatest stresses in the bone at each level and are composed of bending and direct compressive stresses. (A) The leg is represented by a tilted column with an eccentric load. The stress increases uniformly from top to bottom. (B) Three joints appear. The system is stabilized by contraction of the muscles (such as gluteus maximus) round the hip and muscles (such as soleus) round the ankle, and by ligaments (such as the poster capsular ligament) at the back of the knee. The bending stresses in the tibia are now much reduced, there being no bending stresses across the knee joint, only compressive stresses remaining. (C) Biarticular muscles are now included, such as the hamstrings, above, and the gastrocnemius, below. These could reduce still further the bending moments on the tibia and the femur. However, at this stage the structure becomes statically indeterminate, and the magnitude of the effect can only be guessed. (D) The forward bowing of the femur reduces the moment arm of the load about the femur, and so reduces the bending stress in it. This effect is enhanced by a change in shape of the articulation of the femur with the tibia, so their shafts are brought forward a little. (E) The action of another muscle, the short head of the biceps femoris, tends to bend the bottom of the femur in the opposite sense to that produced by the main load, reducing bending stresses still further. (Modified from Pauwels [1980].)

is bowed anteriorly. The middle of the length of the femur is now bent anteriorly by the compressive load acting between its two ends, and posteriorly by the bending moment at the hip. The combined effect is to reduce considerably the maximum tensile stress in the femur.

Pauwels' results and diagrams have a seductive but specious transparency. The actual loads borne by the ligaments and muscles have necessarily been calculated, not measured. They are the loads necessary to produce static equilibrium. However, many of the most interesting diagrams show structures that are statically indeterminate. That is to say, there are redundant members; one could cut one or more of the ties without the structure necessarily collapsing. For example, in figure 9.15B all the muscles and ligaments are needed, but in figure 9.15C it would be possible to cut two muscles without the structure collapsing. So, although in a general way Pauwels' calculations will show what is happening, the actual values will depend on tensions developed in muscles and in ligaments, and this will depend on the activity of the muscles and the stiffness of the ligaments.

Also, the number of muscles acting across joints is large, and it is difficult to know how strongly each muscle is acting at any point in the gait cycle. Workers on this kind of problem, about which there is a large literature, assume some kind of optimization function that the body, if it thought like them, would perform, *and* that is mathematically tractable (e.g., for a lead into the literature, see Rasmussen et al. [2001]). Some have tried the efficacy of neural network models (Sepulveda et al. 1993). An example of the kind of complication necessary is shown by Munih and Kralj (1997), who modeled the effects of muscles in reducing bending moments in bones during the process of standing or gently swaying forward and backward. They had to consider the simultaneous action of 28 muscles.

Another necessary simplification of the Pauwels approach is that the bones are, apart from odd protuberances, uniform in cross-sectional area along their lengths. As a result, the stresses calculated for the bones at any cross section will be directly proportional to the axial loads and bending moments there. In reality, as Pauwels recognizes, the cross sections of bones and their second moments of area vary greatly along their length. It will almost certainly be adaptive to have bones stressed reasonably uniformly (see chapter 4), because this will minimize the mass necessary to bear the load. If these two points are borne in mind, the analyses of Pauwels are very helpful to one's understanding of how bending in bones may, to a large extent, be limited by the soft tissues.

Finally, consider the human foot. In walking, just as the big toe is leaving the ground, the front part of the foot is bearing a load of about 120% of body weight. The foot is loaded like an arch, with the body's inertial weight acting down through the tibia and fibula, and this is opposed by the Achilles tendon acting on the calcaneus and by the ground reaction acting on the metatarsal heads and the toes. The arch is prevented from collapsing by various flexor muscles and their tendons; by ligaments running between the various individual bones, particularly the long plantar ligament; and by the plantar aponeurosis, which overlies the muscles of the foot and runs between the bottom of the calcaneus and the base of the metatarsals and the toes.

Stokes et al. (1979) analyzed the forces in the metatarsals during walking and found that, in fact, the flexor tendons and plantar aponeurosis reduce the bending moment on the metatarsals by only about 10%. The bulk of the resistance to collapse is provided by the ligaments. The values of loading that they calculate are impressive. The first metatarsophalangeal joint has a joint force of about 80% of body weight, and the first metatarsal has a compressive force of 130% of body weight and a shearing force of about 30% of body weight. These values are for walking; they would be much higher in running. This calculation is based on assumptions about the forces in the tendons and aponeurosis that are reasonable, yet it is surprising that their effect in reducing bending moments is not greater. (Ker et al. (1987) showed that the ligaments underneath the foot have an important energy-saving role in running. For every stance phase in running an average man has an energy turnover of about 100 J. If this is not stored somehow the muscles must generate the energy. In fact, about 35 J is stored in the Achilles tendon, and 17 J in the tendons and ligaments of the foot. So, about half the total energy turned over is stored in these two sites.)

In this chapter I have argued that bones should, if possible, be loaded in compression and not in bending. This is not usually possible to arrange because often the *function* of a bone is to produce forces at large angles to its long axis. Also tendons, which could in many circumstances reduce the bending moments on bones, are themselves constrained, by anatomical requirements and by the limited contractility of muscles, to lie very close to the bones.

9.8 Skeletons in General

Bones are arranged together in skeletons, and it would be futile to try to discuss even a tiny proportion of the skeletons that are found in vertebrates. However, some general principles that have to do with mechanical adaptations are worth stating. What follows is, however, the merest taster for the wonderful range of adaptations that we can see around us. If you visit any museum with a collection of skeletons, with your brain in gear, you should leave with lots of new insights and unanswered questions. I shall write very briefly about the dog's skeleton as an example of a mammal, making comparisons between this skeleton and others. An excellent account of the mechanical design of a rather unusual type of skeleton, that of amphisbaenians (legless burrowing lizards), is given by Gans (1974).

9.8.1 Pelvic and Pectoral Girdles

The major parts of the skeleton of the dog are the skull, the vertebral column, which supports the rib cage, and the two limb girdles, which support the fore and hind legs. The pectoral girdle is very much reduced from the condition found in reptiles. In the reptiles from which the mammals arose, and, indeed, in the primitive monotreme mammals, the pectoral girdle is U-shaped and rigid. This results in the head of the humerus being in a fairly constant position in space relative to the vertebral column. In more derived mammals (further removed from the ancestral condition), such as ourselves, the pectoral girdle is reduced to two scapulae and two clavicles, the latter being attached to the sternum. The scapulae are slung in a bag of muscles and have considerable freedom, though the distance from the humeral head to the sternum is held constant by the clavicle. In mammals more derived than ourselves, even this hindrance to movement may be lost. In the dog the clavicle is represented by a little bony plate embedded in a muscle. As a result of this reduction of the pectoral girdle, the forelimb has considerable freedom in relation to the trunk.

The pelvis, by contrast, remains rather more firmly fixed to the sacrum, and little relative movement is possible. Why have these two girdles, which are rather similar in early reptiles, evolved so differently? The center of gravity of the body of the dog, and of most ground-living mammals, is nearer the forelegs than the hind legs, and so when a dog is standing still the forelegs bear a larger share of the body weight. However, this in itself would not account for the evolution of different types of suspension. When the dog is running, each leg, both fore and hind, when it strikes the ground, exerts a force tending to decelerate the body in the direction of motion and then, later in the stride, to accelerate it. However, the effects of the fore and hind legs are different. Jayes and Alexander (1978) have shown that the line of action of the legs tends to act through one point during the time the foot is on the ground. For both fore and hind limbs this point lies above the proximal joint, but is also *behind* it in the forelimb and *in front* of it in the hind limb. The result of this is that, compared with the hind limbs, the forelimbs are more concerned with decelerating the body and less with accelerating it. The forelimbs do not hit the ground while passively stretched out in front of the body; they are rotating backward at the moment of strike. Nevertheless, they do need to have a fairly efficient shock-absorbing mechanism. The hind limbs, on the other hand, have a greater responsibility for accelerating the body and, being more rigidly attached to the spine, they can push the body forward effectively without the need to use muscles to prevent the hip joint from moving around relative to the center of gravity of the animal.

9.8.2 *Limbs*

The proportions of the limb segments differ quite widely between mammals using different kinds of locomotion. Table 9.2 (derived from data in Gambaryan [1974]) shows this. There is a general tendency within groups for fast runners to have distal parts elongated relative to the more proximal parts. Compare the gazelle with the tapir, and the cheetah with the bear. Like the cheetah, the dog (represented in the table by the wolf) is intermediate between the bear and the gazelle, having somewhat elongated metacarpals and metatarsals. The dog is digitigrade, that is, it stands on its toes; the bear is plantigrade, standing on the flat of its feet; and the ungulates are unguligrade, standing on tiptoe.

Associated with this relative elongation of the distal elements is the removal of most muscle mass from the distal parts of the limb. For instance, the widest part of the belly of the gastrocnemius in humans is 54% of the distance down the leg from hip to toe. In the dog it is 43% and in the horse 32%. The muscles of the limbs have, on the whole, the same anatomical relations in a fast-running animal and a more plodding one, yet, because of the stretching out of the distal elements, the muscle mass is more concentrated near the hip. The mechanical advantage of this arrangement is clear. Suppose an animal traveling at a forward speed U has the main joint in the limb a distance h from the ground. As the limb is swung back its extremity is stationary relative to the ground. It therefore has a velocity $-U$ relative to the trunk. It must have an angular velocity $-U/h$. If the moment of inertia of the limb about the shoulder or hip joint is I, then it will have kinetic energy relative to the trunk proportional to $U^2I/2h^2$. The energy expended in the propulsive part of the stride is therefore proportional to I/h^2 (Alexander 1975). It will reduce energy expended, therefore, if legs are long, with a small moment of inertia about the proximal joint. These are, to some extent, contradictory requirements, of course. (I is not constant, since all fast running animals tuck their legs up on the recovery part of the stride; the function of this action is to reduce I.) Other things being equal, I is proportional to length cubed, and therefore longer legs will not be useful from the energy conservation point of view unless I is reduced in some way. In evolution this is done by lengthening the legs while keeping the muscles confined to the proximal part of the leg.

Taylor et al. (1974) compared the cheetah, the gazelle, and the goat, which can be made to run at the same speed and which have similar

TABLE 9.2
Percentage Lengths of the Three Segments of the Hind Limbs of Some Mammals Relative to the Femur Length, Taken as 100%

Species	*Femur*	*Tibia*	*"Foot"*
African elephant	100	61	26
Brown bear	100	83	44
Tapir	100	80	103
Cheetah	100	100	98
Wolf	100	105	100
Quagga	100	106	164
Red Deer	100	120	171
Gazelle	100	121	176

Source. From Gambaryan (1974).

masses but dissimilar leg proportions. They found that the energy consumption of these different animals was very similar and concluded that the effect of moving the muscles proximally was not important. They say, "This suggests that most of the energy expended in running at a constant speed is not used to accelerate the limbs." This result is contrary to experience—Surely anyone who has introspected while sprinting will think that accelerating the limbs requires great effort. It is also not borne out by the later work of Taylor and his group referred to in chapter 7. Perhaps the cheetah, which cannot move its muscles too far proximally, has other adaptations, which overcome the effects of having a larger value of I. One such adaptation is the ability to flex the back much more than can the gazelle or the goat. Steudel (1990a,b) attached weights to dogs, either near their center of mass or to their legs. She found that the cost of locomotion barely increased if the weight were near the center of mass, but that there was a noticeable increase when the weights were attached more distally, to the legs. This certainly suggests that accelerating the legs is an important cost.

9.8.3 *Fusion and Loss of Bones*

In chapter 7, I considered how bones might become adapted so as to perform their functions with minimum mass. In these discussions I assumed that the job had to be done by a single bone. But in skeletons it is commonplace to find a set of bones all doing roughly the same thing, while in other related groups of animals, the number of bones doing the same job is smaller. This reduction is brought about, in evolutionary time, either by the fusion of bones or by their complete loss. A familiar example of this is the reduction of the number of functional metapodial bones in ungulates. Among the artiodactyls, the hippopotamus has four metacarpals, all of them of roughly the same robustness. In the pig and the cow there are also four, of which two are much reduced, while in the llama there are effectively two only. In the perissodactyls, there are four metacarpals in the tapir, three in the rhinoceros, and one in the horse.

There is also a tendency in the mammals to fuse the radius and ulna, and the tibia and fibula, together, so that they become functionally one bone. The radius and ulna are fused in animals as different as the sea cow *Trichecus*, the elephant, and the horse. Birds are well known for the tendency of the bones in the sacral region to become greatly fused, and the sacral region of mammals shows great variation in the number of vertebrae that are fused together. In the dog the radius and ulna are fused along part of their length, and the fibula, besides being very thin, is fused distally to the tibia.

The great advantage in reducing the number of bones to perform a particular function is that the amount of mass necessary for a particular strength or stiffness is also reduced. A simple example shows this. Suppose a cantilever has to support a given load over a given distance and must deflect only by some amount. Suppose also that the cantilever has an overall cross section twice as deep as it is broad. What would be the effect of slitting the cantilever horizontally into two, or into four, separate cantilevers stacked on top of each other?

The separate cantilevers would not support each other, because they would be free to slip past each other in shear (fig. 9.16), and they would, in fact, be equivalent, mechanically, to two or four cantilevers arranged side by side, sharing the load. It is easy to show, given the conditions we have set up, that the mass necessary to produce a given stiffness is proportional to the square of the number of sections into which the cantilever has been sliced, so cutting a cantilever into two horizontally will require it to be four times as heavy if it is to remain as stiff. (Remember, the *overall* cross section remains twice as deep as it is broad.) Slitting the cantilever will leave its resistance to load from the side unaffected, but the resistance to torsion will be reduced, though by how much is too complicated to consider here.

All this shows that if a set of bones are all doing the same thing, then there will be considerable weight advantages in reducing the number of separately moveable bones. But such reduction is advantageous *only* if the concomitant reduction in mobility is acceptable: as the radius becomes fused to the ulna, the mass of bone needed becomes less, but the wrist becomes unable to twist.

A survey of mammalian skeletons shows a great range in the amount

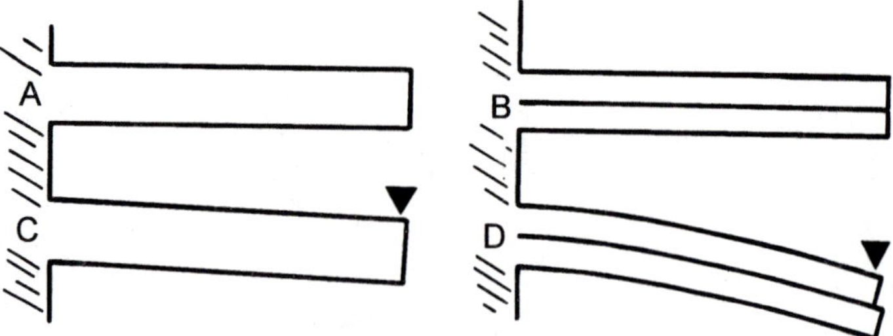

FIG. 9.16 (A) A solid cantilever deflects when it is loaded (C). The same cantilever, if slit in two horizontally (B) will deflect four times as much with the same weight (D).

of mobility allowed between different bones. (Compared with those of mammals the skeletons of birds are extraordinarily uniform, varying mainly in the relative sizes of different bones and little in the amount of fusion or loss.) The primitive state in mammals is to have little fusion or loss of bones. Humans are interesting in that they retain the primitive condition almost entirely.

9.8.4 *The Vertebral Column*

Finally, in this brief survey of the dog's skeleton, we consider the functions of the vertebral column. Of all the main components of the bony skeleton, the vertebral column is perhaps most able to show the division of function between hard and soft tissues, with bone taking the compressive loads and soft tissues the tensile ones. The vertebral column of the mammal has various functions: to support the head, to act as a place of attachment for the limbs, to transfer force from the limbs to the rest of the body, and to support the viscera.

The function of supporting the viscera requires that when a mammal, or any other tetrapod, is standing still, the vertebral column between the limb girdles will be loaded so as to sag. It might be possible to counteract this tendency, without making use of muscles attached to the column itself, by curving the vertebral column into an arch. The column would then be loaded in compression, with the viscera slung from it, and the tendency of the ends of the column to splay apart could be counteracted by the body wall musculature. Figure 9.17 is a diagram showing this. The solid circles represent the main mass to be supported. The head, which is cantilevered out on the end of the neck, is supported by tension in muscles running to the enlarged spinous processes of the anterior thoracic vertebrae. The cervical vertebrae are put into compres-

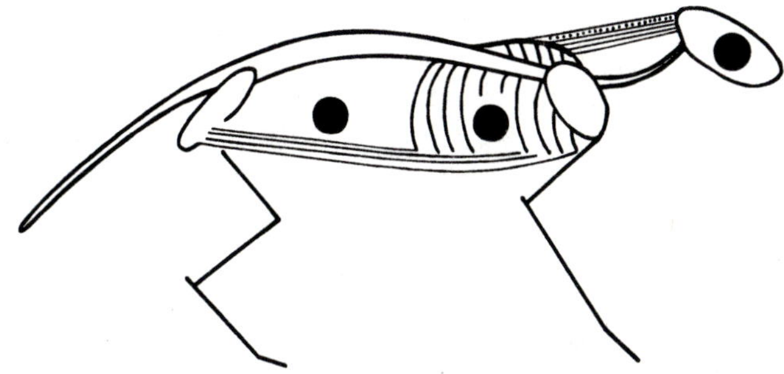

FIG. 9.17 Diagram of a tetrapod showing the complementary effects of the muscles and the spine. Explanation in text.

sion by this, although, as shown here, any real neck is curved and would need intrinsic musculature to stabilize it from buckling.

In mammals such as the cow, which must frequently raise and lower a heavy head, there is a well-developed ligamentum nuchae running along the top of the neck (shown dotted in fig. 9.17). The ligament is composed mainly of elastin, a rubbery protein. When the head is lowered, the ligament is stretched and, unlike muscle, can store the work done in stretching it as strain energy. When the time comes to raise the head and survey the passing scene, the muscles are assisted by the force of the passive contraction of the ligamentum nuchae. In the deer and the sheep the ligament does not entirely balance the weight of the head and neck, and muscles are needed to raise them. However, in the camel the ligament will balance the weight, even when the head is held straight out forward (Dimery et al. 1985). In the trunk, between the fore and hind legs, muscles running from the rib cage to the pelvis act as tension members and the curved spine as the compression member of the system bearing the weight of the viscera.

This description is greatly idealized, because real animals run around, twisting their spine to a greater or lesser amount, and most of the time do not bear much resemblance to the architectural model. In fact, there is usually a great deal of intrinsic musculature associated with the spine itself.

In locomotion different animals flex their spines to greatly differing extents, and figure 9.18 shows the effects of this on lumbar vertebrae. The lumbar vertebrae of the porpoise are very simple. The centra are simple disks, the transverse processes stick out horizontally and directly laterally, and the spinous process, blade-shaped like the transverse processes, has a very simple articulation with the process in front. It is clear that very little relative motion of these vertebrae is possible. Virtually no

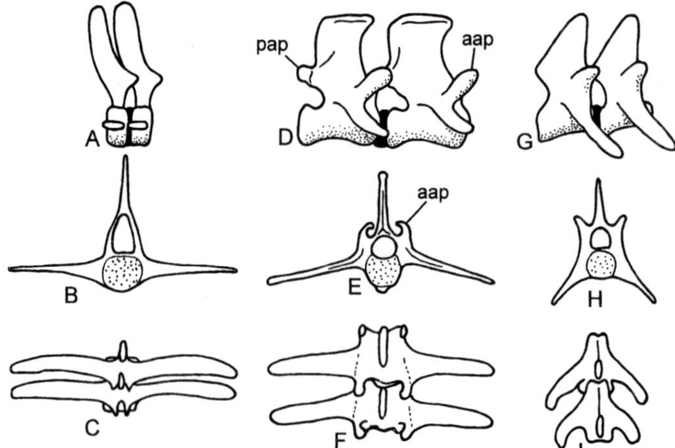

FIG. 9.18 This shows the right-side (A), anterior (B), and dorsal (C) views of lumbar vertebrae of the common porpoise *Phocaena phocaena*; (D–F) are similar views of lumbar vertebrae of the fallow deer *Dama dama*. (G–I) are similar views of lumbar vertebrae of a dog.

flexion in the horizontal plane can occur because the transverse processes would bump into each other. If the vertebrae are to flex ventrally, the movement cannot be brought about by intrinsic muscles attached to the transverse processes, because such muscles would have no turning moment about the center of rotation of the joint. Such a movement could be brought about by blocks of muscles running beneath the vertebral column. In fact, the lumbar region of the porpoise is not capable of much flexion, but the great blocks of muscles running toward the caudal peduncle allow very forceful movement of the tail.

The lumbar region of the fallow deer is somewhat more flexible than that of a porpoise, but, because of this animal's large gut, the spine does not flex much during running. This is shown in the vertebrae. The transverse processes stick out almost laterally and horizontally, and there is not much room for dorsiflexion between the spinous processes. In fact, dorsiflexion is also severely limited by the articular processes. The posterior articular process (pap) of each vertebra fits like a peg into the socket of the overarching anterior articular process (aap).

The lumbar region of the dog, like that of many carnivores, is flexed and extended a great deal in locomotion. This allows the effective length of the stride to be increased beyond that allowed by rotation of the legs. The lumbar vertebrae have much more relative freedom of movement than those of the other two species I have described. There is much more space between the spinous processes. The transverse processes project ventrally, so that muscles running between them can flex the vertebrae, and the spine does not, as it were, have to rely on muscles elsewhere to do its flexion for it. This downward projection of the transverse processes is seen in other noncarnivorous mammals that flex their spines vigorously—in rabbits, for instance. The *anterior* projection of the transverse process is more difficult to explain. Possibly, in vertebrae that are capable of much *lateral* flexion, the overlapping of the lateral processes keeps the moment arm of the muscles joining the processes large at all angles of flexion. They will certainly have this effect, but how important it is, is difficult to say.

A very interesting paper on evolutionary theory has been written about the mammalian vertebral column. There is a tendency to assume that evolution in some way results in progress. Gould (1996) has written half a book on this matter (the other half is about perfection in baseball). He insists that the apparent progress we see is simply a result of the fact that because life started with very simple organisms, some of them, naturally evolved into more complex organisms, and that this in some way might seem to be an advance. However, Gould claims, if one looks at a group of reasonable complexity in the fossil record, one is as likely to see descendants becoming more simple as more complex. McShea (1993) examined the complexity of the mammalian vertebral column and found, in those lineages showing a sufficient fossil record, slightly more lineages becoming less complex than becoming more complex.

9.8.5 The Skull

The skulls of vertebrates are extraordinarily complex and varied, and it would make little sense to try to make any generalizations about them. Volume 3 of *The skull* (Hanken and Hall 1993) is an excellent introduction to the problems of modeling the mechanics of the skull and of attempting to marry the models to what information is available from strain gauging and other techniques (Weishampel 1993). The last section of Russell and Thomason (1993) in this volume is called "The way of the future: finite-element analysis" This is indeed so, but it will probably be some time before much useful and digestible information comes out, though I have mentioned the interesting preliminary results of Rayfield et al. (2001) on the dinosaur skull in section 7.10.

9.9 Conclusion

In this chapter I have tried to describe some of the devices found in animals to take advantage of the different mechanical properties of bone, muscle, and tendon. Perhaps one should say, to mitigate the unsatisfactory features of these tissues. As always, compromises abound, and bone, in particular, is necessarily often stressed in tension, a mode of stressing it is not well able to withstand. As a result, bones often break. In the next chapter we look at the problem of how often they break and whether the safety factors found in bones are appropriate.

Chapter Ten

SAFETY FACTORS AND SCALING EFFECTS IN BONES

IN CHAPTER 7, I discussed the mechanical properties of whole bones. However, there was an important omission in the discussion, which I must now try to fill. We have assumed that natural selection has, for instance, specified some load that must be borne without breaking or too much deflection. But why is this criterion set, rather than some other? In particular, what relationship does the load that the bone is designed to bear have to the loads that are encountered in life? This is a question of safety factors. The chapter also deals with a related question: What kinds of adaptive differences in the general shape of bones are produced by the fact that animals are of different sizes? This is a question of scaling.

10.1 Safety Factors

Every improvement in some mechanical property of a bone will have a cost associated with it. Small changes in the mineralization of bone produce changes in Young's modulus of elasticity and impact energy absorption, which are opposite directions. Similarly, an increase in the stiffness and strength of a bone, given by an increase in mass, will produce the disadvantages of greater mass and the greater metabolic energy, and time, required to produce it.

The most obvious feature about safety factors in bone is that they are insufficient to prevent bones from breaking from time to time. Many readers will have at least cracked a bone during their lifetime. Usually this will have occurred during sport and will have involved a metacarpal, metatarsal, rib, or clavicle. However, more serious fractures are not uncommon and do not all, by any means, result from "unnatural" injuries like automobile or skiing accidents, or bullets. Lovejoy and Heiple (1981), examining a large set of aboriginal Indian skeletons in Ohio, found a 45% chance of a long bone fracture in any individual (mostly adults). Since only 5% of these people lived beyond about 45 years, senile changes were not an important cause of fracture. Bones may break in healthy young people without any apparent accident to cause it. Throwing a ball, hand grenade, or javelin can result in fractures of a bone in the upper arm (Evans et al. 1995). In fact, whether these apparently spontaneous fractures are, in fact, the end result of slowly spreading fatigue fractures is uncertain and irrelevant. Most fatigue fractures, as discussed in chapter 3, are the result of unusual loading, or ordinary loading repeated far more often than usual, but not accidental loading. Even activity only a little more than usual can break bones.

Such apparent weakness is not confined to humans. Running on the flat on grass can cause the fracture of the long bones of horses (Vaughan and Mason 1975; Estberg et al. 1995). Greyhounds are very prone to similar fractures (Devas 1975; Johnson et al. 2000; Muir et al. 1999). Nor is it only highly artificially selected domestic animals that suffer such injury. There is a considerable amount of anecdotal evidence from reliable observers that large wild prey animals, such as zebras and antelope, will sometimes break their legs when accelerating away from a predator.

In wild animals the breaking of a bone, particularly a long bone, is likely to threaten life. It is surprising, therefore, how many animals have healed fractures. A visit to any museum is likely to show a small, but not trivial, proportion of the skeletons with healed fractures. Table 10.1 is a compilation of some of the studies that have been made on healed fractures. In a huge study, Brandwood et al. (1986) examined the incidence of healed fractures in the limb skeleton of various Anatidae (ducks and their relatives), gulls, and pigeons. They examined the incidence in bones, rather than individuals, so it is not possible precisely to say what proportion of individuals had a healed fracture, but it was roughly 5–6% for the Anatidae and gulls, and 3% for the pigeons. A broken bone may be much more serious for birds, than for primates, which show a much higher proportion of injured individuals. Schultz (1939) shot 118 wild gibbons. In these there were 48 healed fractures of the major long bones and 26 healed fractures of other bones. Bramblett (1967) found that among 37 adult baboons *Papio cynocephalus*, only 7 did not have at least 1 fracture. These studies suffer somewhat from the possibility that animals with healed fractures are easier to capture or shoot than others. This problem is absent from the study by Buikstra (1975) in which an entire social group of macaques, *Macaca mulatta*, was culled and the skeletons examined. In the 43 adults there were 17 healed fractures of long bones and clavicles. These were distributed among 11 adults.

These figures show that these primates are quite likely to break an important bone during their lifetime. We do not know the proportion of those who break a bone and survive, not being mortally handicapped. In a small but very interesting study Zihlmann et al. (1990) were able to relate the postmortem state of skeletons of chimpanzees *Pan troglodytes*

TABLE 10.1
Prevalence of Healed Limb Bone Fractures

Group	*Healed bones*	*Total bones*	*Proportion*	*Authors*
Birds				
Vultures	3	17	0.176	Houston (1993)
Laridae	22	5299	0.004	Brandwood et al. (1986)
Anatidae	18	4926	0.004	Brandwood et al. (1986)
Pigeons	5	2635	0.002	Brandwood et al. (1986)
Primates				
Humans	72	2383	0.030	Lovejoy & Heiple (1981)
Baboons	15	732	0.020	Bramblett (1967)
Gibbons	56	2796	0.020	Schultz (1944)
Orang-utans	22	1716	0.013	Schultz (1937, 1941)
Macaques	17	1764	0.010	Buikstra (1975)
Gorillas	18	3912	0.005	Schultz (1937), Randall (1944)
Carnivores				
Viverridae	39	3696	0.011	Taylor (1971)

Note. This is prevalence in bones, not individuals. When the proportion is low, the prevalence of at least one healed fracture in individuals will be, very roughly, 10 to 15 times greater. In the vultures only the ulna was considered.

in the Gombe Reserve in Tanzania to information gleaned from intense observation by Jane Goodall and her co-workers during the animals' lives. One female developed considerable asymmetry in her upper limb skeleton as a result of poliomyelitis at the age of 7 years, but mothered four infants, who did not survive. She died at the age of 19 years. Another female, dying at the ripe age of 43 years, was found to have suffered fractures of a clavicle, an ulna, a metatarsal, and a metacarpal. Despite this she moved around easily. A high-ranking male suffered at least eight fractures, including a very severe crush fracture to the ankle, which resulted in considerable remodeling of the bones. He was seen shortly after he sustained this injury, limping badly, but perked up well, and had no obvious locomotory problems a few months later. Apes are probably much better able to survive bad fractures than many other groups. Being intelligent they have very adaptable ways of living, and being social they get support from their group. In the last case above the authors report that "[He] came into camp limping. The chimpanzees that accompanied him moved slowly to accommodate his impaired gait."

It is difficult to imagine a group of deer or jackdaws being so accommodating. Indeed, the requirements of having an almost perfectly functioning skeleton are probably much more important for many animals than we, living as pampered humans, realize. We are not always being tested by predators, which is the case of, say, wildebeest and gazelles in the Ngorongoro crater, very few of which die of anything except predation, their guts torn out, their testicles mashed, their hoarse cries gradually diminishing as their blood reddens the earth. Hans Kruuk studied these animals for several years (1972). He describes how hyenas attack wildebeest (hyenas are their main predators, and therefore their overwhelmingly important cause of death). "I regularly saw the hunters dash into a large herd, causing the wildebeest to run; then the running hyenas would stop and look at the running animals. . . . Suddenly one of the onlookers might single out a wildebeest from the herd and start a chase. . . . I myself was rarely able to detect any differences in the animals that were chosen . . . they usually seemed to be in perfect condition. But this does not mean, of course, that the hyenas saw no difference; while some people can make a sound judgement about a horse's performance by looking briefly at its gait, I would not be able to do so, and we should not be surprised if a hyena can do even better than a horse dealer."

The predators do not necessarily make the prey run, often enough they just watch them walking. In a world where a slight limp can mean death, the balancing of strength against weight is a very serious matter. Palmqvist et al. (1996) analyzed a Pleistocene bone midden, mainly of herbivores, and showed how the predators took juveniles in preference to adults, but that among the adults, of various species of prey, the prevalence of arthrotic degeneration of the metapodials was far higher than would be expected from an average wild population. These animals would certainly be limping, and would be seen to be the easiest prey.

Teeth and horns (which are really just bones with a protective covering of keratinous horn) are subject to the same kind of constraints as other skeletal elements, and seem to have a variable, but not negligible, prevalence of breakage. Van Valkenburgh (1988) found that large African mammalian predators, lions, hyenas, and the like, had a roughly 0.25 chance of having a tooth broken before death (these animals usually died by being shot), and that the highest prevalence was in the canine teeth. Fenton et al. (1998) found that in 13 species of bats the females had a proportion of 0.05 skulls with at least at one broken or missing tooth, and the males 0.10. Again, the highest prevalence was in the canines. Packer (1983) in a large survey of broken horns in male African antelopes found a species median prevalence of broken horns in living animals of 3.6%. In females the prevalence was much less.

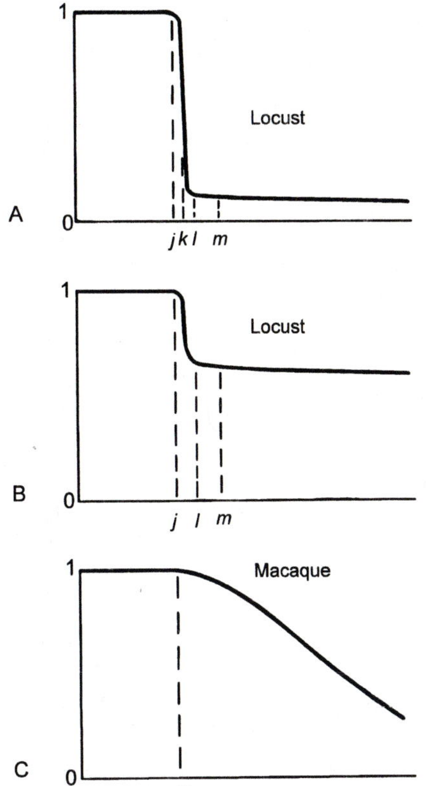

FIG. 10.1. Relationships between the load on a skeletal structure (abscissa) and the probability of such a loading, or less (ordinate). (A) The leg of a locust, with a small probability of predation. (B) The leg of a locust, with a higher probability of predation. (C) Leg bone of a macaque.

We have seen that fracture is not necessarily fatal and that the proportion of fractures in many groups is not trivial. We need now to consider what are the factors affecting the proportion of fractures.

The less variable the loads imposed on a bone during a lifetime, the easier it will be to ensure that the probability of fracture will be small. A nonbony example of this is the jumping leg of the locust. Bennet-Clark (1975) calculated the loads caused by the jump on various structures in the leg, and also the strength of these structures. He showed that the strengths are only about 20% greater than the loads. This is, in engineering terms, a negligently small safety factor. It is made possible, presumably, only because the greatest loads the leg is likely to experience are imposed on it during the jump, and these loads are more or less under the control of the locust itself. Figure 10.1 may help us to visualize the situation. The abscissa is the load on a structure.

The ordinate is the probability of the structure being loaded to that extent, *or less*, during some sensible period, such as from birth to the midpoint of the average reproductive life. In this diagram (fig. 10.1A) the value *j* corresponds to the normal loads imposed during a jump. The locust is *certain* to load its legs to this amount during its life. Because of slight awkwardnesses in the positioning of the leg, the nature of the ground, and so on, the leg will often undergo slightly higher loading than *j* during the jump. The probability decreases rapidly as the loads get higher. The load marked *l* is about the limit of such loads. There is, then, a rather flat tail going toward very high loading, such as would be caused by a predator munching the locust. If the leg is able to bear a load *m*, it will be able to withstand all the loads that are likely to be imposed on it by jumping. Furthermore, increasing the strength beyond *m* will have little effect on the probability of failure, because any loads greater than *l* are quite likely to be *much* greater. A small safety factor *om*/*oj* is probably satisfactory from the point of view of natural selection.

On the other hand, the slightly lower strength *k* would be bad design, because the probability of being loaded beyond this point would be high (50% as drawn here). In other words, a great reduction in the likelihood of failure can be bought for a small increase in skeletal strength. Indeed, this would still be the case if the chances of being untimely eaten by a predator were quite high, as shown in figure 10.1B. If the forces are either highly predictable or irresistible, the safety factor will probably be small.

What is the likely shape of the curve for a macaque? Figure 10.1C shows a possible shape. Up to *n* are the loads imposed by running, jumping, and climbing, and the small falls that all monkeys inevitably suffer. However, above this point there is no sharp decrease in probability, because there will be a gently decreasing probability of ever more disastrous falls. Therefore, in distinction to the locust, there is no dramatic advantage in increasing the skeletal strength a small amount above that necessary to give the strength *n*.

The curves in figure 10.1 are a measure of the benefits associated with various strengths of skeletons. What are the costs? Presumably, the costs are usually some more or less complicated function of the mass of the skeleton. The cost in a currency of metabolic energy, and in a currency of time taken to produce it, will be roughly linear with mass. The *relative* cost of actually producing the skeleton will be markedly different for different vertebrates according to their general level of metabolic activity. For a small endotherm such as a shrew, whose maintenance metabolic activity is a very high proportion of its metabolic rate, the cost of growing the skeleton will be relatively trivial. On the other hand, for an abyssal fish just ticking over on minimal energy intake, the growth of a skeleton could be a sizeable proportion of its total metabolic activity. For all vertebrates the cost of maintaining the bony skeleton will be rather small, because bone probably has a low metabolic and turnover rate compared with other tissues (Thompson and Ballou 1956). Although Schirrmacher et al. (1997) claim that calvarial cells have a high rate of oxygen consumption, it is difficult to know what to make of this claim, because the consumption of bone lining cells is not to be distinguished from that of osteocytes. Even if the individual cells were to have a high metabolic rate, the mass of cells per unit mass of bone is very low (Dodds et al. 1989). However, metabolic rate will certainly be lower in most teleosts than in tetrapods, because teleosts have acellular bone, whose metabolic rate must be very low indeed.

From the mechanical point of view the costs associated with the skeleton's weight or mass are likely to be complicated. Consider a solid bone of constant length whose cross-sectional shape remains constant and whose cross section has some characteristic length L. Suppose we are interested only in the impact strength of the skeleton, which is often likely to be the case. If so, then the resistance to fracture can be considered, for any particular type of bone, to be proportional to mass. This is because shape is rather unimportant in resistance to impact; it is the volume of bone absorbing energy that is important. If, however, bending stiffness were important, then for a solid bone of constant shape and length the stiffness will be proportional to the square of the mass. (Resistance to bending $\propto$ Second moment of area $\propto L^4$; Mass $\propto L^2$.) Other criteria imply other scaling laws, and no bone is likely to be solely adapted to one particular failure mode.

The cost of the skeleton is likely to be very different for different vertebrates according to their mode of life. In fish with swim bladders the component of the cost associated with mass as such will be proportional roughly to mass$^{2/3}$. The reason for this is that the buoyancy will decrease linearly with skeletal mass, and so swim bladder volume will have to increase linearly with skeletal mass. The drag experienced by the fish, and so the power required for swimming, is proportional to cross-sectional area, or to volume$^{2/3}$. Whales and other seagoing air breathers are in a different situation. They need lungs to breathe. If they had no bones, their lungs would make them buoyant. For animals needing to spend much of their time underwater, this would be disadvantageous. Therefore, an increase in skeletal mass would initially be advantageous but would, sooner or later, become disadvantageous. Some seagoing mammals, particularly the Sirenia (the sea cows), which are very slow swimmers and have lungs, which tend to give them unwanted positive buoyancy, have peculiar bones, with small or no marrow cavities and rather dense cancellous bone. This condition, called pachyostosis, is presumably a way of increasing the mass of the skeleton with-

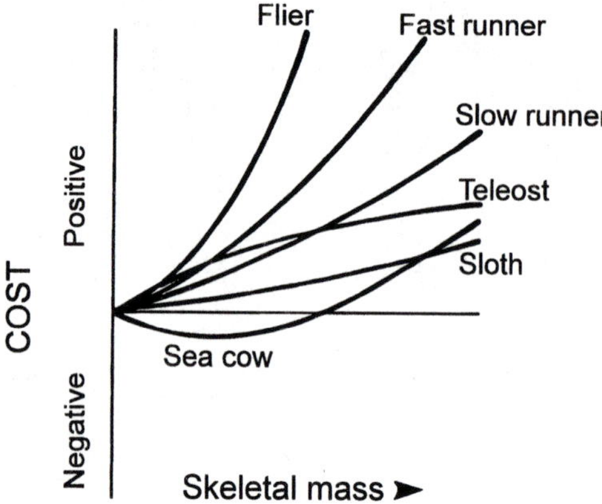

FIG. 10.2. The cost, for different types of vertebrates, of skeletons of different masses. This is very diagrammatic, serving merely to illustrate that the cost of a skeleton will differ very much according to the way of life of the animal.

out much regard for the mechanical consequences. The arrangement of these very dense bones and the soft tissues of the body are well arranged to produce both neutral trim and neutral buoyancy (Domning and de Buffrénil 1991). For nonswimming vertebrates the mechanical component of the cost of increasing skeletal mass can be expected to be greater in the following order: sedentary forms, slow runners, fast runners, fliers (fig. 10.2).

Not only will the cost of increasing the mass of the skeleton, and therefore the safety factor, be greater in more actively locomoting animals, but it will also vary *within* the skeleton. This requirement can be simply shown in the bird. A bird, being a flying animal, will be under considerable selective pressure to have as light a skeleton as possible. Even so, this selective pressure will be greater toward the ends of the wing bones than in the axial skeleton. Consider the wing at the start of the downstroke. To accelerate the wing downward, the pectoralis major muscle will have to exert a torque. Part of this torque will be to overcome the aerodynamic load. This is inescapable, being required for flapping flight. The muscle will also have to exert a torque against the inertial forces caused by the angular acceleration of the wing. These forces will be proportional to the moment of inertia of the wing about the humeroscapular joint. If y_i is the distance of a narrow strip of wing from the joint, and δm is its mass, then the moment of inertia of the whole wing is $\Sigma y_i^2 \delta m$. Any mass of bone, or other tissue, will contribute more to the moment of inertia of the wing if it is far from the shoulder

TABLE 10.2
Distribution of Fractures in Racehorses

Bones involved	*Fall*	*Not in fall*	*% in fall*
1st phalanx	0	14	0
Sesamoid	0	3	0
Metapodial	7	8	47
Carpus, tarsus	3	7	30
Tibia, radius, ulna	11	5	69
Humerus, femur	10	1	91
Pelvis, scapula	6	4	60
Spine	29	0	100

Source. Derived from Vaughan and Mason (1975).

joint than if it is near to it. The greater the moment of inertia, the harder the muscles will have to work. This is a difficulty that any fast-moving tetrapod faces, and it accounts, as we have seen in chapter 6, for the tendency, in fast movers, for the distal parts of the limbs to be lightened as far as possible even, as I mentioned in chapter 4, to the extent of the distal phalanges of a bat being unmineralized (Papadimitriou et al. 1996). The muscles migrate proximally, and the more distal joints are controlled by long tendons.

There will, therefore, be increasing selective pressure along the limb to reduce the margin of safety of the bones. Table 10.2, derived from Vaughan and Mason (1975), is an indication of this. It records the injuries to racehorses in Britain, over a period of about two years, that resulted in the horse's death, either at once, or because it had to be destroyed. I have divided the injuries into those that involved a fall and those that did not, though some of these latter may have occurred when the horse landed after successfully jumping a fence. The loads on the skeleton resulting from a fall are, presumably, not particularly closely related to the loads imposed during ordinary locomotion. There is a clear tendency for the accidents not involving a fall to produce fractures in the distal bones, much more than accidents involving a fall. The only surprise is the four fractures of girdle bones (one of the scapula, three of the pelvis) with no fall. The mathematical basis for such differences in safety factors in different elements in linked systems, such as the limb bones, is developed by Alexander (1997, 1998).

A similar increase in fractures, not involving falls, from proximal to distal is seen in healthy young humans. In athletes and soldiers the tibia is much more likely to develop fatigue fractures than the femur or more proximal bones. The metatarsals in some sports (like dancing) are more prone to fractures than the tibia, in other studies less (Orava et al.

1978). In a study of 114 fatigue fractures in runners and dancers, Brukner et al. (1996) found the following distribution: metatarsals: 27%; ankle: 20%; fibula: 17%; tibia: 26%; femur: 3%; pelvis: 3%; spine: 4%. Fatigue fractures are, as discussed in chapter 3, caused not by very large stresses, but by stresses merely somewhat larger or more frequent than usual. Nor do these stresses act in unusual directions.

This work on two rather different, though domesticated, mammalian species, horses and humans, shows that safety factors in limb bones are probably rather small in relation to the loads imposed during violent, but controlled, locomotion, and that these safety factors decrease distally along limbs, as we might expect. In chapter 7, I discussed the effects of the presence of marrow on the weight advantage of having hollow bones. The presence of fatty marrow makes the advantage small. If long bones were filled with something of greater density than fat, the advantage would be even less. Fat is the least dense tissue the body produces. Hematopoeitic marrow (red marrow) is denser than water. During the ontogeny of mammals about half the red marrow is eventually replaced by yellow marrow. It would obviously be mechanically advantageous to have the denser red marrow that remains concentrated in the more proximal bones. This is what happens (Ascenzi 1976; Piney 1922). Various experiments have been designed to explain this central–peripheral distribution of yellow marrow. These experiments tested the ideas that temperature, or blood supply, may be the controlling factor and ignore the possibility that there is an important functional reason for the distribution of the different types of marrow. The functional, mechanical explanation seems both adequate and satisfactory.

Alexander (1981) has tried to formalize the various factors affecting safety factors in bone and other skeletal materials. He suggests that the safety factor to be found in a bone can be determined by knowing two variables. One is the ratio of the cost of growing and using a bone of a particular "designed" strength, stiffness, or whatever, to the cost to the animal of its failure. The other variable is a combination of the variability of the accidental loading that will be imposed on the bone and of the quality control of the construction of the bone. Before showing how these variables may interact, we must say a little more about them.

If an engineer were to design a structure in, say, an airplane to have a particular strength, it would be fairly easy to quantify its costs. The cost would include the manufacturing cost and the effect that the mass of the structure would have on the performance of the airplane. In general, the greater the strength of the structure, the greater will be its mass, and so the greater the cost, whether measured in dollars or performance. The cost of failure of the structure would not be the same for all structures. If the catch retaining a blanket locker door broke, the cost would be trivial, whereas if the main wing spar were to break, the airplane would probably be destroyed. For a manufacturing company it might be possible to forecast the cost of an airplane crashing, in dollars, but such a quantification would clearly be highly imperfect, depending as it would on such factors as how full the airplane was, the cupidity of lawyers, and so on. When trying to apply concepts such as safety factors to animals, biologists are, in one way, in an easier position than engineers, because the currency of cost is rigidly defined; it is Darwinian fitness.

Suppose a population of animals consists of two types, *A* and *B*, which differ genetically. The average number of offspring produced by an *A* individual (counted from the moment of its own fertilization) is, say, *Ra*, and the average number produced by a *B* individual is *Rb*. Suppose that *Ra* is greater than *Rb*. Then the *fitness* of the two types is defined as *A*: fitness = 1; *B*: fitness = *Rb*/*Ra*. Notice that fitness says nothing about the *number* of offspring successfully produced. The chance of survival to sexual maturity of a fertilized codfish egg might be only 0.0001, while a fertilized mammal egg might have a 0.5 chance of surviving. Fitness, however, will vary between 1 and 0, depending on how well the opposition is doing.

The currency determining whether changes in the safety factor of bones will, or will not, spread through a population is therefore simply whether the change enhances, or does not enhance, the competitive ability, relative to other members of the species, of the bearers of the change to produce offspring in the next generation. All the costs and benefits of a particular safety factor are expressed, in Alexander's 1981 formulation, in the term $(G + U)/F$: the costs of growth and use divided by the cost of failure. In the airplane example above, this could not be readily done, because the different costs are measured in different currencies (though no doubt clever finance directors have come up with ways of doing so). In animals there is, in theory, no problem, because all costs are measured in terms of Darwinian fitness. In practice, actually determining the effects of various characters on Darwinian fitness can be very difficult indeed.

As well as the concepts of cost and benefit, the concept of the safety factor is inevitably also different in biology and in engineering. It is, really, too grand a concept for biology. Human designers may be thought to go about designing in one of two ways. One is to have a rather precise idea of the loads that are to be imposed on the designed structure, and then to design the structure so that, theoretically, it will bear, say, twice that load without failing. Any refining of the process depends on a better understanding of the imposed loads and better stress analysis. The other method is to design something that "looks about right," and then to see how it does in practice. If too many stewards complain that the catch of the locker has broken off, the design can be improved. If, on the other hand, no one *ever* complains, the designer may think that he has produced an object that is too robust and, in particular, too heavy. Weight could be saved, and the performance of the plane (very) marginally improved, by making the catch a little weaker. Natural selection, I am sorry to say, acts in the latter way. Structures will be increased or decreased in strength until an acceptable probability of failure is attained that will maximize Darwinian fitness.

Alexander suggests that the maximum loads imposed on a population of bones, up to some point of interest like production of offspring, can be considered to have a log normal distribution. That is, if the values of load are expressed using a log scale, the distribution of load against probability of that load will have a normal distribution. I use "normal" here in the statistical sense. Alexander recognizes that the main advantage of the log normal distribution is its mathematical convenience, but it does, in fact, describe many loading distributions quite well (Bompas-Smith 1973), though not some that I discussed earlier in the chapter. Alexander defines a safety factor (arbitrarily, for we are dealing with natural selection, not a written specification) as having a value of 1 if the probability of failure is 0.5. If the safety factor defined in this way is called n and the probability of failure with any particular safety factor is called $P(n)$, then

$$\frac{dP(n)}{dn} = \frac{-1}{nV\sqrt{2\pi}} \exp\left[\frac{-(\ln n)^2}{2V^2}\right]$$

where V is a measure of the variability of the loading and of the strength of bony structures of a particular mass. Once this is accepted, it is possible to make calculations concerning the values of the safety factors one should find in bone if natural selection is working in this way. Figure 10.3 shows the results of calculations of safety factors for bones loaded in bending. (The way in which the mass and, therefore, the cost increase with safety factors will vary according to the loading mode that is important; see chapter 4. However, the general shape of the distribution will not be greatly affected by the loading mode.)

Figure 10.3A has several features of interest. As might be expected, the safety factor is high if the cost of $G + U$ is low relative to the cost of failure and, at the same time, the loads imposed are variable. This latter condition is important because, although by definition a safety factor of unity gives a 0.5 chance of survival, a large value for V implies that to have a high, rather than a moderate, chance of survival, the safety factor must be much more than unity.

If the cost of failure is small and the variability of loading is large,

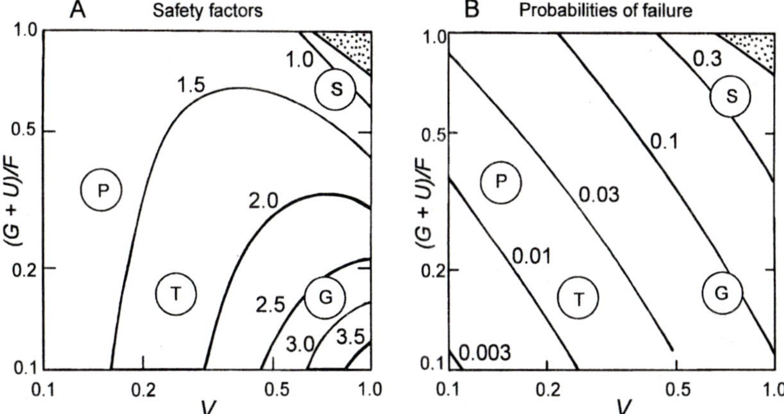

FIG. 10.3 Diagrams, slightly modified from Alexander (1981) showing the safety factors and the probability of accidental fracture that may be expected in different bones as a result of natural selection. Ordinate: (Cost of growth + Use)/(Cost of failure). Abscissa: variability of load and of the bone's response to load. S, stag's antler; G, gibbon's long bone; T, terrestrial animal's long bone; P, pterodactyl's wing bone. The contours represent (A) safety factors and (B) probability of fracture.

then the safety factor should become small. Indeed, the diagram shows a region (dotted) where it is not worthwhile keeping a bone at all, and it should disappear. Figure 10.3B shows the probability of failure with a particular safety factor. Note that having the same safety factor does *not* mean that the probability of failure is the same; that depends on the variability of loading as well. Shown on these diagrams are points that Alexander suggests are roughly where different bones should lie. T refers to the limb bones of terrestrial mammals; G to monkeys and arboreal apes such as gibbons. S refers to the antlers of stags. The stag's antler is not critical for its survival, though it does help it reproduce in any one season. It is a heavy bone, and it has to be re-grown every spring; therefore, its cost of use and production are fairly high compared with the cost of its failure. Since it is used in fights, the variability of loading on it will be very large. These together imply a position on the top right of the diagram. The frequency of fracture in one population per season is known to be about 0.3 (Clutton-Brock et al. 1979), and so it is reasonable to place the stag by the 0.3 chance of failure line.

In terrestrial mammals the cost of growing and using a bone will be about the same as in gibbons, as will the cost of failure, which will be high in each case. For an arboreal animal, the cost of increased mass may be slightly higher than for a ground-hugging one, if only because for animals in trees the ability of the terminal branches to bear the animals' weight without excessive bending may be important. In general, the earth is more unyielding (though my sons when small were better able to traverse bogs than I was). On the other hand, for arboreal animals, which cannot limp from tree to tree while their bones heal, the cost of failure of a long bone is probably higher. Terrestrial mammals and gibbons should, therefore, have about the same position on the ordinate. However, because of falls, the *variability* of loading in an arboreal animal is likely to be much higher than in one living on the ground all the time. Therefore the gibbon should be further to the right on the diagram, and should have a higher value for its safety factor. An important point to notice is that although the gibbon will have a higher safety factor than a land-based mammal, it nevertheless will have higher frequency of failure. This is partly because of the arbitrary definition of safety factor, but it also shows that some ways of life are more hazardous than others, and greater risks have to be accepted.

I have added another point, P, to Alexander's diagram. This represents the limb bones of pterosaurs. In chapter 4, I showed that they were extremely thin-walled, fragile, and pneumatized. Their thin-walledness made them liable to local buckling, and they were obviously very much minimum weight structures. Pterosaurs had large wings and were very light (Bramwell and Whitfield 1974). This very low wing loading was essential to their mode of flight. Therefore, although the cost of breaking a limb for a pterosaur would be as high as, or even higher than, for a gibbon, the cost of making the respective bones a few percent heavier would be much greater for the pterosaur. So, the ratio $(G + U)/F$ would also be greater. On the other hand, the flight behavior of pterosaurs was probably so sedate that the variability of loading would have been less than for a ground living mammal, and much less than for a gibbon. Therefore, it is reasonable to place the point for pterosaurs above, and to the left of, the terrestrial mammals. We shall never know the frequency of fractures in pterosaur bones, but if I have put the point anywhere near the right place, then although the safety factor for the limb bones would have been very low, the probability of fracture would not have been much higher than for a terrestrial mammal.

Vultures might appear to be birds with rather similar habits to pterosaurs. Houston (1993) examined 17 wing skeletons from 14 vultures (he usually could not get both left and right) of two species. He found that of these 17, three had healed fractures of the ulna. It is remarkable that at least 3 out of 14 birds had survived the fracture of a bone vitally important for their flight. Houston points out that such a high proportion of survivors of fractures is not found in other species. (Brandwood et al. in their study of birds [1986] found 3 healed fractures of the ulna in 1802 bones, although the radius was more frequently affected.) This could be caused by a lower proportion of fractures, or a lower ability to survive fractures, or both. In fact, Houston suggests that both are true. Vultures carry considerable reserves of body fat, and could survive unfed for perhaps three weeks, in which time healing could be well advanced. Vultures also have very light and delicate skeletons and are rather squabbly at kills, and therefore they may be more likely to fracture their bones.

The discussion above has concentrated almost entirely on accidental loading of the bones. However, fatigue loading of bones may also be an important source of failure. Alexander (1984) made a heroic attempt to describe the effects on the optimum strength of bones when the possibility of both fatigue and accidental damage are taken into account. The fatigue strength of bone is, of course, less than the strength in one-off accidents, and is more likely to be related to the stresses imposed in normal locomotion. Again, as in Alexander's 1981 formulation, important features of the equations are the cost of growing and maintaining bones versus the cost of fracture, and the variability of the accidental loading. There are, unfortunately, a very large number of parameters in the equations that can be estimated only very roughly. Using sensible values for the parameters, the equations make two predictions, neither of which seems to be very well borne out in life.

The first is that most bones should fail either predominantly in fatigue or predominantly in accidents. Table 10.2 shows that although this seems to be true of the very distal and proximal bones of racehorses, there are many bones, in horses and humans, that fail frequently by both methods. A difficulty here is that a bone may fail in an accident because it was already weakened by fatigue, a complication that Alexander's analysis could not accommodate.

The second prediction is that safety factors (in normal locomotion) should not be less than about 3.4. (This is because fatigue loading would make bones with a lower safety factor very likely to fracture.) Tables 10.4 and 10.5 show a number of bones with safety factors considerably less than this. One possible, quite good, explanation for the discrepancy is that the model does not account for the *selective* removal of fatigue-damaged bone by remodeling. If this occurs, and the discussion in section 11.10.5 will show that it probably does, then the likelihood of fatigue fracture is considerably reduced, and a lower safety factor could be adaptive. Although Alexander's model has a large number of unknowns in it, it is useful in demonstrating rigorously what kinds of considerations need to be taken into account when we think about safety factors.

There is an increasing amount of information available about the magnitude of safety factors found in nature. In what follows, I use a more convenient measure of safety factor than the more universal but essentially unknowable formulation of Alexander. This is the strength of the bone compared with the stresses imposed on it during extreme activity, usually fast locomotion. The technical difficulty of this is not so much that of finding out the strength of bone, but rather of determining the stresses imposed on bones during locomotion. There are two ways of approaching this problem. One is to calculate what stresses should be present in the bone, by making use of force plate or other data external to the bone. Another is to measure the strains in the bones during locomotion directly.

As an example of the first, I take Alexander's (1977) investigation of the stresses found in a 70-kg antelope when it is running flat out. He is able to talk of "an antelope" because the various antelopes that he investigated showed such precise allometric relationships in all the relevant variables that it is not necessary to tie the discussion down to a particular individual or even species. Alexander works from a number of biomechanical certainties, such as that the sum of the vertical forces on the four feet must, over time, average the weight of the animal, and a number of reasonable assumptions, to estimate that the maximum compressive stress is about 50 MPa, the maximum tensile stress about 130 MPa.

Alexander and Vernon (1975) obtained values for the stress in the tibia of a wallaby running on a treadmill and also on a force platform. The calculations are more direct than the previous ones, because the use of the force platform allows the ground reaction force to be determined directly. The values of the greatest stresses in the bones vary from about 60 to 110 MPa in tension and 90 to 150 MPa in compression. The

TABLE 10.3
Calculated Maximum Stresses in Mammal Bones During Strenuous Activity

Animal	*Activity*	*Mass (kg)*	*Tensile stress (MPa)*	*Compressive stress (MPa)*
Dog	Long jump	36	60–80	100
Dog	High jump	36	60	80
Kangaroo	Hopping	7	60	90
Wallaby	Hopping	11	65	90
Antelope	Galloping	70	80–150[a]	
Buffalo	Galloping	500	35–95	60–115
Elephant	Running	2500	45–70	55–85

Source. From Alexander et al. (1979a).

[a]This is the mean of the absolute values of the maximum tensile and compressive stresses, and is not quite comparable with the other values.

TABLE 10.4
Safety Factors in Various Bones and Tendons, Calculated from Force Plate and Similar Data, from Alexander (1981)

		Bone		
Animal	*Activity*	*Tension*	*Compression*	*Tendon Tension*
Dog	Jumping	2–3	2.8	1
Kangaroo	Hopping	3	3.2	1–2
Kangaroo	Pulling on lead	1.6	1.9	—
Buffalo	Galloping	1.8–5	2.5–5	1.8–6
Rhinoceros	Galloping	3.7	4.8	—
Elephant	Running	2.5–4	3.3–5	—
Horse	Galloping	4.8	4.9	—
Horse	Pacing	4.4	4.2	—
Human	Weightlifting		1–1.7	—
Ostrich	Running	2.5	2.6	2.6
Goose	Flying		6	—
Pigeon	Flying	4–5.6	—	1.6
Sunfish	Feeding		8	—

Note. Assumed yield strengths in bone: tension, 172 MPa; compression, 284 MPa; in tendon: tension, 84 MPa. For details of calculations and authors see Alexander's paper. Rhinoceros calculated from Alexander and Pond (1992).

wallaby was running on a rather short treadmill; in the wild it could presumably hop faster. In doing so it would have imposed greater stresses on its bones. Alexander et al. (1979a) have extended this work to mammals of great size, using ciné film, but not, for obvious practical reasons, force platform data. They find that the stresses in the bones during reasonably fast locomotion are about the same in large animals as in small ones. Table 10.3 shows this.

It is interesting that ordinary, though fast, locomotion induces stresses that are roughly the same in animals that differ in mass by a factor of 350. Although we have rather little comparative information, it seems that the mechanical properties of the compact bone of the long bones of different animals do not differ greatly (chapter 3). Alexander (1981), using the kind of analysis discussed above, has tabulated the safety factors for bones involved in a range of strenuous activities (Table 10.4). He also does the same for tendon. The reliability of the results varies greatly. For instance, analyses for the horse, by two different workers, suggest that the safety factor in pacing, in which the legs on one side are in phase, is about 4.3, while in galloping, a much more strenuous gait, there is another estimate of a safety factor of 4.8. However, the experiments of Rubin and Lanyon (1982) show that peak stresses in the radius of a horse become less when the gait changes from the trot to the canter, so the findings about the pace and the gallop may be correct. The general impression is that safety factors vary from the barely adequate to around 5 or 6. The higher apparent values are probably the result of the animals' inability to run hard enough in the experimental conditions.

Kirkpatrick (1994) has estimated the bending and shear stresses, while hovering, in the humeri of 11 species of bird and 7 species of bat. The birds ranged in mass from 0.019 to 2.5 kg, the bats from 0.004 to 0.039 kg. These estimated stresses were produced by calculations using estimates of the lift that must be generated during hovering, determining the center of lift and using the force and moment arm so calculated to produce the stresses in the humeri, whose geometrical properties had also been measured. The safety factors for shear were very high, suggesting that this mode of loading would be most unlikely to break the bones. The average safety factors for bending were 2.2 for birds and 1.4 for bats. I suspect that these safety factors are a little low, because the tensile strengths measured for the bones appear to me to be also rather low. In particular, the value for bats (75 MPa) is too low to be accepted without caution.

Another method of estimating safety factors is to measure the strains in the bones during locomotion. Strain gauges are attached to the bones, the animals are allowed to recover, and they then run while the strains in the bone are measured. Table 10.5 shows the results of such tests. This table is pleasing because it shows results that are reasonably consistent with each other and also with the values tabulated by Alexander. The safety factors for the iguana and alligator shown in table 10.5 are higher than those for the other animals. It is possible that torsion is a more dangerous mode of loading in these animals than is tension (Blob and Biewener 1999).

This fairly close relationship between load and strength almost certainly means that the build of the long bones is actually determined by the forces imposed during locomotion. This statement does not distinguish between determination over evolutionary time and determination during a lifetime. It is probable that the build of bones is roughly established by the genes of an animal, but the precise relationship between build and imposed load is established during life. There is much evidence that the build of a bone can be altered by loads imposed on it, so it is *not* the case that animals have a particular build of their bones and simply limit their own locomotory activity to what their bones are capable of bearing, though obviously elephants do not behave like month-old lambs. I write about this matter in the next section.

It is as if the osteogenic mechanism has a means of measuring strains in the bone, and can increase or decrease the amount of bone to keep reasonable safety factors. However, some bones, such as the human cranium, are capable of bearing loads much greater than those to which

TABLE 10.5
Safety Factors (*S*) Related to Ordinary Activities, Calculated from Strain Gauge Data

Animal	*Bone*	*Activity*	*Strain*	S	*Authors*
Horse (young)	Metacarpus	Gallop	0.0048	1.4	1
Horse (mature)	Metacarpus	Gallop	0.0033	2.1	1
Horse	Radius	Gallop	0.0020	3.4	2
Horse	Tibia	Gallop	0.0031	2.2	2
Horse	Radius	Gallop	0.0022	3.1	3,4
Horse	Metacarpus	Gallop	0.0016	4.2	3,4
Horse	Tibia	Gallop	0.0020	3.4	3,4
Horse	Metatarsus	Gallop	0.0017	4.1	3,4
Horse	Radius	Jump	0.0037	1.9	3,4
Horse	Metacarpus	Jump	0.0024	2.9	3,4
Horse	Tibia	Jump	0.0052	1.4	3,4
Horse	Metatarsus	Jump	0.0040	1.8	3,4
Dog	Radius	Gallop	0.0024	2.9	2
Dog	Tibia	Gallop	0.0020	3.5	2
Goat	Radius	Gallop	0.0018	3.8	5
Goat	Tibia	Gallop	0.0020	3.6	5
Sheep	Radius	Gallop	0.0023	3.0	6
Sheep	Tibia	Trot	0.0021	3.3	7
Rhesus monkey	Mandible	Bite	0.0022	3.2	8
Turkey	Tibiotarsus	Run	0.0024	2.9	9
Chicken	Tibiotarsus	Run	0.0019	3.7	10
Bat	Humerus	Fly	0.0020 (t)	3.5	11
Bat	Radius	Fly	0.0022	3.2	11
Goose	Humerus	Fly	0.0028	2.5	9
Pigeon	Humerus	Fly vertically	0.0023	3.1	12
Human	Tibia	Run	0.0012(t) 0.0020(s)	5.8 10 (s)	13
Human	Tibia	Jump	0.0020	3.5	14
Iguana	Femur	Run	0.00091(t) 0.0016(s)	10.8 4.9(s)	15
Iguana	Tibia	Run	0.0017(t)	5.7	15
Alligator	Femur	Run	0.00071(t) 0.0010(s)	6.3 5.4(s)	15
Alligator	Tibia	Run	0.0012(t) 0.0020(s)	5.5 3.9(s)	15

Note. The safety factors are calculated relative the strain at yield. The strains are the peak strains and are in compression except for those marked "t" (tension) or "s" (shear). Authors: 1, Nunemaker et al. (1990); 2, Rubin and Lanyon (1982); 3, Biewener et al. (1983b); 4, Biewener et al. (1988); 5, Biewener and Taylor (1986); 6, O'Connor et al. (1982); 7, Lanyon and Bourne (1979); 8, Hylander (1979); 9, Rubin and Lanyon (1984); 10, Biewener et al. (1986); 11, Swartz et al. (1992); 12, Biewener and Dial (1992); 13, Burr et al. (1996); 14, Hillam (unpublished observation); 15, Blob and Biewener (1999). I made a number of assumptions and simplifications in constructing this table, and the original papers must be read before grand theories are based on it!

they are normally subjected, and show that this cannot be the whole story. These possible control mechanisms are discussed in chapter 11.

10.2 Size and Shape

10.2.1 *Scaling*

Animals that are closely related to each other, but are of different size, tend also to be of different shapes. This well-known fact, whose significance has been debated for many years, is of importance in the understanding of bones. Suppose animals of different species in a group, such as the quadrupedal mammals, differed only in size, that is, they all were scale models of each other. This is called *geometric* scaling. How might the stresses in the bones differ between the species?

There are various geometrical features of the bone, and the animal, that are of interest, and we can see how they vary in proportion to the length l of the bone. Following Selker and Carter (1989), on which the first part of this section is mainly based, I shall consider the diameter d, cross-sectional area A, second moment of area I, polar moment of inertia J of the bone, and the mass M of the animal. The relationships, assuming geometric scaling are $d \propto l$, $A \propto l^2$, $I \propto l^4$, $J \propto l^4$, and $M \propto l^3$.

Now consider the size of the force F necessary to break the bone in three situations, in which the weight of the animal is doing the loading: (1) when it is loaded axially; (2) when it is loaded in bending, the bending force having a moment arm proportional to the length of the bone; and (3) when it is loaded in torsion in which the torque has a moment arm again proportional to the length of the bone.

For axial loading $F \propto A$. Since $A \propto d^2 \propto l^2$ and $l \propto M^{1/3}$, therefore $F \propto M^{2/3}$

For bending loading $F \propto I/ld$ (this follows from beam theory). Since $I \propto d^4 \infty l^4$, it follows that $F \propto l^2$ and so $F \propto M^{2/3}$.

For torsional loading $F \propto J/ld$ (this also follows from beam theory). Since $J \propto d^4 \propto l^4$, it follows that $F \propto l^2$ and so $F \propto M^{2/3}$.

These calculations imply that if animals are the same shape though of different sizes, and if the *material* properties of the bone are the same, the force required to break a bone would be proportional not to the mass of the animal, but to the ⅔ power of the mass. Putting it the other way round, the maximum stresses in bones should scale $\propto M^{1/3}$ and increase with size. Much of chapters 3 and 4 of this book suggests that ordinary bone of different animals has similar properties, and there is no indication that larger animals increase the strength of their bones by increasing the strength of their bone material.

A ⅓ power law is a very long way from proportionality. The calculations above imply that if one animal is 100 times the mass of another, the stress in the bones, produced by its body mass, is 4.6 times larger. These geometrical and mechanical facts imply that, if they are not to break their bones when going about their business, either animals must change their habits as they get bigger and so reduce the loads on their bones, or their shape must change. In fact, both happen.

There is a technical point that I introduce here, with some hesitation because it confuses things so, before we come to the actual values found in nature. That concerns the fact that we are looking for a *functional relationship* between the two variables, say, bone length and diameter. Most regression analyses, including those discussed below, use a so-called type I regression. This expresses y as a function of x. But length is not a *function* of diameter, or vice versa, they both vary together according to some overarching relationship between them. If a normal algebraic equation relating y to x is $y = ax^b$ then it follows that $x = (y/a)^{1/b}$; the relationships are the inverse of each other. If, however, you take somewhat correlated data and run two ordinary regressions on them, y on x and x on y, you will find that that the two resulting equations are *not* the inverse of each other. If the data are very highly correlated, the difference is small, but as the correlation between them gets less, the difference gets larger. In data such as we are dealing with below, in which both sets of data are logged (and only in this case), a good estimate of the "true" value of the exponent is b/r, where r is the correlation coefficient. (Readers new to this concept may care to run a few regressions on dummy data; I think they will find the results rather surprising.) Fortunately, many of the relationships discussed below are very tight, and the difference in the exponents is small. I shall mention one case below, but, in general, it would be too difficult, in some cases impossible, to translate all the reported exponents into "true" exponents. Nowadays, people concerned with allometry have become aware of this difficulty and allow for it. The earlier literature, however, is more problematical. It will not materially affect the discussion below, but readers should be aware of this lurking problem. The matter is dealt with in many papers. Harvey and Pagel (1991, chapter 6) is a good introduction to the difficult literature.

10.2.2 Elastic Similarity

McMahon (1973) proposed a relationship that he held to be generally true and that had the great virtues of elegance and simplicity. In what follows I shall produce evidence that McMahon's ideas are not generally true, but even if the reader accepts these arguments, this should not detract from the great importance of McMahon's contribution, for it is against his ideas that later ones were be measured. McMahon's basic idea was that organisms are designed so that the *deflections* they undergo are what is controlled, not the stresses they bear. As we saw in chapter 4, a slender bone may collapse by Euler buckling even though its compressive strength is not even approached as it starts to collapse. Take the case of a cylindrical column. How tall can it be made before it collapses by buckling under its own weight? McMahon calculated that the collapse length is $0.79(E/D)^{1/3}d^{2/3}$, where d is the diameter of the column, E is Young's modulus, and D is the density. One can achieve greater lengths than this by, for instance, making the column taper. However, it turns out that the length is always proportional to the ⅔ power of some characteristic diameter, such as the diameter halfway up. To take a related example, suppose we have a branch of a tree that is growing in length without getting any wider. As it grows, it will droop more and more until eventually its end is actually nearer the trunk than when it was shorter. If the proportional drooping, or deflection/l, where l is the length, is to remain constant as the branch grows, the length should increase as diameter to the ⅔ power again or, in other words, the diameter should be proportional to the 1.5 power of the length.

McMahon called structures in which the relationship between the deflections and the length are the same "elastically similar." It turns out to be very generally true that elastically similar structures scale so that $d \propto l^{1.5}$. This is so not only with solid cylinders, but also with more bonelike structures with marrow cavities, and so on. (Remember that if the animal, and its bones, remain the same shape, $d \propto l$.) McMahon argued that it is a general property of organisms that all body segments scale in this way, so that the trunk as well as the limbs will obey $d \propto l^{1.5}$. If McMahon's view is right, bones, and animals, should get stockier as they get larger. This increasing stockiness is seen very clearly in many trees (though often not in saplings, which shoot lankily up in a desperate attempt to get into the light). McMahon gave two examples in animals where this has been shown to be the case, in studies on macaques and Holstein cattle. This idea about the proportions of animals, if true, is, obviously very important.

There have been several attempts to test McMahon's ideas on animals. The first was by himself; he investigated the limb bones of adult ungulates (McMahon 1975). The results are shown in table 10.6. The

TABLE 10.6
Allometric Exponents for Bones

Group	*Humerus*	*Ulna*	*Femur*	*Tibia*	*Metatarsal*	*Forelimb*	*Hind limb*
Bovidae	0.63	0.62	0.76	0.65	0.61	0.64	0.62
Bovidae (modified)	*0.69*	*0.68*	*0.78*	*0.69*	*0.75*	*0.68*	*0.68*
Cervidae	0.79	0.97	0.79	0.83	0.80	0.85	0.76
Suidae	0.54	0.54	0.58	0.41	—	0.52	—
Artiodactyla	0.66	0.63	0.75	0.65	0.68	0.67	0.65
Perissodactyla	0.60	0.20	0.86	0.28	0.53	0.54	0.87
All ungulates	0.65	0.62	0.72	0.60	0.68	0.67	0.60

Source. From McMahon (1975). I have added a line, "Bovidae (modified)," which is the values in the first line divided by the appropriate correlation coefficient.

Note. Values of the exponent b in the equation Length $\propto$ Diameter^b. Expected values: elastic similarity, 0.67; geometric similarity, 1.00.

ungulates as a whole agree rather well with McMahon's predictions, although the perissodactyls (horses, tapirs, etc.) do not. McMahon attributed their lack of agreement, reasonably, to the very different locomotory patterns of the rhinoceroses and the horses. The artiodactyls are more similar in their locomotion. Furthermore, the range of size in the perissodactyls is small. Heinrich and Biknevicius (1998) found that within a set of mustelids (martens, wolverines, etc.), which had a rather similar locomotory style, the bone dimensions fitted elastic scaling reasonably well. This is not the end of the story, however.

10.2.3 Geometric Similarity

If organisms of different size are simply scale models of each other, they are said to show geometric similarity. Alexander (1977) examined the relations between body mass and various bone sizes in bovids. (McMahon had not known the masses of the animals whose skeletons he measured, although he was able to estimate them reasonably well.) Alexander's bovids ranged in size from the dik-dik (4.4 kg) to the oryx (176 kg). The fit in these cases to elastic similarity is good.

However, there are many studies in which the fit is much less good.

Howell and Pylka (1977) found in the femora of bats that elastic similarity would fit in vampires, which shuffle around on the ground a good deal, although the size range was too small for Howell and Pylka to be very confident about this. However, for the great majority of bats, which hang upside down, the femora obey geometric similarity more closely. Howell and Pylka suggest that this is because the femur, being hardly loaded in compression in life, is not loaded in the kind of way considered by McMahon in developing his hypothesis. However, Norberg (1981), in an extensive survey, also considered the wings of bats, and found that more than half the species showed a reasonable fit to geometrical similarity, and that none showed a good fit to elastic similarity.

Alexander and his co-workers made a large survey, the results of which are impossible to reconcile with elastic similarity (Alexander et al. 1979b). McMahon had chosen the Bovidae as a subject for his detailed measurements because they show a good range of size and have a reasonably uniform mode of locomotion. Alexander et al. measured homologous bones from insectivores, primates, lagomorphs, rodents, fissipede carnivores, an African elephant, and various artiodactyls. The overall equation for all species, and, indeed, the equations within most of the orders, have exponents close to those expected if the animals were obeying geometric similarity (fig. 10.4, table 10.7).

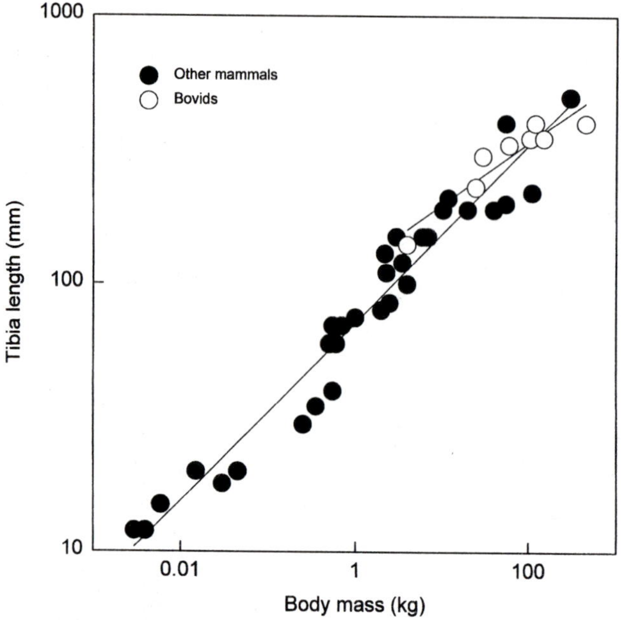

FIG. 10.4. Relationship between the length of the tibia and the body mass of various mammals. The fitted line for all points has a slope of 0.32. The bovids are fitted by a slope of 0.22. (Derived from Alexander et al. [1979b].)

The mammals being considered here have an enormous range of body mass, six orders of magnitude, from the pygmy shrew *Sorex minutus*, 2.9 g, to the elephant *Loxodonta africana*, 2.5 tonnes. The modes of locomotion are very different at the two ends of the mass distribution (the way the limbs are held in the scuttling shrew are quite different from what happens in the ponderous elephant) and so it could be argued that the allometric relations proposed by McMahon could not be expected to hold. In fact, the exponents for bone diameter as a function of body mass agree better with elastic similarity than with geometric similarity. However, the main thrust of the idea of elastic similarity is that bones (and segments of the body as a whole) should get relatively much stockier as body mass increases. In particular, the length of bones should scale to the ⅔ power of bone diameter. Geometric similarity should make the exponent 1. Table 10.7 shows that although bone diameter scales to body mass with an exponent of about 0.38, bone length scales to bone diameter to the 0.82 to 0.97 power (the bovids excepted, as usual).

Prange et al. (1979) examined the scaling in birds' femora and hu-

TABLE 10.7
Allometric Exponents for Bone Dimensions for Mammals as a Function of Body Mass

		Bone length	*Bone diameter*	*b in* L ∝ D^b
Geometric similarity (expected exponent):		0.33	0.33	1.0
Elastic similarity (expected exponent):		0.25	0.38	0.67
Group	*Range of masses (kg)*			
Insectivora	0.003–0.6	0.38	0.39	0.97
Primates	0.3–20	0.31	0.38	0.82
Primates	0.6–60	0.34	0.39	0.87
Rodentia	0.01–2	0.33	0.40	0.82
Carnivora	0.6–150	0.36	0.40	0.90
Bovidae	4–500	0.26	0.36	0.72
All mammals	0.003–2500	0.35	0.36	0.97

Source. From Alexander et al. (1979b).

meri. They measured birds ranging in mass from hummingbirds (0.0036 kg) to the swan (10.8 kg). They found

Humerus length ∝ $\text{Diameter}^{1.24}$
Ulna length ∝ $\text{Diameter}^{1.23}$
Femur length ∝ $\text{Diameter}^{0.88}$

These results do not confirm McMahon's hypothesis. In fact, it is very difficult to find any fairly simple hypothesis that will account for the humerus and ulna results, which show the bones getting relatively *thinner* as they get longer. Prange and his colleagues were at a loss to explain these findings. Olmos et al. (1996) had similar findings, and attributed the anomalies to differences in locomotory pattern, which is the great problem in calculating allometries over large size ranges. Kirkpatrick (1994), in a study of birds and bats over a large size range, calculated that the stresses in the humerus were scale independent. However, the problem in analyzing flying animals is that larger animals have different shapes imposed by aerodynamic factors; in particular, they tend to have relatively smaller wings. This complicates matters greatly, although the selective reason for these differences is at root the same as that producing differences in the shape of terrestrial quadrupedal animals.

These findings together mean the predictive and explanatory power of McMahon's hypothesis is markedly less than originally appeared. Another problem remains, however: it is strange that geometric similarity is adhered to, because this implies that larger bones will have larger stresses imposed on them as explained at the beginning of this section. However, the analysis of Alexander et al. (1979a) in which stresses in the legs of buffaloes and elephants were estimated and compared with those in smaller animals, such as dogs and kangaroos, shows this is not completely so during actual locomotion. The estimated stresses in all animals were roughly similar.

There is *some* general thickening of the skeleton with body size, and this is shown in the exponent for skeletal mass as a function of body mass; this is slightly more than unity (Prange et al. 1979). For the mammals they found an exponent of 1.09, for birds 1.07. Although these are not large exponents, the range of body masses is great. As a result, the skeleton of the shrew is 5% of its body mass, but it is 18% of the mass of a human and 27% of the mass of an elephant (These are measured values, not derived from the regression.)

Rubin and Lanyon have emphasized the role that behavioral adaptations may play; larger animals may, by simply behaving more sedately, not expose their bones to such large forces, relatively, as do small animals (Lanyon and Rubin 1984; Rubin and Lanyon 1982). Biewener has demonstrated how the change in gait with size allows the stresses in different sized animals to be nearly the same, although the size and shape of the bones implies greatly increased stresses with size.

Another general attack on the scaling problem was made by Selker and Carter (1989). They pointed out that for geometrically similar animals the force F required to fracture the bone is given by $F \propto M^{2/3}$, whatever the mode of loading (see section 10.2.1), assuming that the force is proportional to body mass. They also point out that axial loading is not in life a very significant cause of bone fracture and that bending and torsion, the other two principal modes, have the length l as a factor in producing $F \propto M^{2/3}$. They then suggest that one consider a "bone strength index" (S_B), whose value is J/ld or I/ld, according to the mode of loading. The greater the value of S_B the stronger the bone. (Readers may be momentarily surprised that the bone is weaker if d is larger, when they supposed that a large diameter is a good thing in relation to strength. Indeed, large diameter is a good thing, but has already been taken account of in calculating J or I, which has a d^4 term in it.).

Using this theoretical framework, Selker and Carter then examined 93 bones from 12 species of artiodactyls. If S_B were constant with respect to body mass, that is, if the force required to break a bone were proportional to body mass, then the exponent x in the equation $S_B \propto M^x$

would be 1.0. If, on the other hand, the bones were showing geometric similarity, and therefore getting relatively weaker as body mass increased, the exponent would be 0.67. The actual values they found for the exponents for femur, tibia, humerus, and radius/ulna (if fused) were 0.76, 0.76, 0.84, and 0.89, respectively, with an overall value of 0.82. These are the values using resistance to torsion; the values for resistance to bending were very similar. They also, by doing some back-calculation, suggested tentatively that the results of Biewener (1983), who did not make quite the same measurements, but made them on a larger range of quadrupeds, agreed well with their findings.

These findings of Selker and Carter suggest that the shape of the bones of the animals they examined changed with increasing body mass so as to overcome partially the reduction of strength that would be associated with geometric similarity. However, the animals would also to some extent need to load their bones less strongly. Interestingly, the way in which muscle force scales with body mass is such that muscles would not be capable of loading the bones as strongly in larger animals as in smaller ones. Of course, the artiodactyls, on which their study was based, include the bovids, which McMahon studied and which are to some extent atypical. Some of the exponents that Selker and Carter found are closer to the exponents predicted by the elastic similarity hypothesis than the geometric similarity hypothesis.

An attractive theoretical development of Selker and Carter's work has been done by van der Meulen and Carter (1995). They hypothesized that bone responds adaptively to *dynamic* rather than *static* strains. This is reasonable, because dynamic strains are likely to be so much larger than static strains. In general, in static loading, except in pure compression or tension, the loads on the bone are proportional to both the mass of the body and the length of the limb elements, because longer limb elements will allow the mass of the body to have larger turning moments. This was the basis of Selker and Carter's calculations. On the other hand, dynamic loads are more closely related to muscle mass, which is itself likely to be closely proportional to body mass (van der Meulen et al. 1993). In other words, for geometrical similar animals, dynamic loads do not increase as fast as static loads. Incorporating body mass as the driving factor into their computer model, van der Meulen and Carter showed that the result was that bony elements should obey geometric similarity over a very large range of body masses, if the dynamic stresses and strains were to remain the same. Also, they were able to model a proportionally increasing diameter of the marrow cavity using this model. A full description of these models is given in Carter and Beaupré's book (2001).

10.3 Conclusion

This chapter has shown that safety factors seem to fall within a reasonably small range, given the necessary imprecision of the experimental methods available at present. It is not yet clear, despite Alexander's theoretical analysis, why the safety factors we see in life are the appropriate ones. As regards scaling functions, McMahon's concept of elastic similarity appeared, for a few years, to be the basis for a great illumination, but, unfortunately, too many counterexamples have appeared for it to be now any more than a starting point for speculation. The work of Selker and Carter, based on a strength criterion, is a possible way forward. It is clear that vertebrate skeletons are designed in some way to keep the safety factors within bounds. In this chapter I have discussed the relative importance of stiffness, strength, and weight, which is an inevitable concomitant of stiffness, in bones. This is the kind of trade-off between costs and benefits that selection should optimize over evolutionary time. In the next chapter I try to deal with the question of how adaptations may come about during a single lifetime.

Chapter Eleven

MODELING AND RECONSTRUCTION

11.1 The Need for Feedback Control

ONE OF THE most interesting problems in the whole of bone biology and mechanics is how bone responds adaptively to the loads that are imposed on it, and this chapter deals with it. Although it is an extremely interesting problem, it is not so supremely interesting as to justify, in itself, the relative amount of effort that is spent on it compared with all the other questions that can be asked about bone. The reason for the interest is money, of course. The modeling and remodeling ability of bone has important clinical consequences. Two examples: putting prostheses into bones changes the strains in the bone itself, which may have bad consequences, such as the bone melting away; menopausal women may suffer less of the typical bone loss associated with the menopause if they take exercise that loads the bone more heavily than might normally be the case for middle-aged women. Elucidating such features of bone's physiology is so therapeutically important that there is a great deal of money available to help people to do it. Nevertheless, it *is* an interesting problem!

The distinction between modeling and remodeling, which was mentioned in chapter 2, becomes important in this chapter. To recapitulate: in *modeling* the gross shape of the bone may be altered; that is, bone may be added to the periosteal or endosteal surfaces, or it may be taken away from these surfaces. In *remodeling* all surfaces of the bone may be affected, including the vascular cavities. In remodeling the bone involved is usually a small individual packet called a *basic multicellular unit* (BMU), and typically the amount of bone remaining after the remodeling cycle is unchanged; new bone has merely replaced old bone.

In chapter 10, I discussed the ways in which the build of bones and the mechanical properties of bone material may be adapted to the requirements of the animal. I did not deal with how this adaptation was brought about. Bones have quite intricate shapes, yet they also have a general size and build that seems to be adapted to the loads they bear. It is well known that bed rest (Nishimura et al. 1994) and weightlessness, as in space flight (Caillot-Augusseau et al. 1998; Collet et al. 1997; Goodship et al. 1998), lead to a reduction in bone mass and reduced growth rate of bone. The morphological changes resulting from weightlessness reduce the strength, stiffness, and energy-absorbing ability of the bones (Spengler et al. 1979). Similarly, intensive loading of a bone can result in an increase in mass. Middle-aged runners have more mineral in their bones than do sedentary people (Dalén and Olsson 1974), and ballet dancers have disproportionately well-developed leg skeletal mass (Lichtenbelt et al. 1995). In considering such differences in bone mass, one must always be careful to rule out self-selection: people may take up a particular sport *because* they have well-developed legs, or whatever. Where the bones are asymmetrical this is unlikely to be the case. The arm bones of professional or obsessive tennis players are stouter on the racket-holding side than on the other (Jones et al. 1977); the same is true of some Finnish squash players (Haapasalo et al. 1994). Greyhounds, which always sprint the same way round the track, have greater loading on their right than their left ankles. The right central tarsal bone of racing greyhounds has denser bone, with thicker cancellous trabeculae, than the left central tarsal. Retired racers show less asymmetry (Johnson et al. 2000).

Changes in the shapes of bone after severe distortion from accident or disease are often instructive. For instance, Pauwels (1968) shows how rachitic femora, severely bowed by the disease, have cross sections that are strongly elliptical rather than being nearly circular, as is normal. The elliptical shape is clearly adaptive in increasing the second moment of area around the appropriate axis to resist the large bending moments caused by axial loading on the bowed bone.

It is inconceivable, however, that the shapes of bones could result *merely* from mechanical adaptation to loads placed on them. This was shown to be the case by many experiments starting in the 1920s. For instance, Murray and Huxley showed in 1925 that a small fragment of the limb bud of a chick embryo, grafted into the chorioallantois of another, older embryo, developed into a recognizable femur with a head even though there was no pelvis with which it could articulate. Niven (1933) showed in the chick that mesenchyme that would normally develop into a patella would develop into a recognizable patella even in vitro. These and many other very interesting experiments are described by Murray (1936) in a book that is a classic of clear scientific writing.

Bones, therefore, develop at least partly without reference to the load they experience. However, their final build or architecture is dependent in some way on the mechanical environment in which they find themselves, either during development or in maturity. For some years Dennis Carter and co-workers have been developing a theory of bone development that has a rather simple starting point. They are concerned with the question of how, if one starts with a cartilage prefiguration of the bone, one finds some parts becoming bone, some remaining cartilage, some parts producing secondary centers of ossification that then fuse with the main part of the bone and so on. Their starting point is that the development of cartilage is favored when stresses are *hydrostatic* (that is tending to squeeze or expand the material in all directions at once) and the development of bone is favored when stresses are *distortional* (that is, tending to distort the shape). For instance, a compressive stress will tend to make a specimen shorter and fatter and is therefore a distortional stress. It turns out that many things make sense if, along with some refinements, this basic idea is used. Carter and Beaupré (2001) have now gathered their thoughts and experiments into a book where this theory is fully explained. I have no space to do justice to the book here, particularly as there are some aspects with which I am not in agreement. However, I cheerfully commend this book as an example of a good *theory* (which so often nowadays means guess or hunch). It tries to explain a whole set of disparate observations with a few basic postulates, and it makes testable predictions.

I shall barely discuss the intrinsic, genetically determined, development of bone; the subject is too ill-understood. Nevertheless, it must be that the form of bones, lying latent in the genes, is the result of natural selection acting in the past on mechanically functioning skeletons. The question I shall be concerned with is, How is the shape and size of bones affected by the forces acting on them during life? It will become clear that there are many many gaps in our understanding. Of course, there is nothing special about bones in having an interplay between genetic makeup and environmental influences. All aspects of the phenotype are like this. However, the problem can often be stated more starkly in bones, because they are rigid and their shape is easily measurable, and the important mechanical influences acting on them (stresses, strains) can often be well defined.

One may diagram the kind of thing that must happen (fig. 11.1). Loads are applied to a bone and interact with its build to produce strains. These are measured in some way and compared with some allowable or desirable strain. (In this chapter the words *stress* and *strain* are often used almost interchangeably. Usually *strain* is really the appropriate word because stresses are not directly measurable, but it is so intuitive to consider stress to appear in the bone when it is loaded that it often awkward to use the word strain. The context will make it obvious when one needs to know which word is meant.) If the actual strains deviate from the desirable strain, either at all, or by some threshold amount, bone is added to surfaces or removed from them. This reconstruction alters the build of the bone so that the same forces will now result in different and, if the reconstruction has been correct, more appropriate strains in the bone.

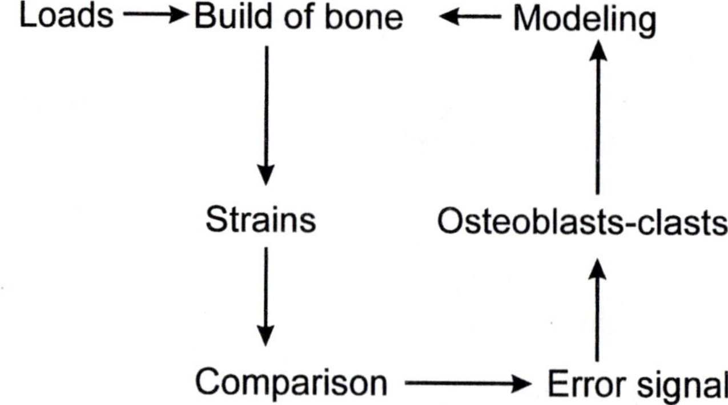

FIG. 11.1 The kind of modeling system that must operate in bones. The strains come from the interaction of the loads applied to the bone, and its build. The error signal stimulates bone cells, and their action results in the modeling of the bone, which changes its build and, in turn, the strains produced by the loads. There is a school of thought that supposes that it is *damage* that is the effective signal. If this is correct, a somewhat different kind of modeling system would be required.

The description of the feedback mechanism above is almost vacuous, and clearly needs to be fleshed out with some facts. An important attempt to do this was by Frost. Harold Frost has, over the years, published an enormous number of papers on the question of bone modeling and remodeling. To some extent this number has been self-defeating, because it is difficult to keep up and to assess the significance of each paper. However, Frost's concept of the "mechanostat" (Frost 1987) has generated considerable interest, and although it can be shown to be faulty in some ways, it has certainly been "seminal" (as Frost rather endearingly claims it to be in his original paper!).

Frost's original idea was this:

- Modeling usually adds bone over a large surface. Therefore, increased modeling increases overall bone mass.
- Remodeling, via BMUs, does not quite replace all the bone taken away. This effect is found in Haversian systems, endosteal cortical bone, and cancellous bone. However, on the periosteal surface bone may actually be added after activation of a BMU. Overall, increased remodeling reduces bone mass.
- Strong activity, leading to large strains in the bone (over a strain of about 0.0015 called the minimum effective strain), increases modeling and decreases remodeling, so bone mass increases. The strain necessary to initiate this activity is the *minimum effective strain.*
- Very low activity prevents modeling, and remodeling activity increases; therefore, bone mass reduces.

- Between these set points is a zone of strains where rather little happens.
- Drugs, hormone imbalances, and so on alter the set points with respect to the strains, so that the bone "feels" according to Frost's proposed feedback mechanism, the mechanostat, to be less (for instance, in the case of fluoride treatment) or more (for instance, in the case of postmenopausal estrogenopenia) robust than it actually is. As a result, inappropriate modeling or remodeling takes place.

A number of bells and whistles can be added to this model, but this is its essence. I introduce it here, before going on to consider a number of other matters. I shall return to the mechanostat idea toward the end of the chapter.

11.2 What Do We Need to Know?

With that general introduction out of the way, it may be helpful to lay out a number of questions that need to be answered, if we are to understand what is going on in bone:

- Does bone model adaptively to its mechanical environment?
- If it does model adaptively, what are the *proximate* stimuli?
- If it does model adaptively, what are the *ultimate* stimuli?
- How do the cells monitor and respond to the proximate stimuli?
- Are adaptive modeling processes capable of being produced by the sum of the actions of individual cells acting on their own, or does it require the cooperative action of cells, which act, in some way, according to what other cells are doing?
- Whatever the answer to the previous point, does adaptive modeling require the cells to "know where they are"?
- Do the cells need to have memory of previous loading situations? If they do, what is the order of magnitude of the time course of the decay of the memory?

The significance and meaning of some of these questions need clarifying. First, it may seem obvious that bone does model adaptively and, most people would think, this is in response to mechanical loading. I have assumed so in section 11.1. However, we must be sure that this is actually the case, and a very useful antidote to a blind acceptance that this is so is the paper by Bertram and Swartz (1991) in which they demonstrate that many phenomena that appear to be adaptive modeling can be explained equally well in other ways. They argue that it is not possible to find a convincing case in the literature of adaptive modeling occurring in *mature* bone, not resulting from trauma or disease, that can unequivocally be shown to be in response to mechanical loading only.

The *proximate* stimuli refer to what the bone cells are reacting to, such as streaming potentials or direct deformation of the cell membrane. The *ultimate* stimuli refer to the significant mechanical features of the strain environment that were being reacted to. This might be the mean strain energy density in a small volume of bone, or it might be the rate of change of the maximum principal strain with distance from the neutral axis, the rate of change as a function of time, the amount of damage in a small volume of bone, or a whole host of other significant stimuli that have been suggested. (Strain energy is the amount of work that could be done by a strained piece of material if it were allowed to relax to the unstrained state. Since there are normal strains, shear strains, bulk strains, and so on, its formulation can be quite complex, and need not concern us. Strain energy density is strain energy per unit volume.)

The question of cells "knowing where they are" is very important. (Of course, cells do not actually *know* where they are; this is merely a shorthand way of saying that the behavior of cells may be affected simply by virtue of where they are in a bone or where that bone is in the body.) For instance, there are bone cells in the distal tibia, the top of the skull, and the auditory ossicles. If bone cells model according to some universal rule about what is an effective strain, then it would seem highly likely that the strain appropriate for the tibia would be completely inappropriate for the top of the skull and the auditory ossicles, which are not loaded to high stresses, but which must not be removed by osteoclasis. (This particular consideration does not apply to algorithms that propose that *damage* is the effective stimulus [Prendergast and Taylor 1994].) I have argued before (1984) that if cells do not "know where they are" then the algorithms required for adaptive modeling are going to be extraordinarily complicated. With the advent of the molecular biology approach to positional signaling (Kay and Smith 1989) the situation is likely to become much more complex in detail, but possibly simpler overall in concept.

The question of memory is important because some hypotheses suppose that bone models if the strains to which it is exposed have an *unusual distribution*. Such an algorithm could be based on the cells "knowing" what kind of bone they were in and determining that the strains were unusual for cells in their position, or it could be based on the cells being agnostic as to their position and instead performing a calculation about the difference between the strain distribution at any particular moment and some "memory-with-decay" of past strains. Implementing an "unusual distribution" algorithm in these two ways would require quite different behavior from the cells.

Clearly, it is important to determine if there is a single mechanism for adaptive modeling. However, we must also bear in mind the words of Frost (1985), quoted by Bertram and Swartz (1991): "Time and again in this field two logical errors have been made by the authorities. On some occasions they assumed that *a* truth was therefore the *whole* truth; so they rejected others. On other occasions they refused to accept a newly perceived truth as genuine simply because it could be shown not to be the whole or only truth. . . ."

It may well be that there are different kinds of proximate stimulus, used concurrently by a single cell, used at different times by the same cell, or used at the same time by different cells according to their situation. It may well be that different ultimate stimuli or different algorithms are used by groups of cells, and that, for instance, in some situations strain energy density is measured, in some way, and at other times or in other places damage is measured in some way. For purposes of actually doing science, it is best to *assume* that there is only a single proximate stimulus, and a single ultimate stimulus, and give up that assumption only when forced to. If one initially assumes that there are many algorithms, then one can always argue that although one's experiments produced the expected results in some cases, the fact that they did not in others was because in the other cases some other, as yet undiscovered, algorithm was appropriate. That is not explaining; it is explaining away.

11.3 Classic Experiments

There were two classic early experiments in which the strain in the living bone was monitored before and after modeling. Both experiments were by Lanyon and his colleagues. Goodship et al. (1979) removed part of the ulna of pigs (fig. 11.2). This removal resulted in roughly a doubling of the peak strain on the radius during walking, as measured by strain gauges attached to the surface of the bone, compared with the strains on the control side. An explosive growth of woven bone followed, the bone became rounded rather than elliptical in cross section, and the difference in peak strain became less. By the end of three months the strain on the operated and unoperated side was about the same. This is a good example of the control system working properly. The other experiment (Lanyon et al. 1982), on the forelegs of sheep, showed that the strain at the end of a year was somewhat lower in the radius whose ulna had been removed than in the radius on the other side. Considerable new growth of bone took place in the radius. These experiments are classic in that later workers (such as Cowin et al. 1985) suggesting algorithms that may produce adaptive modeling of bone have tested them against the Goodship et al. and Lanyon et al. results.

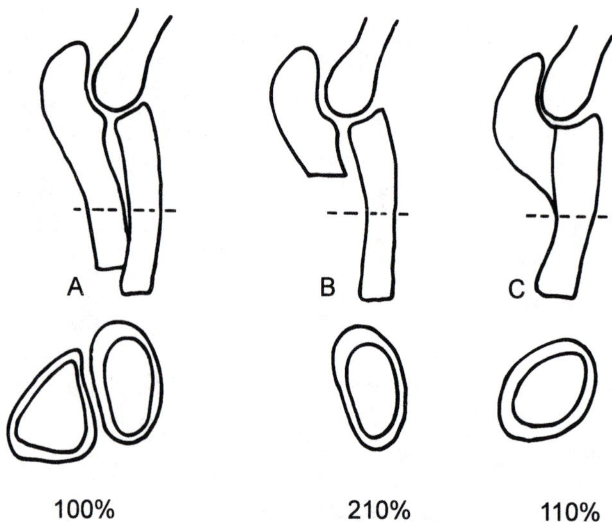

FIG. 11.2 An experiment of Goodship et al. (1979). (A) A normal forelimb of a pig. The ulna is present. The maximum strain on the radius in walking is taken as 100%. Interrupted lines show the plane of the section. (B) Ulna partially removed. The strain is more than doubled. (C) After three months the radius has changed shape, and the strain is about the same as on the control side.

If bones are subjected to an altered loading regime, there are two ways they might respond: by altering the build, or *architecture*, of the bone, or by altering the *quality* of the bone tissue in some way. Woo et al. (1981) have looked for both these responses in bone. In one experiment they studied the effect of saving a dog's femur from strain by screwing a plate to it, so that its overall stiffness was artificially altered. After a year, strips of bone were taken from control animals, as well as from animals that had had attached to them either very stiff plates or flexible plates. The maximum bending moment and the area under the load–deformation curve (a measure of energy absorbed before failure) were measured for strips of standard length and width (their depth, however, was the thickness of the cortex). The control, nonplated strips were stronger and absorbed more energy than strips from the stiff-plated bones, but the strips from the flexible-plated bones were barely affected, behaving just like the controls. However, the Young's modulus and tensile strength of the bone *material* were unaffected whatever the treatment. The cortical thickness had been reduced in the stiff-plated femora, and this accounted completely for the change in the *specimen* properties observed.

The contrary experiment, of increasing the level of activity well above normal, was performed on pigs (Woo et al. 1981). The control animals were allowed to do what pigs like doing, which is nothing much and a lot of eating. The unfortunate experimental animals were trained to trot for about 40 km a week, and this was kept up for 8 months. Again, strips of full-thickness cortex were tested, and marked differences were found in the loads the strips could bear and the energy they were able to absorb, these properties being enhanced in the exercised animals. However, there was again no difference in the Young's modulus, tensile strength, or the mineral and organic constitution of these strips.

In these long-term experiments, therefore, the bone material laid down, if bone mass is increased, or the bone remaining, if bone is removed, is very similar to the original bone. That is what one would expect, because there is no reason for supposing that an altered type of bone material would be more suitable than the original bone for dealing with the altered loading regime. If bone is laid down during modeling, its constitution will at first be different from that of the original bone because new bone is usually woven bone, and, indeed, is often very porous as well. Over a fairly short time it is remodeled to become much more compact.

11.4 The Nature of the Signal

Before adaptive modeling and remodeling can take place, the bone cells must have information concerning the mechanical condition of the bone. Usually it is considered that it is strain that is being measured.

The strain-related messages must be read and acted upon *locally*. It is difficult to imagine how blood-borne factors could be centrally programmed to act at some times more on one bone, and at other times, on another. Blood-borne factors may be necessary for modeling, and, indeed, probably are, but if this is so, their action must be modulated locally. The vast array of cytokines (factors produced by cells that affect the behavior of cells in the vicinity) has various members that have such local effects in bone modeling (Mundy et al. 1996). Loading on bone can also affect cells in other tissues. For instance, Keila et al. (1994) compared marrow from normal and immobilized femora of rats. They found that the osteoblast progenitor cells in the marrow from the unloaded femora had their ability to turn into osteoblasts, and so to form bone, markedly reduced compared with the controls.

11.4.1 Electrical Effects

For many years now electrical effects of some kind have been favored by many investigators as being the likely mode of information transfer between bone strains and the cells in bones. There is, indeed, a great body of evidence that electrical phenomena can alter modeling and fracture healing (Einhorn 1995; Polk 1996; Scott and King 1994). These studies show that the application of potential differences across bone, which causes a current flow, results in new bone formation taking place to a greater extent than in the absence of such potentials. It is also known that electrical potentials can be developed in bone by deforming it. The precise mechanism producing these potentials is still in some doubt. There is a tendency to avoid the issue, if possible, by calling the potentials *stress-generated potentials* (SGP). I shall briefly describe two candidates.

CLASSICAL PIEZOELECTRICITY

A mechanically unstrained crystal, made of anions and cations, will have no net charge, nor will the ends be polarized with respect to each other. However, many crystals have a lattice structure such that, when they are strained, there is a net separation of charges. This results in a potential difference being set up between opposite ends of the crystal. This potential develops because the crystals have no center of symmetry. The effect is called the direct piezoelectric effect. The reverse effect, in which a potential set up across the crystal causes strain in the crystal, also exists, but does not concern us here. Apatite has a center of symmetry and so does not exhibit piezoelectricity. Dry collagen does show it, however, though wet collagen barely does, probably because when wet it has little shearing stiffness. However, it is possible that wet collagen, stiffened by mineral, could be responsible for the SGP of bone.

STREAMING POTENTIALS

If extracellular fluid, which contains mobile ions, is in contact with a solid that has a fixed charge at the surface, then the fluid will have a concentration of oppositely charged ions near the surface. Suppose the solid's charge is negative. There will be a tendency for the fluid side of the interface to be populated predominantly with cations, bound more or less firmly to the surface, while the anions in the fluid will, on average, be more mobile. If the fluid is made to flow with respect to the surface, the less rigidly bound anions will be freer to move than the cations. Such a net movement of charge is an electric current; therefore, a potential difference must appear between the upstream and the downstream parts of the fluid. This potential difference is called a streaming potential. Mak (1990) gives a full description of the general principles, and many useful references. Salzstein et al. (1987) give a theoretical analysis of how SGPs could be produced in bone, and a companion paper (Salzstein and Pollack 1987) gives good experimental confirmation.

In a uniform beam subjected to pure bending the strain increases linearly with distance from the neutral axis. The SGP should be proportional to the difference in strain between any two points. Therefore, for a uniform beam the SGP can be calculated between any two points if the potential difference between the two outermost surfaces can be measured. However, as was foreseen (McElhaney 1967), the situation in bone is complicated because bone cannot be considered to be homogeneous.

Starkebaum et al. (1979), using a microelectrode on human bone at 98% relative humidity, found that the potential distributions, though overall behaving in the expected way, showed great changes over very short distances. Most interestingly, these changes were associated with Haversian systems (fig. 11.3). The specimen was cut in such a way that the histological features were exposed on the lateral surfaces.

The effects of tension or compression on SGPs seemed to be accentuated in Haversian systems compared with the interstitial lamellae. The effects appeared to be actually confined to the Haversian system itself, as if by the cement sheath surrounding the system, even though in life there are some cell–cell connections between Haversian systems and the surrounding interstitial lamellae. The significance of these elegant experiments is not clear. The variations in SGP were recorded from the surface of the test specimens, and it is possible that conditions deep in the bone substance may be different. It may be that these local irregularities in the field, even if they occur in life, are not significant in signaling, because they are confined to the Haversian systems. The morphology of bone is such that the resting cells, lying on the surface, which can become osteoblasts or alternatively can in some way summon up osteoclasts from the circulation and which are, therefore, responsible for initiating modeling, are always outside cement sheaths. If the anomalous effects in Haversian systems are confined to them, the signaling system should be unaffected. Nevertheless, these experiments are important in showing that SGPs are far from being the smoothly varying potentials that it is easy to imagine.

If a specimen of wet bone is loaded in bending, the potential difference that appears is transient. The tension side will become positively charged with respect to the compressive side. The peak difference depends on both the strain and the strain rate and is of the order of 5 to 10 mV (Lanyon and Hartman 1977). When the bone is held at a constant deformation, the potential decays quite quickly, going to near zero in about two seconds (Cochran et al. 1968). If the bone is unloaded, so

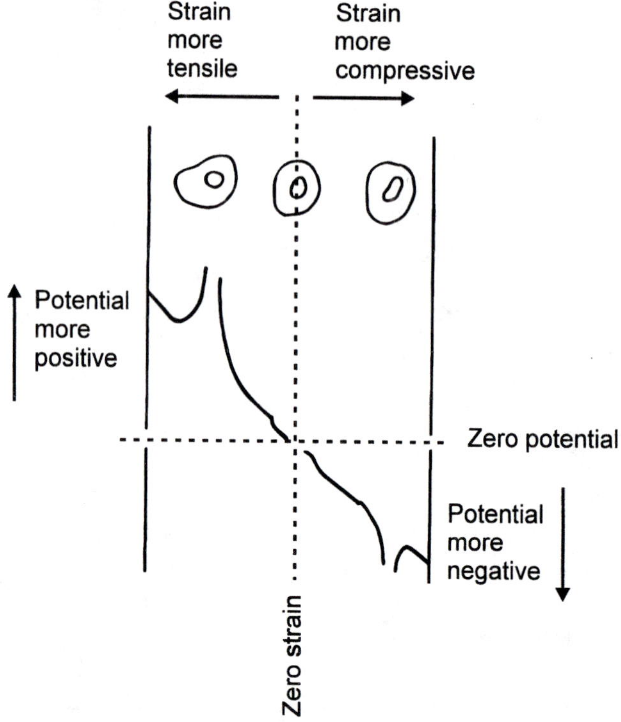

FIG. 11.3. Diagram of the electrical potential gradient in a bending specimen. (*Above*) Three Haversian systems are shown in section. There is a bending moment so that one side is in tension, the other in compression. (*Below*) The potential, shown by the solid diagonal line. Irregularities are associated with Haversian systems. (From Starkebaum et al. [1979].)

that it returns to its original shape, a potential of equal magnitude and opposite sign appears, and decays as quickly.

Cowin and his co-workers have produced a series of extremely elegant papers suggesting that the dimensions of the canaliculi and the physical properties of their contents are just right to produce both in magnitude and time course the kind of potentials that are observed in bone, and that these are streaming potentials (Zeng et al. 1994; Cowin et al. 1995). Rather an alarming number of other assumptions have to be made to arrive at the final conclusion. However, so many features of the analysis are reasonable that one is left with the conclusion that if it is not true, it ought to be. Certainly, the transport of fluids through bones is enhanced by applying bending loads to them (Knothe Tate et al. 1998). An interesting twist to all this is that, as I mentioned in section 1.7, the bone of advanced fish, the teleosts, barely models or remodels at all and most teleost bones have no osteocytes or canaliculi. I do not know whether teleost bone produces streaming potentials.

DAMAGE

One effect of loading bone may be to damage it, and it is possible that damage is an effective stimulus (Prendergast and Taylor 1994). Qiu et al. (1997) have shown that osteocytes near a region of bone that has been damaged by the insertion of a screw undergo *apoptosis*, a kind of programed cell death, quite different in appearance from the ordinary *necrotic* death of cells. Verborgt et al. (2000) showed similar effects in bones loaded excessively. If generally true, this suggests that cells may signal the effect of damage in their vicinity and thereby initiate remodeling. However, it is improbable that things would be arranged so that bone would have to undergo damage, which implies a permanent change in its mechanical properties, to remodel.

11.4.2 Direct Measurement of Strain

BULK STRAIN

There have been suggestions that strain could be measured more directly than through the mediation of electrical effects. For instance, Jendrucko et al. (1976), making some reasonable assumptions about Poisson's ratio for bone and the bulk modulus of water, showed that the hydrostatic strains in the osteocytes could be quite high. They suggest that these transient high pressures may in some way be a stimulus to modeling. Bassett (1976) pointed out that modeling was mediated through cells on free surfaces, not by the osteocytes. This is not to say that hydrostatic strains are not important in the life of osteocytes (indeed, large tensile strains in the bone, applied at a high strain rate, might cause the cell contents to boil). However, if bulk strain is the relevant signal, the enclosed cells must signal to the cells on the surface, which do the modeling.

SHEAR STRAIN

The movement of fluid through the various channels in bone, caused by the deformations of the solid bone, will induce shear strains in the osteocyte cell bodies and their processes. It is possible that it is these shear strains that are measured by the cells, rather than the electric potentials. (There is, for instance, much good evidence that the cells of the endothelium of blood vessels make use of shear strain information both for short-term responses and for long-term "modeling" responses [Davies 1995].) Again, Cowin and his co-workers (Weinbaum et al. 1994) have modeled this process and shown it to be feasible. To complement the hypothesis, Klein-Nulend et al. (1995) have shown that osteocytes are quite sensitive to shear deformation, releasing a specific prostaglandin, but are less sensitive to hydrostatic pressure changes. Chen et al. (2000) show how a set of relevant responses are produced in osteoblasts by shear strain after a cascade of events, including intracellular Ca^{2+} increase and changes in the cytoskeleton, has been initiated by the fluid flow.

11.5 How Does Bone Respond to the Signal?

Suppose, then, that bone cells can in some way determine the local state of strain or damage, or both. How are they to react adaptively to this information? Studies on bone modeling have been bedevilled by *Wolff's law*. Wolff (1892) stated (to paraphrase him) that bony structures modeled so as to fit them to their function. He wrote a great deal more, of course. The unfortunate thing is that, for many workers, it seems only necessary to show that bone is adapting, invoke Wolff's law, and depart, conscious of a day's work well done. No thought is given as to how the bone models in an adaptive fashion.

However, in more recent times people have tried to produce testable models of the mechanisms of adaptive modeling. Many years ago Bassett (1965, 1971) suggested that bone cells lay down bone in regions

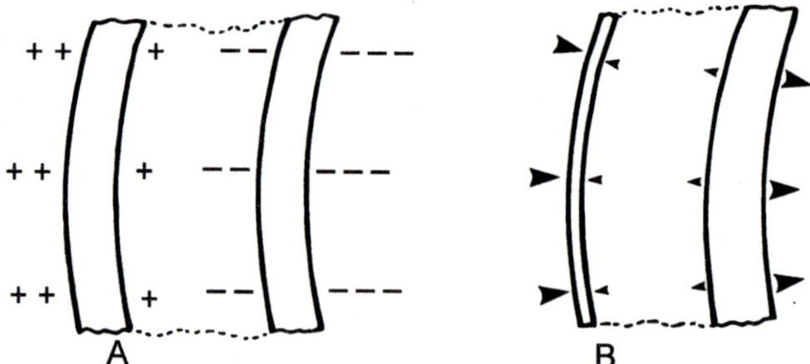

FIG. 11.4 The consequences of Bassett's hypothesis. (A) The strains in an anatomically badly curved, hollow bone, loaded in compression. The strains are progressively less positive from the convex side to the concave one. (B) The arrows show the direction of bone modeling, which will be more intense on the periosteal surfaces than on the endosteal surfaces. The resulting asymmetrical thickness of the cortical walls is shown.

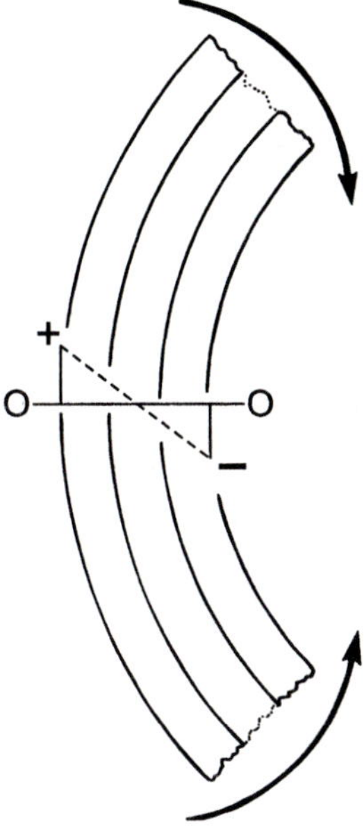

FIG. 11.5 Section of a hollow bone, initially straight, loaded in bending (deformations are grossly exaggerated). The strain in the section O–O is shown by the dotted line. Bones cells might be able to measure the *change* of strain with depth.

of compressive strain and erode it from regions undergoing tensile strain. As it stands, this cannot be correct, because it would lead to a curved hollow long bone that was loaded in compression longitudinally becoming thinner walled on one side of the marrow cavity and thicker walled on the other (fig. 11.4). Nevertheless, this model has been taken as a basis for experimentation and thought.

To overcome this difficulty, in 1964 Frost suggested that bone cells responded to *changes in curvature* of surfaces. If, on the application of load, a surface became more concave, bone would be laid down on it. If it became more convex, it would be eroded. This model does get over the major geometrical problem of Bassett's model, but it has another difficulty. It would be extraordinarily difficult for cells to measure changes of curvature of surfaces. However, with the aid of canaliculi penetrating into the bone substance, it would probably be rather easy for them to measure changes of strain with depth (Currey 1968). If this were so, a bone cell should lay down bone on its local surface if, on application of load, the strains deep to it became more tensile with depth. It should remove bone if the strains became less tensile with depth (fig. 11.5). The model of Frost would work only if the bone were never loaded in net tension—that is to say, if it is loaded in tension as well as being bent or twisted. There seems to be no signaling mechanism, relying solely on *local* information, that can give the right type of

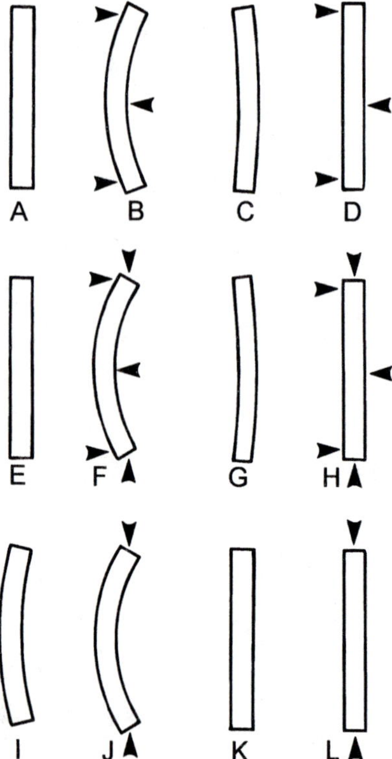

FIG. 11.6 Futile and useful modeling. (A) A straight bone is loaded in bending. (B) Modeling alters its shape (C), so that when it is loaded again (D) it becomes straight. However, this will not materially affect the strains in the bone. (E–H) If an axial load is imposed as well as bending, the resulting modeling is adaptive because, when the bone straightens under the influence of the bending load (H), the axial load will have no bending moment, as it does in (F). Indeed, if the unloaded bone were curved slightly more to the right, the moment produced by the axial load would tend to counterbalance the bending moment induced by the three-point bending. Similarly, it will be advantageous to model an anatomically curved bone (I) subject to an axial load, because in its modeled state (K) it will not be subject to bending when loaded axially (L).

reconstruction if it is possible for bone to be loaded both in net tension and in net compression. In one mode or the other the reconstruction would go in the wrong way, making the situation worse, not better.

Frost has gone on to elaborate his ideas in a series of papers and books. His book of 1973 was a fairly comprehensive account of his theory. (His mechanostat model, which I mentioned above, does not deal with this kind of problem, where geometry is important, at least not explicitly.) One situation in which Frost's model would produce useless modeling is when a bone is loaded in pure bending (fig. 11.6). An

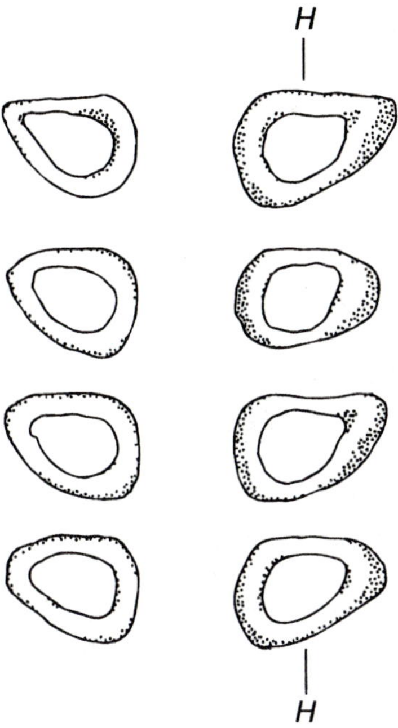

FIG. 11.7 An experiment of Liškova and Heřt (1971). The left-hand sections are the controls; the right-hand sections are the corresponding bones, from the same rabbit, that have been loaded intermittently in bending about the *H–H* axis. New bone is shown by stippling.

originally straight bone would model into an arch, but this would leave the strains almost unchanged. However, this is a small objection because it must be very unusual for a bone to be loaded only in bending. Almost always there will be an axial compressive component as well. The axial component will deflect the bone, and adaptive modeling should act to cancel out the effect.

Such theoretical discussions are all very well, but when we look at what happens in real life, the situation is not very clear cut. Heřt et al. (1969) and Lišková and Heřt (1971) performed some experiments, beautifully designed for their time, in which the tibiae of rabbits were subjected to continuous or to intermittent bending loading. The force on springs producing the bending was occasionally altered so that the maximum bending stress in the bone remained effectively constant, at about 60 MPa, as the bone grew. The experiments showed that the constant stress had very little effect on the bone over many months. In the experiments involving intermittent loading, bending stresses of about the same magnitude as in the static experiments were applied at about 0.5–1 Hz, for from one to three hours a day. The results were quite clear; there was considerable extra bone deposited, both periosteally and endosteally in the experimental bone compared with the control bone (fig. 11.7). Bone was laid down almost equally on the tension and on the compression side. This is not in accord with the predictions of the models discussed above, which are that bone should be lost from the subperiosteal surface that was in tension and from the endosteal surface that was in compression.

I described above the experiment of Goodship and colleagues, which showed modeling reducing strain to acceptable levels. In osteoporosis this does not happen. Osteoporotic vertebrae, in particular, can collapse "spontaneously" when the loading is little, if at all, more than that imposed in day-to-day activities. It would be possible to attribute this lack of adaptive balance between loads and bone erosion to a general decay associated with aging. But bone erosion is an active process, requiring the activity of osteoclasts. It is possible that the feedback system is deranged in some way? This is certainly Frost's view.

Because the SGP declines quite quickly after the bone is loaded, the rate of strain is important in determining the magnitude of the potential evoked. Lanyon and Hartman (1977) investigated the potential produced in the radii of sheep during normal locomotion. Measuring both the strains and the potentials at the same time, they found, as might be expected, that the maximum strain in the bones increased with speed of locomotion. However, they found that the size of the stress-generated potentials increased much more rapidly than the strains themselves with speed. At a brisk trot the strains were obvious, as were the SGPs. However, when the sheep walked slowly, although the strains could be clearly recorded, the SGPs were "small and irregular, with no recognisable strain-linked patterns." In other words, strain *rates* must be high for SGPs to reach reasonably high values.

Lanyon and Hartman suggest that as activity in old people decreases, the strain rates imposed on the bones fall to a level at which effective SGPs are not produced. If, as appears to be the case, bone is removed unless it is actively protected, this level of SGP will result in bone being removed even though the stresses and the strains in the bone become higher and higher as a result. O'Connor et al. (1981) showed that bone modeling is indeed much more sensitive to strain *rate* than to maximum strain. I return to this matter below.

11.6 Postclassical Experiments

There has in recent years been a great flood of experiments designed to test various ideas, hunches, and guesses (often improperly called theories) about adaptive modeling. It is not possible to summarize the findings, because they have been so disparate. A very thoughtful synthesis and analysis of what had been done up to the time of their writing is in

Martin and Burr's book (1989). A brief recent review, emphasizing the interaction of mechanical loading and biological responses, is by Skerry (2000).

The great difficulty, as in any experiments on biological feedback mechanisms, is to "break the loop" and to prevent the homeostatic mechanisms from functioning normally, so that the effect of a particular stimulus is seen on its own. Lanyon and co-workers, particularly Rubin, initiated a whole series of experiments, using the *isolated avian ulna* model (Lanyon and Rubin 1984; Rubin and Lanyon 1984, 1987; Rubin et al. 1995). Part of the shaft of the ulna of a bird, usually a turkey, is isolated mechanically from the rest of the body, but with its blood supply intact, so it remains healthy, and specified loads are imposed on it by means of hydraulic devices. The growth response of the ulna is compared with what happens on the control side.

(It is amusing to go to conferences and hear people muttering under their breaths, "but turkeys aren't *real* animals!" Other animals I have heard classified as unreal are rabbits, rats, battery chickens, and kangaroos. On the other hand, humans are, perversely, considered to be real despite the fact that most work on humans is concentrated on those in the last third of their life span, who in the wild would long ago have left to join their ancestors. As another example of how research is undemocratically targeted, it is probably true that most vertebrate species do not have osteocytes, for most vertebrate species are teleost fishes, and the great majority of these lack osteocytes. From the point of view of bone biology studies, anosteocytic bone might as well not exist. Reality surely lies in the eye of the beholder or, more specifically, the grant giver.)

These avian models quickly showed a number of things: the isolated unstrained bone would develop large cavities; very few cycles a day of physiological loading were necessary to prevent this; if the loading were in an unusual direction extra bone would be laid down in an almost linear dose–response curve; the initial bone deposited was often woven, which was replaced quickly by lamellar bone. More recent work on turkeys has concentrated on, among other things, the effects of strain rate. Rubin and McLeod (1996) have put forward the idea, with some evidence to support it, that high-frequency strains of very low magnitude may have a considerable osteogenic effect. For instance, they suggest that if bone loss is prevented by 100 s of loading, with a maximum strain of 0.001 applied at 1 Hz, the same effect would be produced by 10 min of load at 60 Hz, with a maximum strain of only 0.0002! (Remember, bone yields in tension at a strain of about 0.005–8.). Indeed, Rubin's group has produced evidence that homeopathically small strains (0.000005) at a frequency of 30 Hz can increase markedly the trabecular density in the femora of sheep (Rubin et al. 2001). Old people lose the high-frequency components of muscle tremor, which might account for the failure of their adaptive modeling, rather than their sedater mode of locomotion. These are remarkable results and it will be interesting to see this line of research developing.

It is possible that it is the continuing low amplitude–high frequency of denning bears' muscles that keeps the bone in being. Harlow et al. (2001) have shown that bears retain their muscle tone when they are denning, while barely moving at all. Interestingly, Zhang et al. (1997) produced a model of the behavior of fluid in bone that predicts that a frequency of around 30 Hz is about optimal for bone cells to be electrically stimulated. Another experiment, not using birds, also showed the importance of strain rate. Mosley and Lanyon (1997) showed that in the ulnae of rats, after all other variables had been controlled, loading at a strain rate of $0.1\ s^{-1}$ had produced considerably more new bone than loading at a rate of $0.03\ s^{-1}$.

The osteotomies must be rather traumatic for the bird, and there is always the concern that some or all of the response may be affected by the trauma. Relatively noninvasive methods can be used in certain places. For instance, Hillam and Skerry (1995) loaded the ulnas of anesthetized rats, and were able to inhibit the normal bone resorption that occurs during growth of the ulna. Verborgt et al. (2000) used this system to show the relationship between apoptosis (programmed cell death) and damage to the bone.

Another model that has been used in a rather systematic way is the rat tibia four-point bending model of Turner and his co-workers (Turner et al. 1991; Akhter et al. 1992). The tibia of the living (anesthetized) rat is loaded in four-point bending in a way that allows the strains in various parts of the bone to be well characterized. Using this procedure over a number years these workers reported experiments in which they altered the magnitude of the load, the rate of loading, the duration of loading in a session, the number of sessions, and so on. Among results that have emerged (though not necessarily uniquely) from this system were these:

- There was a threshold strain below which no new bone is laid down; above this threshold it is laid down. Remarkably, the endosteal surface produced lamellar bone, the amount of which was proportional to the "excess" strain, while the periosteal surface produced great quantities of woven bone in what seemed like an all-or-none response. This mass of woven bone was, in time, reduced in size and consolidated.

- Bone formation was stimulated much more by loading at frequencies of 0.5 Hz or greater than at lower frequencies (Turner et al. 1994).
- Increasing the strain is much more effective in producing new bone than increasing the time over which loading takes place.

Many people have adopted, and adapted (Forwood et al. 1998), this model. It has proved useful in quantifying many modeling processes:

- Robling et al. (2000) showed, for instance, that dividing up the daily loading into a series of short bouts was much more effective in inducing bone formation than the same stimulus applied in one burst on one day.
- It has also been used to relate the strain regime in a bone and the expression of cytokines and other biochemicals that may affect the actual bone remodeling process. For instance, Forwood (1996) studied a prostaglandin response. Boppart et al. (1998) used the model, as have others, to determine the time course of cellular events after the initiation of loading.
- Akhter et al. (1998), adapting the model for mice, showed that different inbred strains of mice responded differently. The strain that, in general, had a robust skeleton responded less vigorously to imposed loading than did the strain that in general had a more slender skeleton.

Two of the advantages of the rat tibia model are that it is reasonably easy to carry out and that it is fairly standardized. It is far from being an industry standard but nevertheless it is much more easy to compare experiments that use it than it is to compare completely disparate experiments. It does not "break the loop," like the isolated avian ulna model, but it is much much easier to perform.

11.7 In Search of the Algorithm

There is a great deal of activity at the moment in searching for algorithms, which bone cells could respond to, that will produce adaptive modeling of bone. For instance, one could suppose that cells respond to the axial strain, the strain energy density, local damage, strain rate, and so on. The general procedure adopted is to suppose that one of these variables is the driving variable, to build a finite element model that is driven by the variable, and then to perturb the model and see what happens. What happens is then related to the real-life situation. There are a number of logical and procedural difficulties with this approach, and the subject is well aired in Odgaard and Weinans's book (1995). One of the problems is that many of the algorithms will produce roughly the same answers and experimental results are far too variable to distinguish between them; another is that the algorithms will solve simple problems, but often cannot solve the more complex problems that are likely to be set in any real loading situation.

The general thrust of the work of people like Lanyon and Frost is that bones are adapted to the loads exerted on them, so that the strains they experience are brought to some reasonable level—in fact, that they have some safety factor. If this is accepted, the subtlety of the process becomes apparent, because different bones in the same skeleton will need to have different strains as the "desired" level. This is because the *variability* of loading is different for different bones. I describe here three situations showing that the relationship between the strains experienced by bone and the modeling response cannot be universal.

THE SKULL VAULT

Consider the human skeleton. If the control system in the tibia produces a strain of 0.002 during day-to-day locomotion, a sensible safety factor may well result. But if the bones in the vault of the skull were modeled so that the day-to-day strains were 0.002, what would the effect be? The top of the human skull, which one would suppose is only trivially loaded by muscle action, would have to be very thin indeed if normal activities were to produce strains as great as 0.002 (2×10^{-3}). But a very thin skull would be useless against the occasional sharp blow to which the skull is subjected. In fact, therefore, the vault bones must be in modeling equilibrium while experiencing extremely small strains. In years past I would produce this argument to people who believed in a universal law, only to have them counter with the argument that for all we know the skull vault is as highly strained as the long bones. I thought this unlikely, but there was no hard evidence to refute it.

There is now. Richard Hillam, when a graduate student in Bristol, managed to circumvent ethical committees sufficiently to have two rosette strain gauges implanted on his skull and one on his tibia. He then underwent cruel and unusual punishments such as jumping off ladders, hitting himself on the head with a rubber hammer, and (most cruel) eating a hamburger. The results were unequivocal (table 11.1). The strains on the tibia were quite typical, being like those found in the tibia of Israeli soldiers (Burr et al. 1996). Indeed, the greatest values, found on jumping from a height of 1.7 m, exceeded previous measurements. On the other hand, the strains in the skull could not be made to exceed 2×10^{-4}. The result of this selfless experiment makes one conclude

TABLE 11.1
Maximum Principal Strains in the Skull and Tibia of a Human Subject

	Tensile strains		Compressive strains	
	Skull	*Tibia*	*Skull*	*Tibia*
Mastication	1.3×10^{-4}	—	-1.1×10^{-4}	—
Smiling	10^{-4}	—	-9×10^{-5}	—
Walking	5×10^{-5}	3.5×10^{-4}	-3×10^{-5}	-7.2×10^{-4}
Heading ball	2×10^{-4}	8.4×10^{-4}	-1.5×10^{-4}	-4.2×10^{-4}
Jump from 0.45 m	1.7×10^{-4}	8.5×10^{-4}	-4×10^{-5}	-4×10^{-4}
Jump from 1.7 m	—	2.1×10^{-3}	—	-8.4×10^{-4}

Source. From Hillam (1996).

either that the control system in some bones of the skull has a different set point or gain from that in other bones, or that the bones here do not have a mechanically controlled structure, but instead merely grow to a particular thickness.

An experiment by Lieberman (1996) makes this last conjecture unlikely. Lieberman measured the strains in the skull and tibias of miniature pigs, some of which were made to run, the others allowed to loaf. Running, of course, does induce small strains in the skull, though these are trivial by the standard of the strains developed in the long bones. The running pigs developed thicker tibias, as was to be expected. They also, however, developed somewhat thicker skulls. This does suggest that the control system is different in the two sites, rather than the skull being unresponsive to strains.

EAR BONES

The auditory ossicles are subjected to extremely small forces, and are correspondingly lightly stressed. Yet the control system must ensure that they do not melt away through disuse osteoporosis. Nevertheless, they do erode to some extent (Oxnard et al. 1995).

Sørensen and her co-workers produced a very interesting series of papers (Sørensen 1994, and papers cited therein) showing just how unlike ordinary bone is the bone surrounding the bony labyrinth. The labyrinth is full-sized at birth and does not change shape (Sørensen et al. 1992c). There is a graded increase in internal remodeling as the distance from the inner ear cavity increases (Frisch et al. 2000). In dogs the mount of bone turnover was 0.13% per year very close to the endolymphatic cavity, through to 10% in the periphery. The amount in neighboring skull bones was about 14% per year. If a defect is induced surgically (Sørensen et al. 1992a), it heals in a rather feeble way, and there is little osteonal remodeling in the vicinity, quite unlike the behavior of ordinary bone near a defect.

A quite remarkable observation (Sørensen et al. 1992b) was that most human otic capsules, from mature and older people, contain cracks. These are not artifactual cracks, because they contain thin layers of unmineralized connective tissue. These are presumably fatigue cracks or cracks that have developed after minimal trauma. Such cracks are not found in the perilabyrinthine bone of other species: rats, rabbits, pigs, and dogs. This suggests that the periotic bones in the human skull are subjected to larger stresses than those of the nonhuman animals. Alternatively, their presence may be explained by the fact that they have a much longer time in which to develop. However, the remarkable feature of these human bones is the complete absence of any repair taking place around these cracks. The whole process of remodeling in perilabyrinthine bone is suppressed in some way, and the suppression is very localized, being most profound next to the cavities of the labyrinth, but becoming less, and allowing progressively more remodeling, with distance.

HIBERNATING ANIMALS

The almost universal accompaniment of greatly reduced musculoskeletal activity is increased bone loss, showing itself, among other ways, as an increase in the amount of calcium and phosphate circulating in the blood. Greatly reduced activity is usually the result of illness or extremely peculiar circumstances such as microgravity in space flight. However, hibernating animals reduce their activity spontaneously and naturally. What clues can they give about the processes on bone adaptation? Unfortunately, most hibernating animals are not very good natural experiments because *all* their metabolic processes slow down; their respiration rate falls, as does their core body temperature. In these circumstances it would not be surprising, or particularly significant, if the rate of bone loss was not raised much above normal. Nevertheless, considerable bone loss is reported in some hibernating animals, for instance, in ground squirrels (Haller and Zimny 1977) and hamsters (Kayser and Frank 1963), and this is accompanied by increased activity of osteoclasts.

Bears, however, hibernate in a way that makes them very suitable for seeing the effect of reduced musculoskeletal activity unaccompanied by a general metabolic depression. Black bears *Ursus americanus* undergo an unusual type of hibernation called *denning*. They sleep for 3 to 5 months and remain inactive. They do not eat, drink, urinate, or defecate. Urea is anabolized into protein, and serum urea actually falls during denning. Their serum calcium remains almost unaltered and phosphorus declines slightly. However, the bears' body temperature and cardiac output remain almost normal, and if disturbed they wake up very quickly and, with luck, lumber off. Their muscles remain strong and keep their tone (Harlow et al. 2001). Histomorphometric studies of iliac bone show an increase in the amount of the bone being resorbed compared with summer bears, but there is also an increased amount of bone formation, and, indeed, there is slight net bone formation during the period of denning (Floyd et al. 1990). It is possible that it is a continuing low-strain, high-frequency activity of denning bears' muscles that keeps the bone in being, in the manner suggested by Rubin and McLeod, mentioned above.

The three sets of observations, on the skull, ear bones, and hibernating bears, show that there must be some way of modifying, or even overriding, the control system in particular bones, in different parts of the same bone, and in the same bone at particular times. The system is subtle, indeed.

SOME DIFFICULT PROBLEMS

It is much easier to carry out armchair experiments than to do real ones. However, it really is important to discuss the kind of algorithms that must be available for cells if they are to produce adaptive modeling. I mention here three situations that seem to me *extremely* difficult to accommodate merely through cells making use of local information (Currey 1984).

1. *A cantilever, of minimum weight, has to bear a particular load so that the free end does not deflect more than some given amount.* This rather simple design criterion appears to be impossible to design by making use of only local knowledge. The reason for this is that the deformation of the free end depends much more on the strains near the fixed end than on those near the free end (fig. 11.8). It would, therefore, be adaptive to have quite low strains near the fixed end, while the strains near the load could be higher without affecting the total deformation greatly. For the modeling system to work properly in such a bone would require the local signal producers to have information about the loading system as a whole. It is very difficult to see how this could be effected.

2. *A column of least weight is straight and slender, and is loaded axially. No part of the bone should reach the yield strain.* If a bone is long and slender, it may collapse suddenly by Euler buckling (section 7.7). The problem for a signaling system designed to prevent such buckling is that the strains at any point in the bone give little infor-

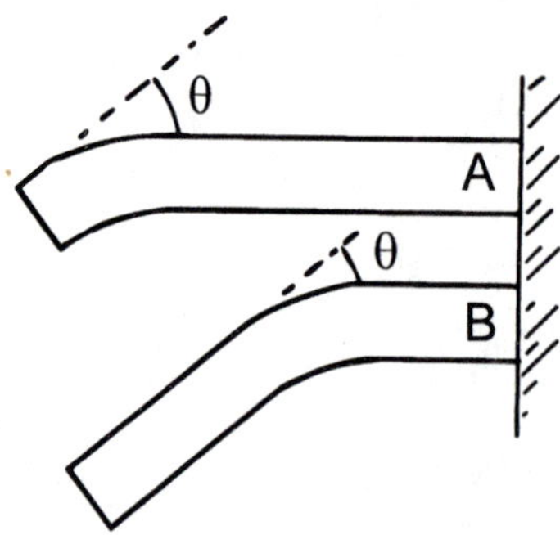

FIG. 11.8 Two cantilevers are both bent though the same angle θ. In A the curvature is distal, in B it is proximal. Identical curvatures (and therefore identical strains) have much more effect on the overall deformation of the cantilever if they occur near the root.

mation about whether the bone is near collapse. In a slender bone loaded very nearly along its long axis (there is bound to be some deviation of the line of action of the force from the long axis), the local stress may become large and give warning of impending doom only just before failure occurs. If, however, a slender bone is slightly curved when unloaded, two things happen: (a) The load that will cause collapse is less, but (b) the greatest stresses in the bone give a better warning of the likelihood of collapse. It is interesting that a bone that is almost certainly loaded in pure compression, the radius of the horse, has a slight anterior curvature. This may not directly help the bone resist the loading, and will indeed make the bones weaker, but may help the modeling system to respond more effectively by giving a greater warning of collapse. (It used to be said that miners preferred wooden pit props to metal ones because, although metal props were stronger, wooden props gave warning of their imminent collapse by creaking and groaning, giving the miners time to escape.) This idea has been developed in relation to bone by Bertram and Biewener (1988).

3. *The hollowness of a tubular bone should be such as to produce a structure of least weight.* As Currey and Alexander (1985) showed (section 7.3.2), the hollowness of tubular bones is appropriate in that, if the mass of the marrow is taken into account, it produces a structure of about least weight. Suppose a bone is too thick-walled, but the overall value of the second moment of area is appropriate in that the bone cells experience appropriate strains. (We have to assume that osteocytes near the endosteum are in equilibrium at lower strains than osteocytes near the periosteum, which again requires that, if modeling is to be adaptive, the cells must "know where they are.") Because the bone is too thick-walled, the endosteal cells should remove bone, even though the strains they measure are "appropriate." All should be well, of course, because the subperiosteal cells will, at

the same time be adding bone even though the strains they are experiencing are also appropriate. (van der Meulen et al. [1993], extended in Carter and Beaupré [2001], have a good try at solving this particular problem.)

These three problems, particularly the cantilever, are somewhat idealized. Nevertheless, adaptive bone modeling requires that the control system be able to solve them or problems like them. One possibility that would *not* require the control system to be so sophisticated, and which I think I favor, is that, in fact, it simply cannot solve the problems, but that some crude rule of thumb is used that produces an answer near enough to the optimum one. It should be realized that making a system more complex has real associated costs; crudely, there are more working parts to go wrong. There will be selection pressures, therefore, preventing any system from becoming too complex. This matter is dealt with theoretically from the population genetics point of view by Orr (2000). The difficulty is that at the moment we really do not know in sufficient detail the loading systems on bones to be able to state with certainty what their ideal construction should be.

(It is worth noting parenthetically that tendons, which show disuse atrophy and use hypertrophy, have a much simpler situation to adapt to. First, tendons are, in the main, loaded only in one direction, in tension along their length, and therefore the appropriate response to changes in loading is to increase or decrease the cross-sectional area. Second, the strains that the tendons habitually undergo, of the order of 4%, are much greater than the strains on bone, and therefore it is easier for the tendon cells to respond directly to strains, through direct strain on the cell membranes [Hsieh et al. 2000].)

In considering the algorithm, many workers have used as a basis Frost's 1987 mechanostat hypothesis. Sometimes, it must be said, this is nothing more than an unthinking obeisance, similar to the remarks people make about Wolff's law. However, unlike Wolff's law, the mechanostat hypothesis does have some testable features, and examining these can sharpen the arguments. For instance, Lanyon (1987) summarized evidence that the minimum effective strain concept needed to be modified to take into account the fact that that not only strain magnitude, but strain distribution and strain rate could be modifiers of the basic response. Lanyon proposed a more general stimulus: the minimum effective strain-related stimulus (MESS), which he said was more accurate and had a more pleasing acronym.

Turner (1999) pointed out that the mechanostat hypothesis predicts that if a bone is completely unloaded, bone mass should decline steadily to zero. This does not happen. (Turner also writes that it is a key assumption of the mechanostat hypothesis that the region of strain where there is no net gain or loss of bone is the same for every bone. I admit that my reading of the Frost 1987 paper does not reveal an explicit statement to this effect, though later papers by Frost do imply a greater universality than the original paper. This illustrates the difficulty of proposing a model and then tweaking it in a series of later publications; everyone is shooting at a different, virtual, target!) Turner's suggestion in relation to the first point is that cells eventually accommodate to a new state of habitual strain, and he produces evidence that this may be the case. Readers may find the spirited correspondence between Frost (2000) and Turner (2000) about the mechanostat entertaining and, possibly, enlightening.

11.8 Precision of Response

It is possible to get a measure of the precision with which adaptive growth and modeling is carried out by examining the asymmetry of bones that ought to be symmetrical. The argument goes like this: we know that paired bones can quite readily become asymmetrical if their habitual use differs. This was shown by the example of the passionate tennis and squash players. Therefore, paired bones are not inherently constrained to be symmetrical. Paired bones will tend to be symmetrical because the construction and reconstruction system in each will be responding to the same kind of loading. Examining the asymmetry of paired bones will give, therefore, a maximum value for the lack of precision of control. It is a maximum value because some of the asymmetry will be the result of different loading histories on the two sides and inevitable experimental error, which will tend to increase the apparent difference between the two sides. Of course, one must also always bear in mind the possibility of the loading on paired bones being asymmetrical, as in the greyhounds I mentioned above.

A group of workers from Leeds, and I, examined the mechanical asymmetry between paired limb bones of the lesser black-backed gull *Larus fuscus* (Alexander et al. 1984). These wild birds (culled on a Manchester rubbish tip) depend a great deal on their flying ability, and there is no reason to think there should be any side-to-side asymmetry in their locomotion.

For each pair of bones we calculated the value $D = 200(L - R)/(L + R)$, where R and L are the values for the property being measured for the right and left bones, respectively. This value essentially gives the difference between the two sides expressed as a percentage of the mean. We tested the humerus, radius, ulna, and tibia, and the properties

TABLE 11.2
Differences in the Mechanical Properties of Left and Right Paired Bones of Black-backed Gulls and Battery Hens

		Gulls		*Hens*	
	Bone	*Mean of* D	*SD*	*Mean of* D	*SD*
Load	Humerus	−1.4	2.1	−1.3	10.8
	Radius	0.9	3.9	2.3	7.6
	Ulna	1.5	3.2	−1.0	5.0
	Tibia	−0.1	3.9	−1.1	5.3
	All	0.24	3.9	−0.23	7.3
Work	Humerus	1.5	5.8	−6.1	33.0
	Radius	−1.6	8.0	3.9	18.4
	Ulna	−0.6	7.1	−0.2	7.7
	Tibia	−4.4	8.1	−1.8	18.7
	All	−1.2	7.4	−0.84	20.6

Note. D is $200(L - R)/(L + R)$, where R and L are the values for the property being measured for the right and left bones, respectively. SD is the standard deviation of D. Despite my strictures in section 3.1, this is one case in which the sample size is not important because the standard error is of no interest.

we examined were the bending load borne in three-point bending and the work shown under the load–deformation trace in this three-point bending. The results are shown in table 11.2.

The mean value for D is always very low, the greatest value for any bone being 4.4.%, with no tendency for the right or left bone to be stronger. Such small differences could well be due to experimental error. This shows that the gulls are not right or left handed or footed. Of more interest are the standard deviations of D. They show the average asymmetry of the bones, without respect to left and right. The overall standard deviation for load is only 3.9%. This implies that in only 1 pair in 20 will the two sides differ by more than 8% Similarly for work, only 1 pair in 20 will differ by more than 14%. These figures are upper values for actual asymmetry, because inevitably our testing procedure introduced some random effects, which will tend to increase the apparent differences between the two sides. These are remarkably low values, and show that for an animal such as a gull the control system determining the build of the bones, and through that their mechanical properties, does indeed exert a very precise control.

We also examined battery hens, and one can see from table 11.1 that in these birds, which are reared in cramped conditions and on which selection for skeletal symmetry must be almost absent, the difference

TABLE 11.3
Standard Deviations of D for Various Bones

Animal	*Bone*	*Test*	*Load or torque*	*Work*	*Authors*
Human	Femur	Bending	6.2	40.5	1
Dog	Fibula	Torsion	16.2	36.6	2
Dog	Tibia	Torsion	4.2	—	3
Rabbit	Humerus	Torsion	34.2	63.0	4
Rabbit	Tibiofibula	Torsion	9.4	17.8	5
Rabbit	Radius	Bending	5.6	—	6
Gull	Various	Bending	3.9	7.4	7
Hen	Various	Bending	7.3	20.6	7

Note. Authors: 1, Mather (1967); 2, Puhl et al. (1972); 3, Strömberg and Dalén (1976); 4, White et al. (1974); 5, Paavolainen (1978); 6, Henry et al. (1968); 7, Alexander et al. (1984).

between members of paired bones is greater. Table 11.3 shows results calculated from the work of other authors. Some of these values are much higher than ours, and may to some extent represent greater experimental error. Anyhow, our results for the gull (the only undomesticated animal tested, on which natural selection is likely to be acting most rigorously) show what the control system is capable of in an animal with strong selection for symmetry.

In flying animals it is highly likely that the symmetry of the arm bones is more important than the symmetry of the leg bones, particularly in bats, in which the leg bones are merely used to hang with. Gummer and Brigham (1995) examined the morphological (not the mechanical) symmetry of the humeri and tibias of the little brown bat *Myotis lucifugus*. They found, as might be expected, that the humeri were more symmetrical than the tibias.

At the other extreme of selection are the Sirenia, the sea cows. These are mammals that live in the sea all their lives and swim around very slowly. As I discussed in chapter 10, in such animals there must be little selection for saving weight in the skeleton, which, indeed, through pachyostosis, acts as ballast to counteract the buoyant effect of the lungs. As a result, there would also be little selection for symmetry. Petit (1955) mentions that the asymmetry of the sirenian skeleton is the most striking thing about it.

The density and structure of cancellous bone varies considerably between different parts of the same bone. Banse et al. (1996) tested the mechanical symmetry of right and left pairs of specimens taken from the human femoral head, making estimates of the stiffness and the strength. They made about 30 tests per pair of heads. They did not express their results in a way that can be related to the other results I report here. However, they showed that there was a fairly tight relationship between the two sides. The correlation coefficient between right and left was 0.89 for strength and 0.85 for stiffness.

Vestigial bones, or bones that are no longer very functional, are also likely to be under relaxed selection and to be variable. Tague (1997) shows that the first metacarpals of two monkeys that have independently developed a very reduced thumb for locomotory reasons are much more variable than the other metacarpals in the hand, and also more variable than the first metacarpal of related species that still have a functional thumb.

A word of warning is needed here. The kind of asymmetry I have been talking about, the asymmetry of structures that have no inherent tendency to be right- or left-handed, is called *fluctuating asymmetry*. It is a matter of intense interest to, and rancorous argument among, people concerned with such matters as developmental stability and the ecological stresses on organisms. What I have discussed is fairly innocuous, but interested readers should consult a reference like Palmer (1994) or Windig and Nylin (2000) to get a flavor of the subject.

11.9 Modeling of Cancellous Bone

In the chapter on cancellous bone I spent some time showing how the general arrangement of cancellous bone was probably well adapted to the loads falling on the bone. We need to consider a little, now, the question as to how cancellous bone models and remodels if the loads falling on it are changed. The extreme view on this is that of Bertram and Swartz (1991), who argue that there is no well-documented case in mature undamaged bone of cancellous bone altering to change its predominant orientation. It must be admitted that they do seem to have a considerable amount of right on their side. Probably the main reason for the lack of evidence is that it would be a very difficult experiment to perform; any intervention harsh enough to alter the loading pattern is likely to cause damage. Some experiments do back up Bertram and Scwhartz's claim. For instance, Biewener et al. (1996) showed that a good relationship existed between the cancellous architecture and the strains imposed on the calcaneus of the potoroo *Potoros tridactylus*, a small marsupial. However, when the calcaneus was freed of significant load by cutting through its tendon, although the number of trabeculae declined and they became thinner, the trabecular pattern did not alter. Whatever was the best orientation of the trabeculae in the tenotomized calcaneus, it surely cannot be that it is best to retain the old arrangement. On the other hand, Kamibayashi et al. (1995) showed that re-alignment of the trabeculae did occur in the human tibia as osteoarthritis developed. There is no doubt that damaged cancellous bone does respond adaptively. Some of the earliest examples, for example, in Jansen (1920), show this very clearly.

Most ideas of modeling assume that bone growth and loss are related in some way to the strains in the bone. If bone is unloaded the strains disappear, and the bone starts to be resorbed. Damage to cancellous bone is interesting in this respect because if a cancellous strut is broken it will in effect be unloaded, its load being taken by other nearby struts. This is unlike the situation in ordinary bones, in which the broken ends will still be in some kind of stressful contact. The different situation in cancellous bone probably accounts for the appearance of *microcalluses*. These are little lumps on the trabeculae, which sometimes may be very irregular in shape, so much so as to be given the name *bird's nest*. In the middle of such bird's nests, which consist of woven bone surrounding the trabecular strut, one very often finds a fractured trabecula (Blackburn et al. 1992; Hahn et al. 1995). Presumably what happens is that the trabecula fractures, so the stresses on it disappear. A race then starts between the osteoclasts, which, if nothing else happens, will erode away the broken unstressed struts, and the repair process, which hastily throws woven bone across the gap, with the result that loading is restored and the strut is saved. Indeed, Hahn et al. (1995) show how the rather disorganized shape of microcalluses can result in new connections being formed, so that after the repair process has settled down there are more trabecular struts than there were previously.

11.10 The Functions of Internal Remodeling

So far in this chapter I have discussed the erosion of bone from, and its deposition onto, the endosteal and periosteal surfaces. I have not discussed internal remodeling, that is, the formation of new Haversian systems. Haversian systems seem to have no straightforward mechanical advantage over primary bone (fig. 11.9). There are, however, some candidate reasons for their formation: they could remove dead bone; they could improve the blood supply; they could have a function in calcium or phosphorus metabolism; they could reorient the grain of the bone if

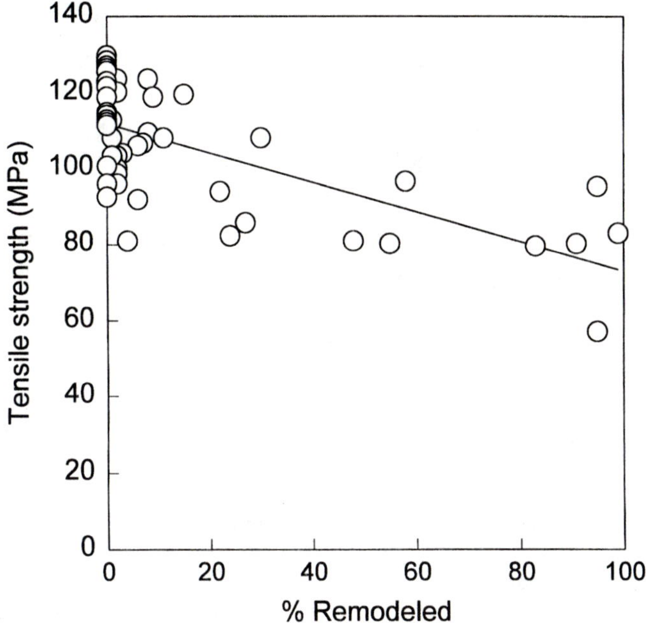

FIG. 11.9 Relationship between tensile strength of bovine femoral specimens and the proportion of the specimen cross section that was occupied by remodeled bone. (From Currey [1959].)

altered loading systems make the present grain inappropriate; they could repair microcracks.

11.10.1 *Removing Dead Bone*

If bone cells die, the bone tissue around them can be considered as dead. It is not at all obvious in what ways dead bone is less effective than living bone. Although there is no doubt (Enlow 1963, 1975; Currey 1968) that Haversian systems quite often form to replace dead bone, this idea of a revitalizing function runs into the problem that the distribution of remodeled bone within bones, between bones in the same animal, and between species is not like the distribution that would result from the death of cells.

Haversian systems are often remarkably symmetrical between right and left bones. Figure 11.10 shows diagrams taken from cross sections of the midshaft of the right and left ulnae of a mature cat. (I chose one section of the pair before the other ulna had been sectioned so as not to

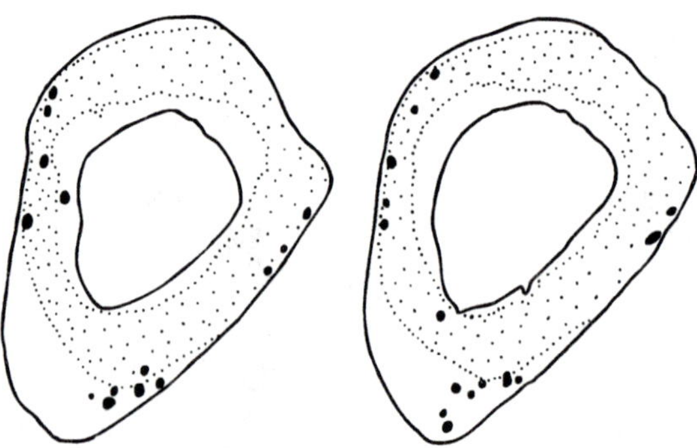

FIG. 11.10 Sections of paired ulnae of a cat. Fine stippling: Haversian bone; black splotches: Haversian systems in process of forming.

compare sections that looked similar. These two sections are, however, typical of several samples I took from cats.) The distribution of Haversian bone in the two sections is very similar, and even some of the erosion cavities appear to be in equivalent places. It is surely stretching things too far to suppose that cell death was occurring with such symmetry in the two bones as to necessitate such symmetrical remodeling. Marotti (1963) also showed in the bones of adult dogs that there is a close similarity in the amount of Haversian remodeling that occurred in equivalent regions of paired bones. Marotti's photographs are worth brooding over. Of course, if the cells are healthy, but are undergoing programmed cell death (apoptosis) then the symmetry is easily explained, but the question is merely pushed back one level—Why are healthy cells dying in this pattern?

11.10.2 Improving the Blood Supply

Another possibility is that remodeling has a physiological function, either in improving the blood supply of bone or by being part of the calcium or phosphate metabolism in the body. The first of this pair of explanations can be readily dismissed. The replacement of fibrolamellar bone by Haversian bone reduces the evenness of the blood supply (Currey 1960; Vasciaveo and Bartoli 1961) (fig. 11.11). Furthermore, it reduces it in a particularly deleterious way: it cuts off the bone in the interstitial lamellae from the nearest blood channels by interposing the cement sheath, across which few canaliculi pass. There is anatomical evidence that the presence of a few Haversian systems in a small volume of bone induces further remodeling because of the death of cells in the interstitial lamellae (Currey 1960).

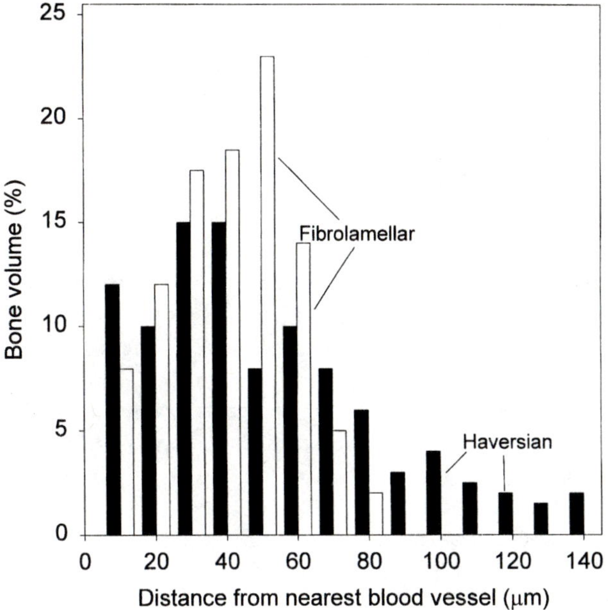

FIG. 11.11 Diagram of the distance of random points from the nearest blood vessel in Haversian and fibrolamellar bone. In Haversian bone the spread is much greater, with more points being close to a blood vessel than in fibrolamellar bone, but many more points being more than 80 μm distant. Osteocytes situated far from the nearest blood vessel are likely to be in metabolic difficulties. (Derived from Currey [1960].)

11.10.3 Mineral Homeostasis

Bone is a convenient source of mineral. Laying birds often develop a mass of bony trabeculae, medullary bone, in the marrow cavity of some of their long bones over a short length of time before egg laying starts. This bone is removed and the calcium is made available for the eggshell. (Because the shell is made of calcium carbonate, the phosphorus in the bone is not needed and must be disposed of.) It would not be sensible to attribute other than an entirely incidental mechanical function to these trabeculae. Ruth (1953) showed that a kind of crude Haversian remodeling could be induced in rats, which do not normally have Haversian systems, by giving lactating mothers a calcium-free diet. The mothers suffer calcium stress, and cavities appear in the bones. When the mothers are later fed a calcium-rich diet, the cavities are filled in with lamellar bone.

A difficulty in accepting calcium regulation as the main function of internal remodeling is the distribution of remodeling between species. Why should young, well-fed human adults, who can surely very rarely have been in negative calcium balance, have considerable Haversian remodeling in their bones, whereas many small mammals and birds, which must be nutritionally near imbalance quite often, do not? Since internal remodeling has quite bad mechanical consequences and since most animals must be in calcium balance in the long run, it is strange, if the function of remodeling is to tide animals over short-term imbalances, that this function is not carried out by mechanically otiose bone, like the medullary spongiosa of laying birds. However, it is also strange that many egg-laying animals, such as the alligators, have not developed the trick of medullary bone, and do undergo considerable bone thinning and increased porosity during egg production (Wink and Elsey 1986).

Burton et al. (1989) point out that Haversian remodeling is going on intensively in the human fetus, by at least the sixth month. Again, it is extraordinarily unlikely that this remodeling is required by the metabolic needs of the fetal body. If the remodeling is to satisfy the *mother's* metabolic requirements, this would be a bizarre case of robbing Peter to pay Pauline so that Peter can be repaid.

11.10.4 Changing the Grain

Where the imposed forces change in direction in relation to the grain of the bone, Haversian remodeling can alter the grain of the bone adaptively. This is seen under large muscle insertions and during fracture repair (Vasciaveo and Bartoli 1961). Enlow (1975) argues persuasively that the Haversian remodeling seen under muscle insertions allows muscles to have firm attachments to the bone even when the muscle insertion is migrating during growth, and also during erosion of the bone surface when the shape of the bone is being altered. Compact coarse-cancellous bone is often badly oriented, and replacing it with well-positioned Haversian systems might be mechanically advantageous. Claes et al. (1995) found that in healing fractures the defect reached almost full density well before it achieved full strength. The final strength was achieved by an intense remodeling that reoriented the bony structure in appropriate direction.

Petrtyl et al. (1996) examined in some detail the orientation of Haversian systems in the human femur. They found they had a gently spiraling course, but the handedness of the spiral was the opposite on the two sides of the bone, so that the two sets of systems met at an angle in the front and the back of the bone. Petrtyl et al. showed that the arrangement of the systems was such that they would be oriented in the direction of the principal stresses if, as well as a large bending moment produced by the offset head of the femur, there was a torque on the femur equal to about one-eighth of the bending stresses. The authors suggest that "new osteons penetrate into the bones in the direction of the first principal dominant stress." This is a very interesting paper, and it would be excellent to see it backed up by research into different bones with different loading systems, the direction of whose principal stresses (strains) in life is known. The grain of the bone often follows the external shape of the bone itself, for instance, in the skull and lower jaw (Tappen 1953, 1970). The struts and buttresses of the skull are themselves oriented in line with the loads in the skull, so it is not possible to say whether the orientation of the grain, usually the secondary osteons, is mechanically adaptive or merely a morphological coincidence. What is particularly interesting about the study of Petrtyl et al. is that the midfemur is nearly cylindrical, yet the Haversian systems do not follow any of the obvious directions of the cylinder (longitudinal, radial, or circumferential).

The work of Riggs et al. reported in chapter 3 does also suggest that remodeling may change the grain of the bone adaptively. Bone in the anterior cortex of horses' radii, which are subjected mainly to tension, was remodeled less, and if it did remodel, the resulting Haversian systems tended to have longitudinally oriented fibers. Unremodeled bone tended to have a longitudinal orientation. Bone in the posterior cortex, subjected mainly to compression, was intensively remodeled, and resulting Haversian systems had more predominantly transverse fibers (Riggs et al. 1993a). This led to mechanical differences in the two cortices, with their strengths each being more appropriate for the main loading situation (Riggs et al. 1993b; see section 3.13). On the other hand, McMahon et al. (1995) could find no relationship between the habitual major stresses in the sheep's calcaneus and the predominant direction of the collagen fibrils. Bone that was, over most of the gait cycle, loaded in compression, showed collagen oriented predominantly longitudinally. McMahon et al. argued that perhaps the small amount of tensile loading may overcome the much greater compressive loading in its histological effects. However, they were clearly not very happy with this interpretation.

Takano et al. (1999) observed changes in the mechanical anisotropy of the collagen in dog's radii that were correlated with the changes in strain induced by osteotomy. Although apparently they did not examine the histology of the bone, and found the structural anisotropy to be too variable to show significant changes, it is not really possible for the mechanical anisotropy to have changed except by secondary remodeling.

11.10.5 Taking out Microcracks

What of the suggestion that Haversian systems form at the end of microcracks and thereby blunt them? Martin and Burr (1982) suggested that when cracks form, inevitably some parts of the bone will be relieved of strain as a result, being sheltered, as it were, by the crack. This volume of bone, being now relieved of strain will send appropriate signals, and as a result the osteoclasts will be recruited and destroy the bone locally—the usual response to a marked reduction of strain in bone. When the cavity becomes filled in, a new Haversian system will have appeared.

It is certainly true that when bone is loaded heavily, it undergoes more remodeling than when it is not (Churches and Howlett 1981; Heřt et al. 1972; Bouvier and Hylander 1981; Bentolila et al. 1997). Indeed, Bentolila et al. (1997) showed that fierce fatigue loading of rats' bone, producing a marked increase in compliance resulting from microcracks, induced internal remodeling in the microcracked regions. Rats normally barely remodel at all. In this case, there can be no doubt that the remodeling was a response to damage. All these experiments accord with the idea that the systems may be repairing microcracks. But it is the distribution of remodeling that argues against this hypothesis. Heřt et al. (1972) loaded rabbit tibiae in bending about the anteroposterior plane and found Haversian remodeling in the regions that had been loaded by bending stresses, the sides, but not in the regions near the neutral axis, the front and back. So far so good. However, Heřt's picture shows that the remodeling was going on not where the stresses induced by bending would be greatest, near the subperiosteal surface, but beneath this, about one-half to three-quarters of the way from the endosteal to the subperiosteal surface.

Burr et al. (1985) showed that if dogs' bones were loaded in fatigue, they showed a large number of microcracks, and also that these microcracks were associated with remodeling resorption cavities far more often than could be accounted for by chance. The obvious explanation for this is that the resorption cavities are there because Haversian remodeling is getting rid of the microcracks. On the other hand, it could be that the resorption cavities are regions of stress concentration, and that microcracks develop, during fatigue, in these regions. A neat experiment by Mori and Burr (1993) showed that the obvious explanation was also much the more likely one. They loaded the left limb of dogs in fatigue for an hour and a half, using a regime that was likely to result in fatigue damage, and then waited eight days. Then they loaded the right limb in the same way, and killed the animals immediately thereafter. During the

TABLE 11.4
Resorption Cavities and Their Relationship to Microcracks

Variable	*Left bone (8 days "recovery")*	*Right bone (killed at once)*
Resorption area (%)	0.075	0.040
Resorption number (N/mm^2)	0.207	0.107
Crack number (N/mm^2)	0.055	0.057
Crack/resorption association (%)	11.1	2.7

Source. Derived from Mori and Burr (1993).
Note. For explanation see text.

period of eight days resorption could start, but would not produce full-sized resorption cavities. They then counted the number and area of resorption cavities and the total number of cracks and the proportion of those cracks associated with a resorption cavity. Table 11.4 is a partial summary of their results.

The number of cracks is about the same after the two treatments, but leaving the bone to recover for eight days has allowed an approximate doubling in the number and total area of resorption cavities. Compared with the bone that was not allowed to do anything after being fatigued, the number of resorption cavities in the left leg that were associated with cracks is much greater (all the differences shown in this table, except in crack number, are highly statistically significant). This does look like good evidence for remodeling taking out microcracks, although the distribution of the microcracking was slightly peculiar, most of it occurring near the neutral axis. This is where shear stresses are largest, and compressive and tensile stresses are smallest.

Verborgt et al. (2000) fatigued rats' bones in vivo and found characteristic microcracking in the heavily strained regions, but not elsewhere. They found that the osteocytes around these microcracked areas were much more likely than control areas to be undergoing apoptosis or to have disappeared, leaving empty lacunae. These differences increased with time over a period of about a week. Furthermore, remodeling cavities were strongly associated with the highly strained/microdamaged areas.

The distribution of remodeling in different species of vertebrates is instructive. In general, in the "lower" vertebrates, that is, all vertebrates except the mammals and birds, there is rather little Haversian remodeling. The dinosaurs and some of the larger mammal-like reptiles do, however, show many Haversian systems. If one were to hold to the idea of Haversian systems functioning by eliminating microcracks, one could explain this distribution of Haversian remodeling either by supposing that reptiles and amphibians do not suffer fatigue cracks or that they are unable, for some physiological reason, to remodel. It is indeed possible that reptiles suffer less microcracking in their bones because, being generally sluggish, the penalties to them for having overdesigned bones with greater safety factors, like the iguana and alligator discussed in the last chapter, would be less than for active mammals or birds.

In the mammals and birds *extensive* internal remodeling goes with size of bone rather than anything else. Biewener (1982, 1993) has shown that there is a reasonable consistency in the greatest strains produced by locomotion and similar activities in bones of very different sizes. So, it would seem that remodeling is a function of bone size, not strain. Small birds have very narrow cortices, and remodeling is almost absent. Ostriches, on the other hand, have extensive Haversian bone. In mammals, the primates and carnivores and artiodactyls show much remodeling, the smaller mammals rather little. The distribution of remodeling in the artiodactyls is interesting because one finds in a cross section of, say, the femur, that most of the bone is still primary fibrolamellar bone, with occasional isolated Haversian systems, but that in one part of the section the bone has been completely remodeled. Usually this remodeling is under a place where large muscles insert. This general distribution of remodeling in birds and mammals does not correspond well with what one feels is likely to be the distribution of more or less highly stressed bone.

Finally, in nearly all tetrapods that show remodeling, the remodeling that does take place is usually more intense toward the marrow cavity than toward the subperiosteal part of the bone (Atkinson and Woodhead 1973; Bouvier and Hylander 1981). This is like the remodeling in Heřt's rabbits, with little remodeling in the most highly stressed regions. Bouvier and Hylander gave monkeys hard or soft diets and observed more Haversian systems in the mandibles of the hard-diet group. The remodeling was sparse in the subperiosteal region, however. In commenting on this they suggest that, because of the way the mandible grows, the subperiosteal bone was younger, and therefore had had less time to develop fatigue cracks and to be remodeled. Against this evidence that internal remodeling is not associated with high stresses is the rather indirect evidence I discuss in chapter 10: that the distribution of remodeling does seem to be related to the safety factors that one would expect in bones in a particular skeleton. Remodeling is less in the distal limb bones, which should have lower safety factors.

In a review paper Burr (1993) attempted to refute the above arguments (essentially reiterated here, with additions, from the first edition of this book) about taking out microcracks, suggesting that I was *against* microdamage being a remodeling stimulus, but was *for* large strains being the stimulus. I was not; I was merely baffled. However, now that I have much more experience in examining the startling arrays of microcracks that can appear when bone reaches the yield region, I am more inclined to think that if Haversian remodeling does *not* have as at least one of its roles that of removing microcracks, it is an opportunity lost. Nevertheless, the arguments above do retain some force.

11.10.6 It's a Pathological Mistake

One of the many strange things about remodeling is that it takes place mainly on the inside of the cortex of the long bones, rather than the outside, even though the strains are, in general, greatest on the outside. This is particularly seen in incipient fatigue fractures, where often quite large cavities open up. Otter et al. (1999) have an interesting hypothesis to account for this. They envisage the following series of events. During locomotion, particularly locomotion of unusual amount, the marrow pressure rises during each stride to greater than the pressure in the arteries supplying the bone via the marrow cavity. The arteries are partially occluded, and this leads to temporary ischemia in the inner part of the cortex (the outer part is usually supplied via the periosteum, so is unaffected). When the locomotion stops, the blood rushes back and damages the somewhat ischemic cells in a process called *reperfusion injury*, which is a quite well attested pathological phenomenon. This damage then results in remodeling, which may in turn weaken the bone and produce fatigue fracture after quite limited amounts of extra exercise. If the ischemia is really severe, as in crush injury, the amount of bone loss is very great indeed (Hsieh et al. 2001), but this is a pathological rather than a supernormal effect.

It is difficult to know what to make of this hypothesis, the evidence for it all being circumstantial. However, it is a useful reminder that bone, although responding to its mechanical environment, is also a biological tissue and therefore prone to various pathologies and derangements of normal physiology.

In summary, there are several explanations for the phenomenon of Haversian remodeling, and there is some evidence for the validity of some of them. Perhaps the three most compelling are those that involve taking out microcracks, changing the grain of the bone, and renewing bone that, simply because of its age, may be becoming weaker (though it may also be developing microcracks). Nevertheless, the way in which remodeling appears to be programmed, particularly on the endosteal side of the cortex of long bones, does make it likely that there is another factor or factors of which we are ignorant.

11.11 Bone Cell Biology

There has recently been a huge increase in knowledge about the molecular and cell biology of bone growth, modeling, remodeling, pathology, and so on. I have ignored this new knowledge in this discussion. Apart from my lack of competence, I think there are two more respectable reasons for doing this. The first is that this chapter is about how bone responds to the mechanical environment. Most of the mechanisms I have discussed relate rather directly to the mechanics. To consider the biology in any detail would have made the chapter much longer than it already is, and might well not have been of great interest to most mechanically minded readers. The second reason is that, although there is an enormous amount of *information* about the molecular biology of bone remodeling, our *understanding* is as yet meager, and it is really too early to attempt a synthesis in a book like this. The splendid books edited by Marcus et al. (1996) and Bilezekian et al. (1996) contain a mass of information, and no doubt there will be many others.

However, the amount of information that is appearing is considerable, and one of the striking things about it is how rapidly measurable responses take place after bone cells in culture, or explants, or whole bones in vivo, are loaded. Table 11.5, modified from Skerry (2000), gives a sampler.

It is clear that mechanically tweaking bone cells in various ways produces immediate, or little-delayed, biological responses. However, understanding such responses is still a million miles away from understanding how the algorithms that must be working to produce adaptive responses are effected. It is going to be fascinating to watch, in the next decade, how these responses, which will certainly become increasingly well characterized, will be married to the complexity of the mechanically induced responses of whole bones, which are also being increasingly well characterized. There is still a huge gap.

11.12 Conclusion

The study of adaptive remodeling, both internal and external, is one of great activity at the moment. Most of the interesting experiments are difficult because they have considerable logistical problems. Imposing excessive loads is technically (and ethically) difficult, and small animals, such as mice, are not usually suitable. But the time scale of growth in bigger animals is quite large, and experiments often have to be carried on for months or even years. There is, furthermore, the difficulty that

TABLE 11.5
Cellular Responses in Bone to Loading

Time after loading	*Response*	*System*	*Authors*
100 ms	Rise in intracellular calcium	Cell culture	2
5 s	Protein kinase C activation	Cell culture	2
5 min	Matrix proteoglycan reorientation	Cell culture	6
		Explants	
5–15 min	Prostaglandin expression	Explants	5
Hours	Gene expression	Cell culture	1
		Explants	7
2 days	Osteoblast formation	Cell culture	4
	Matrix synthesis	Explants	3

Note. Authors: 1, Inaoka et al. (1995); 2, Jones and Bingmann (1991); 3, Lozupone et al. (1992); 4, Pead et al. (1988); 5, Rawlinson et al. (1991); 6, Skerry et al. (1990); 7, Zaman et al. (1992).

the operation to impose the loading pin, plate, or whatever is to be used to overload or underload the bone will itself tend to produce a reactive bone growth, and controls have to be extremely carefully thought out. As a result of these difficulties, it is perhaps not surprising that apparently similar experiments have quite dissimilar results. I have not emphasized these differences in this chapter because they are at the moment merely confusing, and we may possibly see a consensus soon. However, even leaving aside the Greek chorus of Bertram and Swartz (1991) (and we should *not* leave it aside, because they make many excellent points), a reading of the papers in Odgaard and Weinans (1995) will show that some considerable reconciliation must be brought about before such consensus emerges!

A final point must be emphasized: the interaction between the genetic endowment of the cells concerned with remodeling and the strain imposed on the bone must be complex. It must be complex because, in the mature skeleton, the kinds of stresses imposed on bones will differ from place to place. Some bones, such as vertebral centra, function to bear compressive loads. Some, such as long limb bones, function through resistance to bending. Other bones, like much of the vault of the human skull, normally bear very little stress but must be strong in an accident. Some bones, like long bones, must be straight as they bear bending loads, while still others, like the lower jaw, must be complex and curved to bear bending loads. To subsume such different relations between stress and function under any very simple law seems a futile activity.

Chapter Twelve

SUMMING UP

THERE IS NOTHING strange about bone. It may be difficult to understand; indeed, its hierarchical structure makes it particularly difficult to analyze. However, in its hierarchical structure lies its strength, literally, for it is this that makes it difficult for cracks to travel through. I am sure that, in not too many years' time, we shall have an understanding of how bone tissue's mechanical properties are determined by its structure.

We see in bone starkly the compromise that any structure, be it natural or produced by humans must confront. The compromise is not always the same, but it is always there. In the case of bone, the compromise is usually between stiffness and toughness, both qualities that any bone would, were it possible, carry to an extreme. But stiffness requires mineral, and mineral produces brittleness.

So, too, we often see in the design of whole bones answers to problems that an engineer might set the system, say, "Produce a structure of least weight that is stiff enough to do such and such." But any deeper analysis is extraordinarily complicated by the fact that we have no real idea of how costs, in molecules of ATP, in time to produce, in genetic complexity, in weight to carry around, and many other things, should be set off against the various mechanical properties we *can* measure.

Natural selection, which has over the last few hundreds of millions of years produced the bone material and bones that we see now, has been able to work only by choosing what happens to survive. This is an extraordinarily inefficient way to proceed, but that's all there is; there is no other way.

Probably for a biologist the most awkward thing about bone and bones is to work out what it, and they, are designed to do. A clinician, or sports scientist, even an engineer, may be able to take bone and bones as they are, and then decide what they are capable of doing and try to aid them in doing this. For a biologist, however, there is the ever-nagging question: Why this, why not something different? I end this book with the question of Rik Huiskes that I quoted many, many words ago: "If bone is the answer, then what is the question?"

REFERENCES

Abbott, B.C. and Wilkie, D.R. 1953. The relation between velocity of shortening and the tension–length curve of skeletal muscle. *Journal of zoology, London* **120**:214–23.

Ahmed, A.M. and Burke, D.L. 1983. In vitro measurements of static pressure distribution in synovial joints, part 1: tibial surface of the knee. *Journal of biomedical engineering* **105**:216–25.

Akhter, M.P., Cullen, D.M., Pedersen, E.A., Kimmel, D.B. and Reeker, R.R. 1998. Bone response to in vivo mechanical loading in two breeds of mice. *Calcified tissue international* **63**:442–49.

Akhter, M.P., Raab, D.M., Turner, C.H., Kimmel, D.B. and Recker, R.R. 1992. Characterization of in vivo strain in the rat tibia during external application of four-point bending load. *Journal of biomechanics* **25**:1241–46.

Akkus, O. and Rimnac, C.M. 2001 Cortical bone tissue resists fatigue fracture by deceleration and arrest of microcrack growth. *Journal of biomechanics* **34**:757–64.

Akkus, O., Jepsen, K.J. and Rimnac, C.M. 2000 Microstructural aspects of the fracture process in human cortical bone. *Journal of materials science* **35**:6065–74.

Alexander, R.McN. 1975. *Biomechanics*. London, Chapman & Hall.

———1977. Allometry of the limbs of antelopes (Bovidae). *Journal of zoology, London* **183**:125–46.

———1981. Factors of safety in the structure of mammals. *Science Progress* **67**:109–30.

———1982. *Locomotion of animals*. Glasgow, Blackie.

———1983. *Animal mechanics*, 2d ed. Oxford, UK, Blackwell Scientific.

———1984. Optimum strengths for bones liable to fatigue and accidental damage. *Journal of theoretical biology* **109**:621–6.

———1988. *Elastic mechanisms in animal movement*. Cambridge, UK, Cambridge University Press.

———1993. Optimization of structure and movement of the legs of animals. *Journal of biomechanics* **26** (Suppl.1):1–6.

———1997. A theory of mixed chains applied to safety factors in biological systems. *Journal of theoretical biology* **184**:247–52

———1998. Symmorphosis and safety factors. In: *Principles of animal design: the optimization and symmorphosis debate*. E.R. Weibel, C.R. Taylor, and L. Bolis (eds.). Cambridge, UK, Cambridge University Press. pp. 28–35.

Alexander, R.McN. and Bennet-Clark, H. C. 1977. Storage of elastic strain energy in muscle and other tissues. *Nature* **265**:114–17.

Alexander, R.McN and Bennett, M.B. 1987. Some principles of ligament function, with examples from the tarsal joints of the sheep (*Ovis aries*). *Journal of zoology, London* **211**:487–504.

Alexander, R.McN. and Dimery, N.J. 1985. The significance of sesamoids and retro-articular processes for the mechanics of joints. *Journal of zoology, London* **205**:357–71.

Alexander, R.McN. and Jayes, A.S 1978. Optimum walking techniques for idealized animals. *Journal of zoology, London* **186**:61–81.

Alexander, R.McN. and Pond, C.M. 1992. Locomotion and bone strength of the white rhinoceros, *Ceratotherium simum. Journal of zoology, London.* **227**:63–69.

Alexander, R.McN. and Vernon, A. 1975. The mechanics of hopping by kangaroos (Macropodidae). *Journal of zoology, London* **177**:265–303.

Alexander, R.McN., Maloiy, G.M.O., Hunter, B., Jayes, A.S. and Nturibi, I. 1979a. Mechanical stresses in fast locomotion of buffalo (*Sincerus caffer*) and elephant (*Loxodonta africana*). *Journal of zoology, London* **189**:135–44.

Alexander, R.McN., Jayes, A.S., Maloiy C.M.O. and Wathuta, E.M. 1979b. Allometry of the limb bones of mammals from shrews (*Sorex*) to elephant (*Loxodonta*). *Journal of zoology, London* **189**:305–14.

Alexander, R.McN., Brandwood, A., Currey, J.D. and Jayes, A.S. 1984. Symmetry and precision of control of strength in limb bones of birds. *Journal of zoology, London* **203**:135–43.

Alexander, R.McN., Ker, R.F. and Bennett, M.B. 1990. Optimum stiffness for leg bones. *Journal of zoology, London.* **222**:471–8.

Alexander, R.McN., Fariña, R.A. and Vizcaíno, S.F. 1999. Tail blow energy and carapace fractures in a large glyptodont (Mammalia, Xenarthra). *Zoological journal of the Linnean society* **126**:41–49.

Amprino, R. 1948. A contribution to the functional meaning of the substitution of primary by secondary bone tissue. *Acta anatomica* **5**:291–300.

An, Y.H. 2000. Mechanical properties of bone. In: An and Draughn (2000), pp. 41–63.

An, Y.H. and Draughn, R.A. (eds.) 2000. *The mechanical testing of bone and the bone–implant interface.* Boca Raton, FL, CRC Press.

Archer, C.W., Caterson, B., Benjamin, M. and Ralphs, J.R. 1999. *Biology of the synovial joint.* Amsterdam, Harwood.

Aronsson, D.D. and Loder, R.T. 1996. Treatment of the unstable (acute) slipped capital femoral epiphysis. *Clinical orthopaedics and related research* **322**:99–110.

Arramon, Y.P. and Cowin, S.C. 1997. Hydraulic strengthening of cancellous bone. *Forma* **12**:209–21.

Ascenzi, A. 1976. Physiological relationship and pathological interferences between bone tissue and marrow. In: *The biochemistry and physiology of bone*, Vol. 4. Bourne, G.H. (ed.). New York, Academic Press, pp. 403–44.

Ascenzi, A. and Bonucci, E. 1967. The tensile properties of single osteons. *Anatomical record* **158**:375–86.

———1968. The compressive properties of single osteons. *Anatomical record* **161**:377–88.

———1976. Mechanical similarities between alternate osteons and cross-ply laminates. *Journal of biomechanics* **9**:65–71.

Ascenzi, A., François, C. and Bocciarelli, D. S. 1963. On the bone induced by estrogens in birds. *Journal of ultrastructure research* **8**:491–505.

Ascenzi, A., Bonucci, E. and Bocciarelli, D.S. 1965. An electron microscope study of osteon calcification *Journal of ultrastructure research* **12**:287–303.

Ascenzi, A., Bonucci, E. and Bocciarelli, D.S. 1967. An electron microscope study on primary periosteal bone. *Journal of ultrastructure research* **18**:605–18.

Ascenzi, A., Bonucci, E., Ripamonti, A. and Roveri, N. 1978. X-ray diffraction and electron microscope study of osteons during calcification. *Calcified tissue research* **25**:133–43.

Ascenzi, A., Baschieri, P. and Benvenuti, A. 1990. The bending properties of single osteons. *Journal of biomechanics* **23**:763–71.

Ascenzi, A., Baschieri, P. and Benvenuti, A. 1994. The torsional properties of single selected osteons. *Journal of biomechanics* **27**:875–84.

Ascenzi, M-G., Benvenuti, A. and Ascenzi, A. 2000. Single osteon micromechanical testing. In: An and Draughan (2000), pp. 271–90.

Ashby, M.F. 1999. *Materials selection in mechanical design*. Oxford, UK, Butterworth Heinemann.

Ashman, R.B. and Rho, J.Y. 1988. Elastic modulus of trabecular bone material. *Journal of biomechanics* **21**:177–81.

Ashman, R.B., Cowin, S.C., Van Buskirk, W.C. and Rice, J.C. 1984. A continuous wave technique for the measurement of the elastic properties of cortical bone. *Journal of biomechanics* **17**:349–61.

Atkinson, P. J. and Woodhead, C. 1973. The development of osteoporosis: a hypothesis based on a study of human bone structure. *Clinical orthopaedics and related research* **90**:217–28.

Bai, T. Pollard, D.D. and Gao, H. 2000. Explanation for fracture spacing in layered materials. *Nature* **403**:753–56.

Banse, X., Delloye, C., Cornu, O. and Bourgois, R. 1996. Comparative left–right mechanical testing of cancellous bone from normal femoral heads. *Journal of biomechanics* **29**:1247–53.

Bassett, C.A.L. 1965. Electrical effects in bone. *Scientific American* **213** (October):18–25.

———1971. Biophysical principles affecting bone structure. In: *The Biochemistry and physiology of bone*, Vol. 3. G. H. Bourne (ed.). New York, Academic Press, pp. 1–76.

———1976. Comment on theoretical evidence for the generation of high pressure in bone cells by R. J. Jendrucko et al. *Journal of biomechanics* **9**:485.

Batson, E.L., Reilly, G.C., Currey, J.D. and Balderson, D.S. 2000. Postexercise and positional variation in mechanical properties of the radius in young horses. *Equine veterinary journal* **32**:95–100.

Beaupré, G.S. and Carter, D.R. 1992. Finite element analysis in biomechanics. In: Biewener (1992a), pp. 149–74.

Behiri, J. C. and Bonfield, W. 1980. Crack velocity dependence of longitudinal fracture in bone. *Journal of materials science* **15**:1841–49.

———1984. Fracture mechanics of bone—the effects of density, specimen thickness and crack velocity on longitudinal fracture. *Journal of biomechanics* **17**:25–34.

———1989. Orientation dependence of the fracture mechanics of bone. *Journal of biomechanics* **22**:863–72.

Bell, E.C. and Gosline, J.M. 1996. Mechanical design of mussel byssus: material yield enhances attachment strength. *Journal of experimental biology* **199**:1005–17.

Bell, K.L. Loveridge, N., Reeve, J., Thomas, D., Feik, S. and Clement, J. 2000. Cortical remodelling clusters (super-osteons) in the human femoral shaft. *Journal of bone and mineral research* Suppl.1 **15**:SA013.

Benjamin, M. and Ralphs, J.R. 1999. The attachment of tendons and ligaments to bone. In: Archer et al. (1999), pp. 361–371.

Benjamin, M. and Ralphs, J.R. 2000. The cell and developmental biology of tendons and ligaments. *International review of cytology: a survey of cell biology* **196**:85–130.

Benjamin, M., Qin, S. and Ralphs, J.R. 1995. Fibrocartilage associated with human tendons and their pulleys. *Journal of anatomy* **187**:625–33.

Bennet-Clark, H.C. 1975. The energetics of the jump of the locust *Schistocerca gregaria. Journal of experimental biology* **63**:53–83.

Bennett, M.B. and Stafford, J.A. 1988. Tensile properties of calcified and uncalcified avian tendons. *Journal of zoology, London* **214**:343–51.

Bennett, M.B., Ker, R.F., Dimery, N.J. and Alexander, R.McN. 1986. Mechanical properties of various mammalian tendons. *Journal of zoology, London* **209**:537–48.

Bentolila, V., Hillam, R.A., Skerry, T.M., Boyce, T.M., Fyrhie, D.P. and Schaffler, M.B. 1997. Activation of intracortical remodeling in adult rat long bones by fatigue loading. *43rd meeting, Orthopaedic Research Society, San Francisco*, p.578.

Berthet-Colominas, C., Miller, A. and White, S.W. 1979. Structural study of the calcifying collagen in turkey leg tendons. *Journal of molecular biology* **134**:431–445.

Bertram, J.E.A. and Biewener, A.A. 1988. Bone curvature: sacrificing strength for load predictability? *Journal of theoretical biology* **131**:75–92.

Bertram, J.E. and Swartz, S.M. 1991. The "law of bone transformation": a case of crying Wolff? *Biological reviews* **66**:245–73.

Biewener, A.A. 1982. Bone strength in small mammals and bipedal birds: do safety factors change with body size? *Journal of experimental biology* **98**:289–301.

———1983. Allometry of quadrupedal locomotion: the scaling of duty factor, bone curvature and limb orientation to body size. *Journal of experimental biology* **105**:147–71.

———(ed.). 1992a. *Biomechanics (structures and systems) a practical approach.* Oxford, UK, IRL Press.

———1992b. In vivo measurement of bone strain and tendon force. In: Biewener (1992a), pp. 124–47.

———1993. Safety factors in bone strength. *Calcified tissue international* **53**:(Suppl I) S68–S74.

Biewener, A.A. and Dial, K.P. 1992. In vivo strain in the pigeon humerus during flight. *American journal of zoology* **32**:155A.

Biewener, A.A. and Taylor, C.R. 1986. Bone strain: a determinant of gait and speed? *Journal of experimental biology* **123**:383–400.

Biewener, A.A., Thomason, J., Goodship, A. and Lanyon, L.E. 1983a. Bone stress in the horse forelimb during locomotion at different gaits: a comparison of two experimental methods. *Journal of biomechanics* **16**:565–76.

Biewener, A.A., Thomason, J. and Lanyon L.E. 1983b. Mechanics of locomotion and jumping in the forelimb of the horse (*Equus*): in vivo stress developed in the radius and metacarpus. *Journal of zoology, London.* **201**:67–82.

Biewener, A.A., Swartz, S.M. and Bertram, J.E.A. 1986. Bone modeling during growth: dynamic strain equilibrium in the chick tibiotarsus. *Calcified tissue international* **39**:390–95.

Biewener, A.A., Thomason, J.J. and Lanyon L.E. 1988. Mechanics of locomotion and jumping in the horse (*Equus*): in vivo stress in the tibia and metatarsus. *Journal of zoology, London* **214**:547–65.

Biewener, A.A., Fazzalari, N.L., Konieczynski, D.D. and Baudinette, R.V. 1996. Adaptive changes in trabecular architecture in relation to functional strain patterns and disuse. *Bone* **19**:1–8.

Bigot, G., Bouzidi, A., Rumelhart. C. and MartinRosset, W. 1996. Evolution during growth of the mechanical properties of the cortical bone in equine cannon-bones. *Medical engineering and physics* **18**:79–87.

Bilezikian, J.P., Raisz, L.G. and Rodan, G.A. 1996. *Principles of bone biology.* San Diego, CA, Academic Press.

Blackburn, J., Hodgskinson, R., Currey, J.D. and Mason, J.E. 1992. Mechanical properties of microcallus in human cancellous bone. *Journal of orthopaedic research* **10**:237–46.

Bledsoe, A.H., Raikow, R.J. and Glasgow, A.G. 1993. Evolution and functional significance of tendon ossification in woodcreepers (Aves: Passeriformes: Dendrocalaptinae). *Journal of morphology* **215**:289–300.

Blob, R.W. and Biewener, A.A. 1999. In vivo locomotor strain in the hindlimb bones of *Alligator mississipiensis* and *Iguana iguana*: implications for the evolution of limb bone safety factor and non-sprawling limb posture. *Journal of experimental biology* **202**:1023–46.

Bompas-Smith, J.H. 1973. *Mechanical survival: the use of reliability data.* London, McGraw-Hill.

Bonar, L.C., Lees, S. and Mook, H.A. 1985. Neutron diffraction studies of collagen in fully mineralised bone. *Journal of molecular biology* **181**:265–70.

Bonfield, W. 1987. Advances in the fracture mechanics of cortical bone. *Journal of biomechanics* **20**:1071–81.

Bonfield, W. and Behiri, J.C. 1983. Fracture mechanics of bone—evaluation by the compact tension method. In: *Biomedical engineering, I: Recent developments.* S. Saha (ed.). Elmsford, NY, Pergamon, pp. 343–47.

Bonfield, W. and Datta, P.K. 1976. Fracture toughness of compact bone. *Journal of biomechanics* **9**:131–34.

Bonfield, W. and Grynpas, M.D. 1977. Anisotropy of Young's modulus of bone. *Nature* **270**:453–54.

Bonfield, W., Grynpas, M.D. and Young, R.J. 1978. Crack velocity and the fracture of bone. *Journal of biomechanics* **11**:473–79.

Bonfield, W., Behiri, J.C. and Charambilides, B. 1984. Orientation and age-related dependence of the fracture toughness of cortical bone. *Biomechanics: current interdisciplinary research*. S.M.Perrin and E. Schneider (ed.). Dordrecht, Martinus Nijhoff, pp. 185–189.

Bonucci, E. and Gherardi, G. 1975. Histochemical and electron microscope investigations on medullary bone. *Cell and tissue research* **163**:81–97.

Borden IV, S. 1974. Traumatic bowing of the forearm in children. *Journal of bone and joint surgery* **56-A**:611–6.

Boppart, M.D., Kimmel, D.B., Yee, J.A. and Cullen, D.M. 1998. Time course of osteoblast appearance after in vivo mechanical loading. *Bone* **23**:409–15.

Boskey, A.L. 2001. Bone mineralization. In: Cowin (2001a), Chapter 5.

Bosshardt, D.D. and Selvig, K.A. 1997. Dental cementum: the dynamic tissue covering of the root. *Periodontology 2000* **13**:41–75.

Bouvier, M. and Hylander, W.L. 1981. Effect of bone strain on cortical bone structure in macaques (*Macaca mulatta*). *Journal of morphology* **167**:1–12.

Bowen, R.L. and Rodriguez, M.S. 1962. Tensile strength and modulus of elasticity of tooth structure and several restorative materials. *Journal of the American dental association* **64**:378–87.

Bowman, S.M., Keaveny, T.M., Gibson, L.J., Hayes, W.C. and McMahon, T.A. 1994. Compressive creep behavior of bovine trabecular bone. *Journal of biomechanics* **27**:301–10.

Boyde, A. 1972. Scanning electron microscope studies of bone. In: *The biochemistry and physiology of bone*, Vol. 1. G. H. Bourne (ed.). New York, Academic Press, pp. 259–310.

———1980. Electron microscopy of the mineralizing front. *Metabolic bone disease and related research* **2** (Suppl.): 69–78.

———1989. Enamel. In: *Teeth*. A. Oksche and L. Vollrath (eds.). Berlin, Springer Verlag. pp. 309–473.

Boyde, A. and Hobdell, M. H. 1969. Scanning electron microscopy of lamellar bone. *Zeitschrift für Zellforschung* **93**:213–31.

Boyde, A. and Jones, S. 1998. Aspects of anatomy and development of bone: the nm, μm and mm hierarchy. *Advances in organ biology* **5A**:3–44.

Boyde, A., Haroon, Y., Jones, S.J. and Riggs, C.M. 1999. Three dimensional structure of the distal condyles of the third metacarpal bone of the horse. *Equine veterinary journal* **31**:122–29.

Bramblett, C.A. 1967. Pathology of the Darajani baboon. *American journal of physical anthropology* **26**:331–40.

Bramwell, C.D., and Whitfield, C.R. 1974. Biomechanics of *Pteranodon*. *Philosophical transactions of the Royal Society of London* **267–B**:503–81.

Brandwood, A., Jayes, A.S. and Alexander, R.McN. 1986. Incidence of healed fracture in the skeletons of birds, molluscs and primates *Journal of zoology, London* **208**:55–62.

Brazier, L.G. 1927. On the flexure of thin cylindrical shells and other 'thin' sections. *Proceedings of the Royal Society* **116–A**:104–14.

Brear, K., Currey, J.D, and Pond, C.M. 1990a. Ontogenetic changes in the mechanical properties of the femur of the polar bear *Ursus maritimus*. *Journal of zoology, London* **222**:49–58.

Brear, K., Currey, J.D., Pond, C.M. and Ramsay, M.A. 1990b. The mechanical properties of the dentine and cement of the tusk of the narwhal *Monodon monoceros* compared with those of other mineralised tissues. *Archives of oral biology* 35:615–21.

Brear, K., Currey, J.D., Kingsley, M.C.S. and Ramsey M. 1993. The mechanical design of the tusk of the narwhal (*Monodon monoceros*: Cetacea). *Journal of zoology, London* **230**:411–23.

Bright, R.W., Burstein, A.H. and Elmore, S.M. 1974. Epiphyseal-plate cartilage: a biomechanical and histological analysis of failure modes. *Journal of bone and joint surgery* **56–A**:688–703.

Brodt, M.D., Ellis, C. and Silva, M.J. 1999. Growing C57B1/6 mice increase whole bone mechanical properties by increasing geometric and material properties. *Journal of bone and mineral research* **14**:2159–66.

Brookes, M. and Revell, W.J. 1998. *Blood supply of bone: scientific aspects.* London, Springer Verlag.

Brukner, P., Bradshaw, C., Khan, K.M., White, S. and Crossley, K. 1996. Stress fractures: a review of 180 cases. *Clinical journal of sport medicine.* **6**:85–89.

Bryant, J. D. 1983. The effect of impact on the marrow pressure of long bones in vitro. *Journal of biomechanics* **16**:659–65.

Bryant, J.D. 1988. On the mechanical function of marrow in long bones. *Engineering in medicine* **17**:55–58.

Bryant, J.D. 1995. Hydraulic strengthening of bones. *Journal of biomechanics* **28**:353–54.

Bühler, P. 1972. Sandwich structures in the skull capsules of various birds—the principle of lightweight structures in organisms. *Information of the Institute for Lightweight Structures, Stuttgart* **4**:39–50.

Bühler, P. 1992. Light bones in birds. *Los Angeles County Museum of Natural History, science series* **36**:385–93.

Buikstra, J.E. 1975. Healed fractures in *Macaca mulatta*: age, sex, and symmetry. *Folia primatologica* **23**:140–48.

Burgin, L.V. and Aspden, R.M. 2000. Force transmission and dissipation in articular cartilage and bone following an impact load. In: *Proceedings of the 12th Conference of Biomechanics.* P.J. Prendergast, T.C. Lee, and A.J. Carr (eds.). Dublin, Royal Academy of Medicine in Ireland, p. 418.

Burr, D.B. 1993. Remodeling and the repair of fatigue damage. *Calcified tissue international* **53** (Suppl 1):S75–81.

Burr, D.B. and Milgrom, C. (eds.) 2001. *Musculoskeletal fatigue and stress fractures.* Boca Raton, FL, CRC Press.

Burr, D.B. and Stafford, T. 1990. Validity of the bulk-staining technique to separate artifactual from in vivo bone microdamage *Clinical orthopaedics and related research* **260**:305–8.

Burr, D.B., Martin, R.B., Schaffler, M.B. and Radin, E.L. 1985. Bone remodeling in response to in vivo fatigue microdamage. *Journal of biomechanics* **18**:189–200.

Burr, D.B., Milgrom, C., Fyrhie, D., Forwood, M., Nyska, M., Finestone, A., Hoshaw, S., Saiag, E. and Simkin, A. 1996. In vivo measurement of human tibial strains during vigorous activity. *Bone* **18**:405–10.

Burstein, A.H. Currey, J.D., Frankel, V.H. and Reilly, D.T. 1972. The ultimate properties of bone tissue: the effects of yielding. *Journal of biomechanics* 5:35–44.

Burton, P., Nyssen-Behets, C. and Dhem, A. 1989. Haversian bone remodeling in the human fetus. *Acta anatomica* **135**:171–75.

Byock, J.L. 1995. Egil's bones. *Scientific American* **272** (1):62–67.

Caillot-Augusseau, A., Lafage-Proust, M.H., Soler, C., Pernod, J., Dubois, F. and Alexandre, C. 1998. Bone formation and resorption biological markers in cosmonauts during and after a 180-day space flight. (Euromir 95). *Clinical chemistry* **44**:578–85.

Caler, W.E. and Carter, D.R. 1989. Bone creep-fatigue damage accumulation. *Journal of biomechanics*. **22**:625–35.

Calow, L. I. and Alexander, R. M. 1973. A mechanical analysis of a hind leg of a frog (*Rana temporaria*). *Journal of zoology, London* **171**:293–321.

Carrier, D. and Leon, L.R. 1990. Skeletal growth and function in the California gull (*Larus californicus*). *Journal of zoology, London* **222**:375–89.

Carter, D.R. and Beaupré, G.S. 2001. *Skeletal function and form*. Cambridge, UK, Cambridge University Press.

Carter, D.R. and Caler, W.E. 1983. Cycle-dependent and time-dependent bone fracture with repeated loading. *Journal of biomechanical engineering* **105**:166–70.

———1985. A cumulative damage model for bone fracture. *Journal of orthopaedic research*. **3**:84–90.

Carter, D.R. and Hayes, W.C. 1976a. Fatigue life of compact bone, I: effects of stress amplitude, temperature and density. *Journal of biomechanics* **9**:27–34.

———1976b. Bone compressive strength: the influence of density and strain rate. *Science* **194**:1174–76.

———1977a. The compressive behavior of bone as a two-phase porous structure. *Journal of bone and joint surgery* **59–A**:954–62.

———1977b. Compact bone fatigue damage, I: residual strength and stiffness. *Journal of biomechanics* **10**:325–37.

———1977c. Compact bone fatigue damage—a microscopic examination. *Clinical orthopaedics and related research* **127**:265–74.

Carter, D.R., Hayes, W.C. and Schurman, D.J. 1976. Fatigue life of compact bone, 2: effects of microstructure and density. *Journal of biomechanics* **9**:211–18.

Carter, D.R., Schwab, G.H. and Spengler, D.M. 1980. Tensile fracture of cancellous bone. *Acta orthopaedica Scandinavica* **51**:733–41.

Carter, D.R., Caler, W.E., Spengler, D.M. and Frankel, V.H. 1981a. Fatigue behavior of adult cortical bone: the influence of mean strain and strain range. *Acta orthopaedica Scandinavica* **52**:481–90.

———1981b. Uniaxial fatigue of human cortical bone: the influence of tissue physical characteristics. *Journal of biomechanics* **14**:461–70.

Carter, D.R., Harris, W.H., Vasu, R. and Caler, W.E. 1981c. In: Cowin (1981), pp. 81–92.

Carter, J.G. 1990. *Skeletal biomineralization: patterns, processes and evolutionary trends*. New York Van Nostrand Reinhold, Vol. 1, pp. 1–832; Vol. 2, pp. 1–99.

Cassidy, J.J., Hiltner, A. and Baer, E. 1989. Hierarchical structure of the intervertebral disc. *Connective tissue research* 23:75–88.

Castanet, J., Curry Rogers, K, Cubo, J. and Boisard, J-J. 2000. Periosteal bone growth rates in extant ratites (ostriche [sic] and emu): implications for assessing growth in dinosaurs. *Comptes Rendus de l'Academie des Sciences, Serie III, Sciences de la Vie* 323:543–50.

Castanet, J., Grandin, A., Abourachid, A. and De Riqlès, A. 1996. Expression de la dynamique de croissance dans la structure de l'os périostique chez *Anas platyrhyncos. Comptes rendus de l'Academie de Sciences, Paris, Sciences de la vie.* 319:301–8.

Cavagna, G.A., Saibene, F.P. and Margaria, R. 1964. Mechanical work in running. *Journal of applied physiology* 19:249–56.

Cezayirlioglu, H., Bahniuk, E., Davy, D.T. and Heiple, K.G. 1985. Anisotropic yield behavior of bone under combined axial force and torque. *Journal of biomechanics* 18:61–69.

Chamay, A. 1970. Mechanical and morphological aspects of experimental overload and fatigue in bone. *Journal of biomechanics* 3:263–370.

Chen, N.X., Ryder, K.D., Pavalko, F.M., Turner, C.H., Burr, D.B. Qiu, J.Y. and Duncan, R.L. 2000. Ca^{2+} regulates fluid shear-induced cytoskeletal reorganization and gene expression in osteoblasts. *American journal of physiology–cell physiology.* 278:C989–97.

Chiappe, L.M. and Chinsamy, A. 1996. *Pterodaustro*'s [sic] true teeth. *Nature* 379:211–12.

Choi, K. and Goldstein, S.A. 1992. A comparison of the fatigue behavior of human trabecular and cortical bone tissue. *Journal of biomechanics* 25:1371–81.

Choi, K., Kuhn, J.L., Ciarelli, M.J. and Goldstein, S.A. 1990. The elastic moduli of human subchondral, trabecular, and cortical bone tissue and the size-dependency of cortical bone modulus. *Journal of biomechanics* 23:1103–13.

Churches, A. E. and Howlett, C. R. 1981. The response of mature cortical bone to controlled time-varying loading. In: Cowin (1981), pp. 69–80.

Claes, L.E., Wilke, H-J. and Kiefer, H. 1995. Osteonal structure better predicts tensile strength of healing bone than volume fraction. *Journal of biomechanics* 28:1377–90.

Clark, J. and Stechschulte, D.J, Jr. 1998. The interface between bone and tendon at an insertion site: a study of the quadriceps tendon insertion. *Journal of anatomy* 193:605–16.

Clutton-Brock, T.H., Albon, S.D., Gibson, R.M. and Guinness, F.E. 1979. The logical stag: adaptive aspects of fighting in red deer (*Cervus elaphus* L.). *Animal behaviour* 27:211–25.

Cochran, G.V.B., Pawluk, R.I. and Bassett, C.A.L. 1968. Electromechanical characteristics of bone under physiologic moisture conditions. *Clinical orthopaedics and related research* 58:249–70.

Collet, P. Uebelhart, D., Vico, L., Moro, L. Hartmann, D. Roth, M. and Alexandre, C. 1997. Effects of 1- and 6-month spaceflight on bone mass and biochemistry in two humans. *Bone* 20:547–51.

Cooper, R.R. and Misol, S. 1970. Tendon and ligament insertion: a light and electron microscopic study. *Journal of bone and joint surgery* 52-A:1–20.

Cooper, R.R., Milgram, J.W. and Robinson, R.A. 1966. Morphology of the osteon: an electron microscopic study. *Journal of bone and joint surgery* **48-A**:1239–71.

Cowin, S.C. (ed.) 1981. *Mechanical properties of bone*, AMD, Vol. 45. New York, American Society of Mechanical Engineers.

———1997. The false premise of Wolff's law. *Forma* **12**:247–62.

———(ed.) 2001a. *Bone mechanics handbook*, 2d ed. Boca Raton, FL, CRC Press.

———2001b. Mechanics of materials. In: Cowin (2001a), Chapter 6.

———2001c. The false premise of Wolff's law. In: Cowin (2001a), Chapter 30.

Cowin, S.C., Hart, R.T. and Balser, J.R. 1985. Functional adaptation in long bones: establishing in vivo values for surface remodeling rate coefficients. *Journal of biomechanics* **18**:665–84.

Cowin, S.C., Weinbaum, S. and Zeng, Y. 1995. A case for bone canaliculi as the anatomical site of strain generated potentials. *Journal of biomechanics* **28**:1281–97.

Craig, R.G., Peyton, F.A. and Johnson, D.W. 1961. Compressive properties of enamel, dental cements and gold. *Journal of dental research* **40**:936–40.

Crofts, R.D., Boyce, T.M. and Bloebaum, R.D. 1994. Aging changes in osteon mineralization in the human femoral neck. *Bone* **15**:147–52.

Cubo, J. and Casinos, A. 2000. Incidence and mechanical significance of pneumatization in the long bones of birds. *Zoological journal of the Linnean society* **130**:499–510.

Currey, J. D. 1959. Differences in the tensile strength of bone of different histological types. *Journal of anatomy* **93**:87–95.

———1960. Differences in the blood-supply of bone of different histological types. *Quarterly journal of microscopical science* **101**:351–70.

———1962a. The histology of the bone of a prosauropod dinosaur. *Palaeontology* **5**:238–46.

———1962b. Stress concentrations in bone. *Quarterly journal of microscopical science* **103**:111–33.

———1964a. Three analogies to explain the mechanical properties of bone. *Biorheology* **2**:1–10.

———1964b. Metabolic starvation as a factor in bone reconstruction. *Acta anatomica* **59**:77–83.

———1965. Anelasticity in bone and echinoderm skeletons. *Journal of experimental biology* **43**:279–92.

———1967. The failure of exoskeletons and endoskeletons. *Journal of morphology* **123**:1–16.

———1968. Adaptation of bones to stress. *Journal of theoretical biology* **20**:91–106.

———1969. The mechanical consequences of variation in the mineral content of bone. *Journal of biomechanics* **2**:1–11.

———1975. The effect of strain rate, reconstruction and mineral content on some mechanical properties of bovine bone. *Journal of biomechanics* **8**:81–86.

———1979a. Mechanical properties of bone with greatly differing functions. *Journal of biomechanics* **12**:313–19.

———1979b. Changes in the impact energy absorption of bone with age. *Journal of biomechanics* **12**:459–69.

———1984. Can strains give adequate information for adaptive bone remodeling? *Calcified tissue international* **36**:118–22.

———1987. The evolution of the mechanical properties of amniote bone. *Journal of biomechanics* **20**:1035–44.

———1988. Strain rate and mineral content in fracture models of bone. *Journal of orthopaedic research* **6**:32–38.

———1989. Strain rate dependence of the mechanical properties of reindeer antler and the cumulative damage model of bone fracture *Journal of biomechanics* **22**:469–75.

———1990. Physical characteristics affecting the tensile failure properties of compact bone. *Journal of biomechanics* **23**:837–44.

———1999. What determines the bending strength of compact bone? *Journal of experimental biology* **202**:2495–503.

Currey, J.D. and Alexander, R.McN. 1985. The thickness of the walls of tubular bones. *Journal of zoology, London* **206A**:453–68.

Currey, J.D. and Brear, K. 1974. Tensile yield in bone. *Calcified tissue research* **15**:173–79.

Currey, J.D. and Butler, G. 1975. The mechanical properties of bone tissue in children. *Journal of bone and joint surgery* **57–A**:810–14.

Currey, J.D. and Kohn A.J. 1976. Fracture in the crossed-lamellar structure of *Conus* shells. *Journal of materials science* **11**:1615–23.

Currey, J.D. and Pond, C.M. 1989. Mechanical properties of very young bone in the axis deer (*Axis axis*) and humans. *Journal of zoology, London* **218**:59–67.

Currey, J.D., Brear, K. and Zioupos, P. 1994. Dependence of mechanical properties on fibre angle in narwhal tusk, a highly oriented biological composite. *Journal of biomechanics* **27**:885–97.

Curtis, T.A., Ashgrafi, S.H. and Weber, D.F. 1985. Canalicular communication in the cortices of human long bones. *Anatomical record* **212**:336–44.

Dacke, C.G., Arkle, S., Cook, D.J., Wormstone, I.M., Jones, S., Zaidi, M. and Bascal, Z.A. 1993. Medullary bone and avian calcium regulation. *Journal of experimental biology* **184**:63–88.

Dalén, N. and Olsson, K.E. 1974. Bone mineral content and physical activity. *Acta orthopaedica Scandinavica* **45**:170–74.

Dalstra, M., Huiskes, R., Odgaard, A. and van Erning, L. 1993. Mechanical and textural properties of pelvic trabecular bone. *Journal of biomechanics* **26**:523–25.

Davies, P.F. 1995. Flow-mediated endothelial mechanotransduction. *Physiological reviews* **75**:519–60.

Day, W.H., Swanson, S.A.V. and Freeman, M.A.R. 1975. Contact pressures in the loaded human cadaver hip. *Journal of bone and joint surgery* **57–A**:302–13.

de Buffrénil, V. and Francillon-Vieillot, H. 2001. Ontogenetic changes in bone compactness in male and female Nile monitors (*Varanus niloticus*). *Journal of zoology* **254**:539–46.

de Buffrénil, V. and Mazin, J-M. 1990. Bone histology of the ichthyosaurs: comparative data and functional interpretation. *Paleobiology* **16**:435–47.

de Buffrénil, V. and Schoevaert, D. 1988. On how the periosteal bone of the delphinid humerus becomes cancellous: ontogeny of a histological specialization. *Journal of morphology* **198**:149–64.

de Ricqlès, A. 1977. Recherches paléohistologiques sur les os longs des tétrapodes VII (deuxième partie, fin). *Annales de paléontologie* **63**:133–60.

de Ricqlès, A., Padian, K., Horner, J.R. and Francillon-Vieillot, H. 2000. Palaeohistology of the bones of pterosaurs (Reptilia: Archosauria): anatomy, ontogeny and biomechanical implications. *Zoological journal of the Linnean society* **129**:349–85.

De Santis, R., Anderson, P., Tanner, K.E., Ambrosio, L., Nicolais, L., Bonfield, W. and Davis, G.R. 2000. Bone fracture analysis on the short rod chevron-notch specimens using the X-ray computer micro-tomography. *Journal of materials science: materials in medicine.* **11**:629–36.

Devas, M. 1975. *Stress fractures*. Edinburgh, Churchill Livingstone.

Dimery, N.J., Alexander, R.M. and Deyst, K.A. 1985. Mechanics of the ligamentum nuchae of some artiodactyls *Journal of zoology, London* **206**:341–51.

Dingerkus, G., Séret, B. and Guilbert, E. 1991. Multiple prismatic calcium phosphate layers in the jaws of present-day sharks (Chondrichthyes; Selachii) *Experientia* **47**:38–40.

Dodds, R.A., Emery, R.J.H., Klenerman, L., Chayen, J. and Bitensky, L. 1989. Comparative metabolic enzymatic activity in trabecular as against cortical osteoblasts. *Bone* **10**:251–54.

Domning, D.P. and de Buffrénil, V. 1991. Hydrostasis in the Sirenia: quantitative data and functional interpretations. *Marine mammal science* **7**:331–68.

Duncan, J.L. 1992. Strain measurement by thermoelastic emission. In: Miles and Tanner (1992), pp. 156–68.

Eckstein, F., Löhe, F., Hillebrand, S., Bergmann, M., Schulte, E., Milz, S. and Putz, R. 1995a. Morphomechanics of the humero-ulnar joint, I: joint space width and contact areas as a function of load and flexion angle. *Anatomical record* **243**:318–26.

Eckstein, F., Merz, B., Müller-Gerbl, M., Holznecht, N., Pleier, M. and Putz, R. 1995b. Morphomechanics of the humero-ulnar joint, II: concave incongruity determines the distribution of load and subchondral mineralization. *Anatomical record* **243**:327–35.

Einhorn, T.A. 1995. Enhancement of fracture-healing. *Journal of bone and joint surgery* **77-A** 940–56.

Ekanayake, S. and Hall, B.K. 1988. Ultrastructure of the osteogenesis of acellular vertebral bone in the Japanese medaka, *Oryzias latipes* (Teleostei, Cyprinodontidae). *American journal of anatomy* **182**:241–49.

Elliott, D.H. 1965. Structure and function of mammalian tendon. *Biological reviews* **40**:392–421.

Ellis, M.I., Seedhom, B.B., Wright, V. and Dowson, D. 1980. An evaluation of the ratio between the tensions along the quadriceps tendon and the patellar ligament. *Engineering in medicine* **9**:189–94.

Enlow, D.H. 1963. *Principles of bone remodeling*. Springfield, IL, Charles C. Thomas.

———1969. The bone of reptiles. In: *Biology of the reptilia*, Vol. 1. C. Gans (ed.). New York, Academic Press. pp. 45–80.

———1975. *A handbook of facial growth*. Philadelphia, W. B. Saunders.

Enlow, D.H. and Brown, S.O. A comparative histological study of fossil and recent bone tissues. *Texas journal of science* Part I 1956 **8**:405–43; Part II 1957 **9**:186–214; Part III 1958 **10**:187–230.

Erts, D., Gathercole, L.J. and Atkins, E.D.T. 1994. Scanning probe microscopy of intrafibrillar crystallites in calcified collagen. *Journal of materials science: materials in medicine* **5**:200–206.

Estberg, L., Gardner, I.A., Stover, S.M., Johnson, B.J., Case, J.T. and Ardans, A. 1995. Cumulative racing-speed exercise distance cluster as a risk factor for fatal musculoskeletal injury in thoroughbred racehorses in California. *Preventive veterinary medicine* **24**:253–63.

Evans, E.J., Benjamin, M. and Pemberton, D.J. 1990. Fibrocartilage in the attachment zones of the quadriceps tendon and patellar ligament of man. *Journal of anatomy* **171**:155–62

Evans, F.G. 1973. *Mechanical properties of bone*. Springfield, IL, Charles C. Thomas.

Evans, F.G. and Bang, S. 1967. Differences and relationships between the physical properties and the microscopic structure of human femoral, tibial and fibular cortical bone. *American journal of anatomy* **120**:78–88.

Evans, F.G. and Lebow, M. 1957. Strength of human compact bone under repetitive loading. *Journal of applied physiology* **10**:127–30.

Evans, P.A., Farnell, R.D., Moalypour, S. and McKeever, J.A. 1995. Thrower's fracture: a comparison of two presentations of a rare fracture. *Journal of accident and emergency medicine* **12**:222–24.

Fazzalari, N.L., Forwood, M.R., Manthey, B.A., Smith, K. and Kolesik, P. 1998. Three-dimensional confocal images of microdamage in cancellous bone. *Bone* **23**:373–78.

Fedak, M.A., Heglund, N.C. and Taylor, C.R. 1982. Energetics and mechanics of terrestrial locomotion, 2: kinetic energy changes of the limbs and body as a function of speed and body size in birds and mammals. *Journal of experimental biology* **79**:23–40.

Fenton, M.B., Waterman, J.M., Roth, J.D., Lopez, E. and Fienberg, S.E. 1998. Tooth breakage and diet: a comparison of bats and carnivorans. *Journal of zoology, London* **246**:83–88.

Fiala, P and Heřt, J. 1993. Principal types of functional architecture of cancellous bone in man. *Functional and developmental morphology* **3**:91–99.

Fink, R.J. and Corn, R.C. 1982. Fracture of an ossified Achilles tendon. *Clinical orthopaedics and related research* **169**:148–50.

Floyd, T., Nelson, R.A. and Wynne, G.F. 1990. Calcium and bone metabolic homeostasis in active and denning black bears (*Ursus americanus*). *Clinical orthopaedics and related research*. **255**:301–9.

Fondrk, M., Bahniuk, E., Davy, D.T. and Michaels, C. 1988. Some viscoplastic characteristics of bovine and human cortical bone. *Journal of biomechanics* **21**:623–30.

Fondrk, M.T., Bahniuk, E.H. and Davy, D.T. 1999a. A damage model for nonlinear tensile behavior of cortical bone. *Journal of biomechanical engineering—transactions of the ASME* **121**:533–41.

———1999b. Inelastic strain accumulation in cortical bone during rapid transient tensile loading. *Journal of biomechanical engineering—transactions of the ASME* **121**:616–21.

Fong, H., Sarikaya, M., White, S.N. and Snead, M.L. 2000. Nano-mechanical properties profiles across dentin–enamel junction of human incisor teeth. *Materials science and engineering C Biomimetic and supramolecular systems* **7**:119–28.

Forwood, M.R. 1996. Inducible cyclo-oxygenase (COX-2) mediates the induction of bone formation by mechanical loading *in vivo*. *Journal of bone and mineral research* **11**:1688–93.

Forwood, M.R., Bennett, M.B., Blowers, A.R. and Nadorfi, R.L. 1998. Modification of the in vivo four-point loading model for studying mechanically induced bone adaptation. *Bone* **23**:307–10.

Francillon-Vieillot, H., de Buffrénil, V., Castanet, J., Géraudie, J., Meunier, F.J., Sire, J.Y., Zylberberg, L. and de Ricqlès, A. 1990. Microstructure and mineralization of vertebrate skeletal tissues. In: Carter (1990) Vol. 1, pp. 471–530.

Frank, R.M. and Nalbandian, J. 1989. Structure and ultrastructure of dentine. In: *Teeth*. A. Oksche and L. Vollrath (eds.). Berlin, Springer Verlag, pp. 173–247.

Frasca, P. 1981. Scanning-electron microscopy studies of "ground substance" in the cement lines, resting lines, hypercalcified rings and reversal lines of human cortical bone. *Acta anatomica* **109**:114–21.

Frasca, P., Harper, R. and Katz, J. L. 1977. Collagen fiber orientations of human secondary osteons. *Acta anatomica* **98**:1–13.

———1981. Strain and frequency-dependence of shear storage modulus for human single osteons and cortical bone microsamples—size and hydration effects. *Journal of biomechanics* **14**:679–90.

Fraser, F.C. and Purves, P.E. 1960. Hearing in cetaceans: evolution of the accessory air sacs and the structure and function of the outer and middle ear in recent cetaceans. *Bulletin of the British museum of natural history* 7:1–140.

Fratzl, P., Groschner, M., Vogl, G., Plenk, H., Eschberger, J., Fratzl-Zelman, N., Koller, K. and Klaushofer, K. 1992. Mineral crystals in calcified tissues: a comparative study by SAXS. *Journal of bone and mineral research* 7:329–34.

Freeman, M.A.R. and Kempson, G.E. 1973. Load carriage. In: *Adult articular cartilage*. M.A.R. Freeman (ed.). London, Pitman Medical, pp. 228–46.

Frey, E., Sues, H.-D. and Munk, W. 1997. Gliding mechanism in the late Permian reptile *Coelurosauravus*. *Science* **275**:1450–52.

Frisch, T., Overgaard, S., Sørensen, M.S. and Bretlau, P. 2000. Estimation of volume referent bone turnover in the otic capsule after sequential point labeling. *Annals of otology, rhinology and laryngology* **109**:33–9.

Frost, H.M. 1960. Presence of microscopic cracks in vivo in bone. *Henry Ford Hospital bulletin* 8:25–35.

———1964. *The laws of bone structure*. Springfield, IL, Charles C. Thomas.

———1973. *Bone modeling and skeletal modeling errors*. Springfield, IL, Charles C. Thomas.

———1985. The pathomechanics of osteoporoses. *Clinical orthopaedics and related research* 200:198–225.

———1987. Bone "mass" and the "mechanostat": a proposal. *Anatomical record* **219**:1–9.

———2000. Toward a mathematical description of bone biology: the principle of cellular accommodation. *Calcified tissue international* **67**:184–85.

Fukuda, S. 1988. Calcification and fracture of costal cartilage in beagles. *Japanese journal of veterinary science* **50**:1009–16.

Fukuda, Y., Takai, S., Yoshino, N., Murase, K., Tsutsumi, S., Ikeuchi, K. and Hirasawa, Y. 2000. Impact load transmission of the knee joint—influence of leg alignment and the role of meniscus and articular cartilage. *Clinical biomechanics* **15**:516–21.

Fung, Y.C. 1993. *Biomechanics: mechanical properties of living tissues* New York, Springer Verlag.

Fyhrie, D.P. and Schaffler, M.B. 1994. Failure mechanisms in human vertebral cancellous bone. *Bone* **15**:105–9.

Galante, J., Rostoker, W. and Ray, R. D. 1970. Physical properties of trabecular bone. *Calcified tissue research* **5**:236–46.

Galton, P.M. 1971. A primitive dome-headed dinosaur (Ornithischia: Pachycephalosauridae) from the lower Cretaceous of England, and the function of the dome in the pachycehphalosaurids. *Journal of paleontology* **45**:40–47.

Gambaryan, P.P. 1974. *How mammals run: anatomical adaptations*. New York, Halsted Press.

Gans, C. 1974. *Biomechanics: an approach to vertebrate biology*. Philadelphia, Lippincott.

Ganss, B., Kim, R.H. and Sidek, J. 1999. Bone sialoprotein. *Critical reviews in oral biology and medicine*. **10**:79–98.

Gerstenfield, L.C. 1999. Osteopontin in skeletal tissue homeostasis: an emerging picture of the autocrine/paracrine functions of the extracellular matrix. *Journal of bone and mineral research* **14**:850–55.

Gibson, L.J. 1985. The mechanical behavior of cancellous bone. *Journal of biomechanics* **18**:341–49.

Gibson, L.J. and Ashby, M.F. 1997. *Cellular solids: structure and properties* Cambridge, UK, Cambridge University Press.

Gibson, V.A., Stover, S.M., Martin, R.B., Gibeling, J.C., Willits, N.H., Gustafson, M.B. and Griffin, L.V. 1995. Fatigue behavior of the equine third metacarpus: mechanical property analysis. *Journal of orthopaedic research* **13**:861–68.

Gillette, E.P. 1872. Des os sesamoides chez l'homme. *Journal d'anatomie et physiologie* **8**:506–38.

Gilmore, R.S., Pollack, R.P. and Katz, J.L. 1969. Elastic properties of bovine dentine and enamel. *Archives of oral biology* **15**:787–96.

Giraud-Guille. M.M. 1988. Twisted plywood architecture of collagen fibrils in human compact bone osteons. *Calcified tissue international* **42**:167–80.

Glowacki, J., Cox, K.A., O'Sullivan, J., Wilkie, D. and Deftos, L.J. 1986. Osteoclasts can be induced in fish having an acellular bony skeleton. *Proceedings of the national academy of sciences* **83**:4104–7.

Goodship. A.E., Lanyon, L.E. and McFie, H. 1979. Functional adaptation of bone to increased stress. *Journal of bone and joint surgery* **61-A**:539–46.

Goodship, A.E., Cunningham, J.L. Oganov, V., Darling, J., Miles, A.W. and Owen, G.W. 1998. Bone loss during long term space flight is prevented by the application of a short term impulsive mechanical stimulus. *Acta astronautica* **43**:65–75.

Gordon, A.M., Huxley, A.F. and Julian, F.J. 1966. The variation in isometric tension with sarcomere length in vertebrate muscle fibres. *Journal of physiology* **184**:170–92.

Gorski, J.P. 1998. Is all bone the same? Distinctive distributions and properties of noncollagenous matrix proteins in lamellar vs. woven bone imply the existence of different underlying osteogenic mechanisms. *Critical reviews in oral biology and medicine*. **9**:201–23.

Gould, S.J. 1996. *Full house*. New York, Harmony Books.

Gould, S.J. and Lewontin, R.C. 1979. The spandrels of San Marco and the Panglossian paradigm: a critique of the adaptationist programme. *Proceedings of the Royal Society of London* **205B**:581–98.

Goulet, R.W., Goldstein, S.A., Ciarelli, M.J., Kuhn, J.L., Brown, M.B. and Feldkamp, L.A. 1994. The relationship between the structural and orthogonal compressive properties of trabecular bone. *Journal of biomechanics* **27**:375–89.

Granados, H. 1986. Physiological role of the pigment of the enamel in the rodent incisors: its importance in experimental biology and medicine. *Archivos de investigacion medica (Mexico)* **17**:37–54.

Gray, R.J. and Korbacher, G.K. 1974. Compressive fatigue behaviour of bovine compact bone. *Journal of biomechanics* **7**:287–92.

Green, A.E. and Taylor, G.I. 1945. Stress systems in aelotropic plates, III. *Proceedings of the Royal Society of London* **184A**:181–95.

Griffith, A.A. 1920. The phenomena of rupture and flow in solids. *Philosophical transactions of the Royal Society of London* **221A**, 163–97.

Gummer, D.L. and Brigham, R.M. 1995. Does fluctuating asymmetry reflect the importance of traits in little brown bats (*Myotis lucifugus*)? *Canadian journal of zoology* **73**:990–92.

Guo, X.E. and Goldstein, S.A. 1997. Is trabecular bone tissue different from cortical bone tissue? *Forma* **12**:185–96.

Haapasalo, H., Kannus, P., Sievanen, H., Heinonen, A., Oja, P. and Vuori, I. 1994. Long-term unilateral loading and bone-mineral density and content in female squash players. *Calcified tissue international* **54**:249–55.

Habelitz, S., Marshall, S.J., Marshall G.W.Jr and Balooch, M. 2001. Mechanical properties of human dental enamel on the nanometre scale. *Archives of oral biology* **46**:173–83.

Hahn, M., Vogel, M., Amling, M., Ritzel, H. and Delling, G. 1995. Microcallus formations of the cancellous bone: a quantitative analysis of the human spine. *Journal of bone and mineral research* **10**:1410–16.

Haines, R.W. 1969. Epiphyses and sesamoids. In: *Biology of the reptilia*, Vol. 1. C. Gans (ed.). New York, Academic Press, pp. 81–115.

Haire, T.J. and Langton, C.M. 1999. Biot theory: a review of its application to ultrasound propagation through cancellous bone. *Bone* **24**:291–95.

Haldane, J.B.S. 1953. Animal populations and their regulation. *New biology* **15**:9–24.

Haller, A.C. and Zimny, M.L. 1977. Effects of hibernation on interradicular alveolar bone. *Journal of dental research* **56**:1552–57.

Hancox, N.M. 1972. *Biology of bone*. Cambridge, UK, Cambridge University Press.

Hanken, J. and Hall, B.K. (eds.). 1993. *The skull* Vol. 3: *functional and evolutionary mechanisms*. Chicago, University of Chicago Press.

Harlow, H.J., Lohuis, T., Beck, T.D.I. and Iaizzo, P.A. 2001. Muscle strength in overwintering bears. *Nature* **409**:997.

Harner, J.P.III and Wilson, J.H. 1985. Bone strength statistical distribution functions for broilers. *Poultry science* **64**:585–87.

Harvey, P.H. and Pagel, M.D. 1991. *The comparative method in evolutionary biology*. Oxford, UK, Oxford University Press.

Hawkins, S.A., Schroeder, E.T., Wiswell, R.A., Jaque, S.V., Marcell, T.J. and Costa, K. 1999. Eccentric muscle action increases site-specific osteogenic response. *Medicine and science in sports and exercise* **31**:1287–92.

Hazlehurst, G.A. and Rayner, J.M.V. 1992. Flight characteristics of Triassic and Jurassic Pterosauria: an appraisal based on wing shape. *Paleobiology* **18**:447–63.

Heglund, N.C., Cavagna, G.A. and Taylor, C.R. 1982a. Energetics and mechanics of terrestrial locomotion, 3: energy changes of the centre of mass as a function of speed and body size in birds and mammals. *Journal of experimental biology* **97**:41–56.

Heglund, N.C., Fedak, M.A., Taylor, C.R. and Cavagna, G.A. 1982b. Energetics and mechanics of terrestrial locomotion, 4: total mechanical energy changes as a function of speed and body size in birds and mammals. *Journal of experimental biology* **97**:57–66.

Heinrich, R.E. and Biknevicius, A.R. 1998. Skeletal allometry and interlimb scaling pattern in mustelid carnivorans. *Journal of morphology* **235**:121–34.

Heinrich, R.E., Ruff, C.B. and Adamczewski, J.Z. 1999. Ontogenetic changes in mineralization and bone geometry in the femur of muskoxen (*Ovibos moschatus*). *Journal of zoology, London* **247**:215–23.

Henry, A.N. Freeman, M.A.R. and Swanson, S.A.V. 1968. Studies of the mechanical properties of healing experimental fractures. *Proceedings of the Royal Society of Medicine* **61**:902–6.

Hernandez, C.J., Beaupré, G.S. Keller, T.S. and Carter, D.R. 2001. The influence of bone volume fraction and ash fraction on bone strength and modulus. *Bone* **29**:74–8.

Herring, S.W. and Mucci, R.J. 1991. In vivo strain in cranial sutures: the zygomatic arch. *Journal of morphology* **207**:225–39.

Heřt, J. 1994. A new attempt at the interpretation of the functional architecture of the cancellous bone. *Journal of biomechanics* **27**:239–42.

Heřt, J., Kučera, P., Vávra, M. and Voleník, V. 1965. Comparison of the mechanical properties of both the primary and haversian bone tissue. *Acta anatomica* **61**:412–23.

Heřt, J., Lišková, M. and Landgrot, B. 1969. Influence of the long term, continuous bending on the bone. *Folia morphologica* **17**:389–99.

Heřt, J., Přibylová E. and Lišková, M. 1972. Microstructure of compact bone of rabbit tibia after intermittent loading. *Acta anatomica* **82**:218–30.

Hill, A.V. 1938. The heat of shortening and the dynamic constants of muscle. *Proceedings of the Royal Society of London* **126B**:136–95.

Hillam, R.A. 1996. *Response of bone to mechanical load and alterations in circulating hormones*. Thesis, University of Bristol, UK, pp. 160–206.

Hillam, R.A. and Skerry, T.M. 1995. Inhibition of periosteal bone resorption and stimulation of formation by in vivo mechanical loading of the modeling rat ulna. *Journal of bone and mineral research* **10**:683–89.

Höcker, K 1995. Die traumatische Knochenverbiegung im Kindes-, Jugend- und frühen Erwachsenalter—Pathomechanik und Literaturübersicht. *Unfallchirurg* **98**:540–44.

Hodge, A.J. and Petruska, J.A. 1963. Recent studies with the electron microscope on ordered aggregates of the tropocollagen molecule. In: *Aspects of protein structure*. G.N. Ramachandran (ed.). New York, Academic Press, pp. 269–300.

Hodgskinson, R. and Currey, J.D. 1990a. Effects of structural variation on the Young's modulus of non-human cancellous bone. *Engineering in medicine* **204**:43–52.

———1990b. The effect of variation in structure on the Young's modulus of cancellous bone: a comparison of human and non-human material. *Engineering in medicine* **204**:115–21.

———1992. Young's modulus, density and material properties in cancellous bone over a large density range. *Journal of materials science: materials in medicine.* **3**:377–81.

Hodgskinson, R., Currey, J.D. and Evans, G.P. 1989. Hardness, an indicator of the mechanical competence of cancellous bone. *Journal of orthopaedic research* 7:754–58.

Horsman, A. and Currey, J.D. 1983. Estimation of mechanical properties of the distal radius from bone mineral content and cortical width. *Clinical orthopaedics and related research* **176**:298–304.

Houston, D.C 1993. The incidence of healed fractures to wing bones of white-backed and Rüppell's griffon vultures *Gyps africanus* and *G. rueppellii* and other birds. *Ibis* **135**:468–69.

Howell, D.J. and Pylka, J. 1977. Why bats hang upside down: a biomechanical hypothesis. *Journal of theoretical biology* **69**:625–31

Hsieh, A.H., Tsai, C.M.H., Ma, Q.J., Lin, T., Banes, A.J., Villarreal, F.J., Akeson, W.H. and Sung, K.L.P. 2000. Time-dependent increases in type-III collagen gene expression in medial collateral ligament fibroblasts under cyclic strain. *Journal of orthopaedic research* **18**:220–27.

Hsieh, A.S., Winet, H., Boa, J.Y., Glas, H. and Plenk, H. 2001. Evidence for reperfusion injury in cortical bone as a function of crush injury ischemia duration: a rabbit bone chamber study. *Bone* **28**:94–103.

Huang, T-J.G, Schilder, H. and Nathanson, D. 1992. Effects of moisture content and endodontic treatment on some mechanical properties of human dentin. *Journal of endodontics* **18**:209–15.

Hughes. J., Paul, J. P. and Kenedi, R. M. 1970. Control and movement of the lower limbs. In: *Modern trends in biomechanics*. D. C. Simpson (ed.). London, Butterworths, pp. 147–79.

Huiskes, R. 2000. If bone is the answer, then what is the question? *Journal of anatomy* **197**:145–56.

Huiskes, R., Janssen, J.D. and Sloof, T.J. 1981. A detailed comparison of experimental and theoretical stress-analysis of a human femur. In: Cowin (1981), pp. 211–34.

Huja, S.S., Hasan. M.S., Pidaparti, R., Turner, C.H., Garetto, L.P. and Burr, D.B. 1999. Development of a fluorescent light technique for evaluating microdamage in bone subjected to fatigue loading. *Journal of biomechanics* **32**:1243–49.

Hukins, D.W.L. 1978. Bone stiffness explained by the liquid crystal model for the collagen fibril. *Journal of theoretical biology* **71**:661–67.

Hull, D. and Clyne, T.W. 1996. *An introduction to composite materials*. Cambridge, UK, Cambridge University Press.

Hulmes, D.J.S., Wess, T.J., Prockop, D.J. and Fratzl, P. 1995. Radial packing, order and disorder in collagen fibrils. *Biophysical journal* **68**:1661–70.

Hunt, J.D. 1982. *The wider sea*. London, Dent.

Hylander, W.M. 1979. Mandibular function in *Galago crassicaudatus* and *Macaca fascicularis*: an in vivo approach to stress analysis of the mandible. *Journal of morphology* **159**:253–96.

Inaoka, T., Lean, J.M., Bessho, T., Chow, J.W.M., Mackay, A., Kokubo, T. and Chambers, T.J. 1995. Sequential analysis of gene-expression after an osteogeneic stimulus—c-fos expression is induced in osteocytes. *Biochemical and biophysical research communications* **217**:264–70.

Inglis, C.E. 1913. Stresses in a plate due to the presence of cracks and sharp corners. *Transactions of the institute of naval architects* **55**:219–30.

Ishiyama, M. and Teraki, Y. 2000. The fine structure and formation of hypermineralized petrodentine in the tooth plate of extant lungfish (*Lepidosiren paradoxa* and *Protopterus* sp.) *Archives of histology and cytology* **53**: 307–21.

Jacobs, C.R., Davis, B.R., Rieger, C.J., Francis, J.J., Saad, M. and Fyrhie, D.P. 1999. The impact of boundary conditions and mesh size on the accuracy of cancellous bone tissue modulus determination using large-scale finite-element modeling. *Journal of biomechanics* **32**:1159–64.

Jäger, I. and Fratzl, P. 2000. Mineralized collagen fibrils: a mechanical model with a staggered arrangement of mineral particles. *Biophysical journal* **79**:1737–46.

Jameson, M.W., Hood, J.A.A. and Tidmarsh, B.G. 1993. The effects of dehydration and rehydration on some mechanical properties of human dentine. *Journal of biomechanics* **26**:1055–65.

Janis, C.M. and Fortelius, M. 1988. On the means whereby mammals achieve increased functional durability of their dentitions, with special reference to limiting factors. *Biological reviews* **63**:197–230.

Jansen, M. 1920. *On boneformation*: [sic] *its relation to tension and pressure.* Manchester, UK, Manchester University Press.

Jaslow, C.R. 1990. Mechanical properties of cranial sutures. *Journal of biomechanics* **23**:313–21.

Jaslow, C.R. and Biewener, A.A. 1995. Strain patterns in the horncores, cranial bones and sutures of goats (*Capra hircus*) during impact loading. *Journal of zoology, London* **235**:193–210.

Jayes, A.S. and Alexander, R.McN. 1978. Mechanics of locomotion of dogs (*Canis familiaris*) and sheep (*Ovis aries*). *Journal of zoology, London* **185**:89–108.

Jecker, P. and Hartwein, J. 1992. Ossification of the stapedial tendon—a rare cause of conductive hearing-loss *Laryngo-rhino-otologie* **71**:344–346 (in German).

Jendrucko, R.J., Hyman, W.A., Newell, P.H. and Chakraborty, B. K. 1976. Theoretical evidence for the generation of high pressure in bone cells. *Journal of biomechanics* **9**:87–91.

Jenkins, F.A., Dial, K.P. and Goslow, G.E. 1988. A cineradiographic analysis of bird flight: the wishbone in starlings is a spring. *Science* **241**:1495–98.

Jensen, K.S., Mosekilde, L. and Mosekilde, L. 1990. A model of vertebral trabecular architecture and its mechanical properties. *Bone* **11**:417–23.

Jepsen, K.J., Davy, D.T. and Krzypow, D.J. 1999. The role of the lamellar interface during torsional yielding of human cortical bone. *Journal of biomechanics* **32**:303–10.

Jeronimides, G. 1980. Wood, one of nature's challenging composites. In: *The mechanical properties of biological materials.* J.F.V. Vincent and J.D. Currey (eds). Symposia of the Society for Experimental Biology, no. 34. Cambridge, UK, Cambridge University Press, pp. 169–83.

Jeyasuria, P. and Lewis, J.L. 1987. Mechanical properties of the axial skeleton in gorgonians. *Coral reefs* **5**:213–39.

Johnson, K.A., Muir, P., Nicoll, R.G. and Roush, J.K. 2000. Asymmetric adaptive modeling of central tarsal bones in racing greyhounds. *Bone* **27**:257–63.

Jones, D.B. and Bingmann, D. 1991. How do osteoblasts respond to mechanical stimulation. *Cells and materials.* **1**:329–40.

Jones, H.H., Priest, J.D., Hayes, W.C., Tichenor, C.C. and Nagel, D.A. 1977. Humeral hypertrophy in response to exercise. *Journal of bone and joint surgery* **59-A**:204–8.

Kachanov, M. 1994. Elastic solids with many cracks and related problems. *Advances in applied mechanics* **30**:259–445.

Kadler, K. (ed.). 1994. Extracellular matrix, 1: fibril-forming proteins. In: *Protein profile.* P. Sheterline (ed.). London, Academic Press, pp. 517–638.

Kafka, V. 1993. On hydraulic strengthening of bones. *Journal of biomechanics* **26**:761–62.

Kamibayashi, L., Wyss, U.P., Cooke, T.D.V. and Zee, B. 1995. Changes in mean trabecular orientation in the medial condyle of the proximal tibia in osteoarthritis. *Calcified tissue international* **57**:69–73.

Katz, J.L. 1971. Hard tissue as a composite material, I: bounds on the elastic behavior. *Journal of biomechanics* **4**:455–73.

———1981. Composite material models for cortical bone. In: Cowin (1981), pp. 171–84.

Kay, R. and Smith, J. (eds.). 1989. The molecular basis of positional signalling. *Development*. Supplement, pp. 1–186.

Kayser, C. and Frank, R.M. 1963. Comportement des tissus calcifiés du hamster d'Europe *Cricetus cricetus* au cours de l'hibernation. *Archives of oral biology* 8:703–13.

Keaveny, T.M., Borchers, R.E., Gibson, L.J. and Hayes, W.C. 1993. Theoretical analysis of the experimental artifact in trabecular bone compressive modulus. *Journal of biomechanics* 26:599–607.

Keaveny, T.M., Guo, X.E., Wachtel, E.F., McMahon, T.A. and Hayes, W.C. 1994a. Trabecular bone exhibits fully linear elastic behavior and yields at low strains. *Journal of biomechanics* 27:1127–36.

Keaveny, T.M., Wachtel, E.F., Ford, C.M and Hayes, W.C. 1994b. Differences between the tensile and compressive strengths of bovine tibial trabecular bone depend on modulus. *Journal of biomechanics* 27:1137–46.

Keila, S., Pitaru, S., Grosskopf, A. and Weinreb, M. 1994. Bone marrow from mechanically unloaded rat bones expresses reduced osteogenetic capacity in vitro. *Journal of bone and mineral research* 9:321–27.

Keller, J.B. 1960. The shape of the strongest column. *Archive for rational mechanics and analysis* 5:275–85.

Keller, T.S. 1994. Predicting the compressive mechanical behavior of bone. *Journal of biomechanics* 27:1159–68

Keller, T.S., Spengler, D.M. and Carter, D.R. 1986. Geometric, elastic and structural properties of maturing rat femora. *Journal of orthopaedic research* 4:57–67.

Kempson, G.E. 1991. Age-related changes in the tensile properties of human articular cartilage: a comparative study between the femoral head of the hip joint and the talus of the ankle joint. *Biochimica et biophysica acta* 1075:223–30.

Ker, R.F. 1999. The design of soft collagenous load-bearing tissues. *Journal of experimental biology* 202:3315–24.

Ker, R.F., Bennett, M.B., Bibby, S.R., Kester, R.C. and Alexander, R.M. 1987. The spring in the arch of the human foot. *Nature* 325:147–49.

Ker, R.F., Alexander, R.McN. and Bennett, M.B. 1988. Why are mammalian tendons so thick? *Journal of zoology, London* 216:309–24.

Ker, R.F., Wang, X.T. and Pike, A.V.L. 2000. Fatigue quality of mammalian tendons. *Journal of experimental biology* 203:1317–27.

Kerin, A.J., Wisnom, M.R. and Adams, M.A. 1998. The compressive strength of articular cartilage. *Proceedings of the institution of mechanical engineers Part H—Journal of engineering in medicine* 212:273–80.

Kim, H.D. and Walsh, W.R. 1992. Mechanical and ultrasonic characterization of cortical bone. *Biomimetics* 1:293–310.

Kim, H.M, Rey, C. and Glimcher, M.J. 1995. Isolation of calcium–phosphate crystals of bone by nonaqueous methods at low temperature. *Journal of bone and mineral research*. 10:1589–601.

King, A.I. and Evans, F.G. 1967. Analysis of fatigue strength of human compact bone by the Weibull method. In: *Digest of the 7th international conference on medical and biological engineering.* B. Jacobson (ed.). Stockholm, The organizing committee, p. 514.

Kingdon, J. 1979. *East African mammals,* Vol. III, Part B: *large mammals.* London, Academic Press, p.14.

Kingsmill, V.J., Jones, S.J. and Boyde, A. 1998. Mineralization density of aged human mandibular bone. In: *Biological mechanisms of tooth eruption, resorption and replacement by implants.* Z. Davidovitch and J. Mah (eds.). Boston, Harvard Society for the advancement of orthodontics, pp. 167–71.

Kinney, J.H., Balooch, M., Marshall, G.W. and Marshall, S.J. 1999. A micromechanics model of the elastic properties of human dentine. *Archives of oral biology* **44**:813–22.

Kirkpatrick, S.J. 1994. Scale effects on the stresses and safety factors in the wing bones of birds and bats. *Journal of experimental biology* **190**:195–215.

Klein-Nulend, J., Van der Plas, A., Semeins, C.M., Ajubi, N.E., Frangos, J.A., Nijweide, P.J. and Burger, E.H 1995. Sensitivity of osteocytes to biomechanical stress in vitro. *The FASEB journal* **9**:441–45.

Klisch, S.M. and Lotz, J.C. 1999. Application of a fiber-reinforced continuum theory to multiple deformations of the annulus fibrosus. *Journal of biomechanics* **32**:1027–36.

Knothe Tate, M.L., Knothe, U. and Niederer, P. 1998. Experimental elucidation of mechanical load-induced fluid flow and its potential role in bone metabolism and functional adaptation. *American journal of the medical sciences* **316**:189–95.

Kodaka, T., Debari, K. and Yamada, M. 1991. Physicochemical and morphological studies of horse dentin. *Journal of electron microscopy* **40**:385–91.

Krenchel, H. 1964. *Fibre reinforcement.* Copenhagen, Akademisk Forlag.

Kriewall, T.J., McPherson, G.K. and Tsai, A.C. 1981. Bending properties and ash content of fetal cranial bone. *Journal of biomechanics* **14**:73–79.

Kruuk, H. 1972. *The spotted hyena: a study of predation and social behavior.* Chicago, University of Chicago Press.

Kuhn, J.L., Goldstein, S.A., Choi, K., London, M., Feldkamp, L.A. and Matthews, L.S. 1989. Comparison of the trabecular and cortical tissue moduli from human iliac crests. *Journal of orthopaedic research.* **7**:876–84.

Lafferty, J. F. 1978. Analytical model of the fatigue characteristics of bone. *Aviation space and environmental medicine* **49**:170–74.

Lakes, R. 1987. Foam structures with negative Poisson's ratio. *Science* **235**:1038–40.

Lakes, R. 1995. On the torsional properties of single osteons. *Journal of biomechanics* **28**:1409–10.

Lakes, R. and Saha, S. 1979. Cement line motion in bone. *Science* **204**:501–3.

Landis, W.J. 1995. The strength of calcified tissue depends in part on the molecular structure and organization of its constituent mineral crystals in their organic matrix. *Bone* **16**:533–44.

Landis, W.J., Song, M.J., Leith, A., McEwen, L. and McEwen, B.F. 1993. Mineral and organic matrix interaction in normally calcifying tendon visualised in three dimensions by high-voltage electron microscopic tomography and by graphic image reconstruction. *Journal of structural biology* **110**:39–54.

Landis, W.J., Hodgens, K.J., Arena, J., Song, M.J. and McEwen, B.F. 1996. Structural relations between collagen and mineral in bone as determined by high voltage electron microscopic tomography. *Microscopy research and technique* **33**:192–202.

Lanyon, L.E. 1974. Experimental support for the trajectorial theory of bone structure. *Journal of bone and joint surgery* **56-B**:160–66.

———1987. Functional strain as an objective, and controlling stimulus for adaptive bone remodelling. *Journal of biomechanics* **20**:1083–95.

Lanyon, L.E. and Baggott, D. G. 1976. Mechanical function as an influence on the structure and form of bone. *Journal of bone and joint surgery* **58-B**:436–43.

Lanyon, L.E. and Bourne, S. 1979. The influence of mechanical function on the development of remodeling of the tibia: an experimental study in sheep. *Journal of bone and joint surgery* **61-A**:263–73.

Lanyon, L.E. and Hartman, W. 1977. Strain related electrical potentials recorded in vitro and in vivo. *Calcified tissue research* **22**:315–27.

Lanyon, L.E. and Rubin, C.T. 1984. Static vs dynamic loads as an influence on bone remodelling. *Journal of biomechanics* **17**:897–905.

Lanyon, L.E., Goodship, A.E., Pye, C.J. and MacFie, J.H. 1982. Mechanically adaptive bone remodelling. *Journal of biomechanics* **15**:141–54.

Lawn, B. 1993. *Fracture of brittle solids*. Cambridge, UK, Cambridge University Press.

Lease, G.O. and Evans, F.G. 1959. Strength of human metatarsal bones under repetitive loading. *Journal of applied physiology* **14**:49–51.

Lee, D.D. and Glimcher M.J. 1991. Three-dimensional relationship between the collagen fibrils and the inorganic calcium phosphate crystals of pickerel (*Americanus americanus*) and herring (*Clupea harengus*) bone. *Journal of molecular biology* **217**:487–501.

Lees, S. 1968. Specific impedance of enamel and dentine. *Archives of oral biology* **13**:1491–500.

Lees, S. and Davidson, C.L. 1977. The role of collagen in the elastic properties of calcified tissues. *Journal of biomechanics* **10**:473–86.

Lees, S., Tao, N.-J. and Lindsay, S.M. 1990. Studies of compact hard tissues and collagen by means of Brillouin light scattering. *Connective tissue research* **24**:187–205.

Lees, S., Prostak, K.S., Ingle, V.K. and Kjoller, K. 1994. The loci of mineral in turkey leg tendon as seen by atomic force microscope and electron microscopy. *Calcified tissue international* **55**:180–89.

Lees, S., Hanson, D.B. and Page, E.A. 1996. Some acoustical properties of the otic bone of a fin whale. *Journal of the acoustical society of America* **99**:2421–27.

Leguillon, D., Lacroix, C. and Martin, E. 2000. Interface debonding ahead of a primary crack. *Journal of the mechanics and physics of solids* **48**:2137–61,

Lehman, M.I. 1967. Tensile strength of human dentin. *Journal of dental research* **46**:197–201.

Les, C.M., Stover, S.M., Keyak, J.H., Taylor, K.T. and Kaneps, A.J. 1995. Stiff and strong material properties are associated with brittle post-yield behavior in cortical bone. *Abstract, 41st Annual Meeting, Orthopaedic Research Society*. 131–132.

Lian, J.B. and Stein, G.S. 1996. Osteoblast biology. In: Marcus et al. (1996), pp. 23–59.

Lichtenbelt, W.D.V., Fogelholm, M., Ottenheijm, R. and Westerterp, K.R. 1995. Physical activity, body composition and bone density in ballet dancers. *British journal of nutrition* **74**:439–51.

Lieberman, D.E. 1996. How and why humans grow thin skulls: experimental evidence for systemic cortical robusticity. *American journal of physical anthropology* **101**:217–36.

Linde, F., Pongsoipetch, B., Frich, L.H. and Hvid, H. 1990. Three-axial strain controlled testing applied to bone specimens from the proximal tibial epiphysis. *Journal of biomechanics* **23**:1167–72.

Linde, F., Hvid, I. and Madsen, F. 1992. The effect of specimen geometry on the mechanical behaviour of trabecular bone specimens. *Journal of biomechanics* **25**:359–68.

Lišková, M. and Heřt, J. 1971. Reaction of bone to mechanical stimuli, part 2: periosteal and endosteal reaction of tibial diaphysis in rabbit to intermittent loading. *Folia morphologica* **19**:301–17.

Liu, D., Wagner, H.D. and Weiner, S. 2000. Bending and fracture of compact circumferential and osteonal lamellar bone of the baboon tibia. *Journal of materials science—materials in medicine* **11**:49–60.

Lord, M.J., Ha, K.I. and Song, K.S. 1996. Stress fractures of the ribs of golfers. *American journal of sports medicine* **24**:118–22.

Lovejoy, C.O. and Heiple, K. G. 1981. The analysis of fractures in skeletal populations with an example from the Libben site, Ottawa County, Ohio. *American journal of physical anthropology* **55**:529–41.

Lowenstam, H.A. and Weiner, S. 1989. *On biomineralization*. New York, Oxford University Press.

Lozupone, E. and Favia, A. 1990. The structure of the trabeculae of cancellous bone, 2: long bones and mastoid. *Calcified tissue international* **46**:367–72.

Mack, R.W. 1964. Bone—a natural two-phase material: a study of the relative strength and elasticity of the organic and mineral components of bone. *Technical memorandum of the biomechanics laboratory*, University of California at Berkeley, pp. 1–36.

Mahoney, E., Holt, A., Swain, M. and Kilpatrick, N. 2000. The hardness and modulus of elasticity of primary molar teeth: an ultra-micro-indentation study. *Journal of dentistry* **28**:589–94.

Maitland, M.E. and Arsenault, A.L. 1991. A correlation between the distribution of biological apatite and amino acid sequence of type I collagen. *Calcified tissue international*. **48**:341–52.

Mak, A.F.T. 1990. Streaming potential in bone. In: Mow et al. (1990), Vol. II, pp. 175–94.

Marcus, R., Feldman, D. and Kelsey, J. (eds.). 1996. *Osteoporosis*. San Diego, CA, Academic Press.

Marotti, G. 1963. Quantitative studies of bone reconstruction, 1: the reconstruction in homotypic shaft bones. *Acta anatomica* **52**:291–333.

Marotti, G. 1993. A new theory of bone lamellation. *Calcified tissue international* **53** (Suppl. 1):S47–55.

Marotti, G., Ferretti, M., Remaggi, F. and Palumbo, C. 1995. Quantitative evaluation on [sic] osteocyte canalicular density in human secondary osteons. *Bone* **16**:125–28.

Marotti, G. and Zallone, A.Z. 1980. Changes in the vascular network during the formation of Haversian systems. *Acta anatomica* **106**:84–100.

Martin, R.B. and Boardman, D.L. 1993. The effects of collagen fiber orientation, porosity, density, and mineralization on bovine cortical bone bending properties. *Journal of biomechanics* **26**:1047–54.

Martin, R.B. and Burr, D.B. 1982. A hypothetical mechanism for the stimulation of osteonal remodeling by fatigue damage. *Journal of biomechanics* **15**:137–39.

———1989. *Structure, function and adaptation of compact bone.* New York, Raven Press.

Mather, B.S. 1967. The symmetry of the mechanical properties of the human femur. *Journal of surgical research* **7**:222–25.

Matyas, J.R., Anton, M.G., Shrive, N.G. and Frank. C.B. 1995. Stress governs tissue phenotype at the femoral insertion of the rabbit MCL. *Journal of biomechanics* **28**:147–57.

Mauch, M., Currey, J.D. and Sedman, A.J. 1992. Creep fracture in bones with different stiffnesses. *Journal of biomechanics* **25**:11–16.

McAlister, G.B. and Moyle, D.D. 1983. Some mechanical properties of goose femoral cortical bone. *Journal of biomechanics* **16**:577–89.

McConnell, D. 1962. The crystal structure of bone. *Clinical orthopaedics* **23**:253–68.

McCutchen, C.W 1975. Do mineral crystals stiffen bone by straightjacketing its collagen? *Journal of theoretical biology* **51**:51–58.

McElhaney, J. H. 1966. Dynamic response of bone and muscle tissue. *Journal of applied physiology* **21**:1231–36.

———1967. The charge distribution on the human femur due to load. *Journal of bone and joint surgery* **49-A**:1561–71.

McKee, M.D. and Nanci, A. 1996. Osteopontin at mineralized tissue interfaces in bone, teeth and osseointegrated implants: ultrastructural distribution and implications for mineralized tissue formation, turnover, and repair. *Microscopy research and technique* **33**:141–64.

McMahon, J.M., Boyde, A. and Bromage, T.G. 1995. Pattern of collagen fiber orientation in the ovine calcaneal shaft and its relation to locomotor-induced strain. *Anatomical record* **242**:147–58.

McMahon, T.A. 1973. Size and shape in biology. *Science* **179**:1201–04.

———1975. Allometry and biomechanics: limb bones in adult ungulates. *American naturalist* **109**:547–63.

McNally, D.S. and Arridge, R.G.C. 1995. An analytical model of intervertebral disc mechanics. *Journal of biomechanics* **28**:53–68.

McPherson, G.K. and Kriewall, T.J. 1980a. The elastic modulus of fetal cranial bone: a first step towards an understanding of the biomechanics of fetal head molding. *Journal of biomechanics* **13**:9–16.

McPherson, G.K. and Kriewall, T.J. 1980b. Fetal head molding: an investigation utilizing a finite element model of the fetal parietal bone. *Journal of biomechanics* **13**:17–26.

McShea, D.W. 1993. Evolutionary change in the morphological complexity of the mammalian vertebral column. *Evolution* **47**:730–40.

Melvin, J.W. 1993. Fracture mechanics of bone. *Journal of biomechanical engineering* **115**:549–54.

Melvin, J.W. and Evans, F.G. 1973. Crack propagation in bone. *Biomechanics Symposium ASME*, New York, pp. 87–88.

Mente, P.L. 2000. Micromechanical testing of single trabeculae. In: An and Draughn (2000), pp. 291–304.

Mente, P.L. and Lewis, J.L. 1989. Experimental method for the measurement of the elastic modulus of trabecular bone tissue. *Journal of orthopaedic research* **7**:456–61.

Mente, P.L. and Lewis, J.L. 1994. Elastic modulus of calcified cartilage is an order of magnitude less than that of subchondral bone. *Journal of orthopaedic research* **12**:637–47.

Meredith, N., Sherriff, M., Setchell, D.J. and Swanson, S.A.V. 1996. Measurement of the microhardness and Young's modulus of human enamel and dentin using an indentation technique. *Archives of oral biology* **41**:539–45.

Merrilees M.J, and Flint, M.H. 1980. Ultrastructural study of tension and pressure zones in a rabbit flexor tendon. *American journal of anatomy* **157**:87–106.

Meunier, F.J. 1987. Nouvelles données sur l'organisation spatiale des fibres de collagène de la plaque basale des écailles des téléostéens *Annales des sciences naturelles, zoologie, Paris* 13th Series **9**:113–21.

Meunier, F.J. and Huysseune, A. 1992. The concept of bone tissue in osteichthyes. *Netherlands journal of zoology* **42**:445–58.

Michel, M.C., Guo, X-D.E., Gibson, L.J., McMahon, T.A. and Hayes, W.C. 1993. Compressive behavior of bovine trabecular bone. *Journal of biomechanics* **26**:453–63.

Miles, A.W. and Tanner, K.E. (eds.) 1992. *Strain measurement in biomechanics.* London, Chapman & Hall.

Miller, A. 1984. Collagen: the organic matrix of bone. *Philosophical transactions of the Royal Society of London* **304B**:455–77.

Miller, S.C., de Saint-Georges, L., Bowman, B.M. and Jee, W.S.S. 1988. Bone lining cells: structure and function. *Scanning microscopy* **3**:953–60.

Mori, S. and Burr, D.B. 1993. Increased intracortical remodeling following fatigue damage. *Bone* **14**:103–9.

Mosley, J.R. and Lanyon, L.E. 1997. The influence of strain rate on the adaptive response to in vivo loading of the growing rat ulna. *43rd meeting, Orthopaedic Research Society, San Francisco*, p. 329.

Moss, M.L. 1961a. Osteogenesis of acellular teleost fish bone. *American journal of anatomy* **108**:99–109.

Moss, M.L. 1961b. Studies of the acellular bone of teleost fish, 1: morphological and systematic variations. *Acta anatomica* **46**:343–62.

Moss, M.L. 1977. Skeletal tissues in sharks. *American zoologist* **17**:335–342.

Mow, V.C., Ratcliffe, A. and Woo, S.L-Y. (eds.) 1990. *Biomechanics of diarthrodial joints.* Two volumes. New York, Springer Verlag.

Moyle, D.D. and Gavens, A.J. 1986. Fracture properties of bovine tibial bone. *Journal of biomechanics* **19**:919–27.

Muir, P., Johnson, K.A. and Ruaux-Mason, C.P. 1999. In vivo matrix microdamage in naturally occurring canine fatigue fracture. *Bone* **25**:571–76.

Mullender, M.G., Huiskes, R., Versleyen, H. and Buma, P. 1996. Osteocyte density and micromorphometric parameters in cancellous bone of the proximal femur in five mammalian species. *Journal of orthopaedic research* **14**:972–79.

Mundy, G.R., Boyce, B.F., Yoneda, T., Bonewald, L.F. and Roodman, G.D. 1996. Cytokines and bone remodeling. In: Marcus et al. (1996), pp. 301–13.

Munih, M. and Kralj, A. 1997. Modelling muscle activity in standing with considerations for bone safety. *Journal of biomechanics* **30**:49–56.

Murray, P.D.F. 1936. *Bones, a study of the development and structure of the vertebrate skeleton.* Cambridge, UK, Cambridge University Press.

Murray P.D.F. and Huxley, J.S. 1925. Self-differentiation in the grafted limb bud of the chick. *Journal of anatomy* **59**:379–84.

Muzik, K. and Wainwright, S.A. 1977. Morphology and habitat of five Fijian sea fans. *Bulletin of marine science* **27**:308–37.

Nachtigall, W. 1971. *Biotechnik: Statische Konstruktionen in der Natur.* Heidelberg, Quelle & Meyer.

Nanci, A 1999. Content and distribution of noncollagenous matrix proteins in bone and cementum: relationship to speed of formation and collagen packing density. *Journal of structural biology* **126**:256–69.

Nesbitt, S.A. and Horton, M.A. 1997. Trafficking of matrix collagens through bone-resorbing osteoclasts. *Science* **276**:266–69.

Netz, P., Eriksson, K. and Stromberg, L. 1980. Material reaction of diaphyseal bone under torsion. *Acta orthopaedica Scandinavica* **51**:223–29.

Neville, A.C. 1993. *Biology of fibrous composites: development beyond the cell membrane.* Cambridge, UK, Cambridge University Press.

Nichols D. and Currey, J.D. 1968. The secretion, structure, and strength of echinoderm calcite. In: *Cell structure and its interpretation.* S. M. McGee Russell and K.F.A. Ross (eds.). London, Edward Arnold. pp. 251–61.

Nishimura, Y., Fukuoka, H., Kiriyama, M., Suzuki, Y., Oyama, K., Ikawa, S., Higurashi, M. and Gunji, A. 1994. Bone turnover and calcium metabolism during 20 days bed rest in young healthy males and females. *Acta physiologica Scandinavica.* **150** (S616):27–35.

Niven, J.S.F. 1933. The development in vivo and in vitro of the avian patella. *Roux' Archiv* **128**:480–501.

Norberg, U.M. 1981. Allometry of bat wings and legs and comparison with bird wings. *Philosophical transactions of the Royal Society* **292B**:359–98.

Norman, T.L., Vashishth, D. and Burr, D.B. 1991. Mode I fracture toughness of human bone. In: *Advances in bioengineering*, Vol. 20. R. Vanderby (ed.). New York, BED, ASME, 361–64.

———1992. Effect of groove on bone fracture toughness. *Journal of biomechanics* **25**:1489–92.

———1995. Fracture toughness of human bone under tension. *Journal of biomechanics* **28**:309–20.

Nunemaker, D.M., Butterweck, D.M. and Provost, M.T. 1990. Fatigue fractures in thoroughbred racehorses: relationships with age, peak bone strain and training. *Journal of orthopaedic research* **8**:604–11.

O'Connor, J.A., Goodship, A.E., Rubin, C.T. and Lanyon, L.E. 1981. The effect of externally applied loads on bone remodeling in the radius of the sheep. In: *Mechanical factors and the skeleton*. I.A.F. Stokes (ed.). London, John Libbey, pp. 83–90.

O'Connor, J.A., Lanyon, L.E. and MacFie, H. 1982. The influence of strain rate on adaptive remodelling. *Journal of biomechanics* **15**:767–81.

O'Connor, J.J., Lu, T.-W., Leardini, A., Wilson, D.R., Feikes, J., Gill, H.S. and Zavatsky, A.B. (1999). Principles of joint mechanics. In: Archer et al. (1999), pp. 373–402.

Odgaard, A. 2001. Quantification of cancellous bone architecture. In: Cowin (2001a), pp.14-1 to 14-19.

Odgaard, A. and Gundersen, H.J.G. 1993. Quantification of connectivity in cancellous bone, with special emphasis on 3-D reconstructions. *Bone* **14**:173–82.

Odgaard, A. and Linde, F. 1991. The underestimation of Young's modulus in compressive testing of cancellous bone specimens. *Journal of biomechanics* **24**:691–98.

Odgaard, A. and Weinans, H. (eds.) 1995. *Bone structure and remodeling*. Singapore, World Scientific Publishing.

Odgaard, A., Hvid, I. and Linde, F. 1989. Compressive axial strain distributions in cancellous bone specimens. *Journal of biomechanics* **22**:829–35.

Odgaard, A., Kabel, J., Van Rietbergen, B. and Huiskes, R. 1999. Architectural 3-D parameters and anisotropic elastic properties of cancellous bone. In: *IUTAM symposium on synthesis in bio solid mechanics*. P. Pedersen and M.P. Bendsøe (eds.). Dordrecht, Kluwer, pp. 33–42.

Ogawa, K. and Ui, M. 1996. Fracture-separation of the medial humeral epicondyle caused by arm wrestling. *Journal of trauma—injury infection and critical care* **41**:494–97.

Oliver, W.C. and Pharr, G.M. 1992. An improved technique for determining hardness and elastic modulus using load and displacement sensing indentation experiments. *Journal of materials research* **4**:1564–83.

Olmos, M., Casinos, A. and Cubo, J. 1996. Limb allometry in birds. *Annales des sciences naturelles, zoologie, Paris*. **Series 13 17**:39–49

Olsen, B.R. and Ninomiya, Y. 1993. Collagens. In: *Guidebook to the extracellular matrix and adhesion proteins*; T. Kreis and R. Vale (eds.). Oxford, UK, Oxford University Press, pp. 32–48.

Orava, S., Puranen, J. and Ala-Ketola, L. 1978. Stress fractures caused by physical exercise. *Acta orthopaedica Scandinavica* **49**:19–27.

Orr, H.A. 2000. Adaptation and the cost of complexity. *Evolution* **54**:13–20.

Ørvig, T. 1967. Phylogeny of tooth tissues: evolution of some calcified tissues in early vertebrates. In: *Structural and chemical organization of teeth*, Vol. 1. A.E.W. Miles (ed.). New York, Academic Press, pp. 45–110.

Otter, M.W., Qin, Y.X., Rubin, C.T. and McLeod, K.J. 1999. Does bone perfusion/reperfusion initiate bone remodeling and the stress fracture syndrome? *Medical hypotheses* **53**:363–68.

Ou-Yang, H., Paschalis, E.P., Mayo, W.E., Boskey, A.L. and Mendelsohn, R. 2001. Infrared microscopic imaging of bone: spatial distribution of CO_3^{2-}. *Journal of bone and mineral research* **16**:893–900.

Oxnard, C.E. 1971. Tensile forces in skeletal structures. *Journal of morphology* **134**:425–36.

Oxnard, C.E. 1993. Bone and bones, architecture and stress, fossils and osteoporosis. *Journal of biomechanics* **26** (Suppl. 1):63–79.

Oxnard, C.E., Lannigan, F. and O'Higgins, P. 1995. The mechanism of bone adaptation: tension and resorption in the human incus. In: Odgaard and Weinans (1995), pp. 105–25.

Paavolainen, P. 1978. Studies on mechanical strength of bone, I: torsional strength of normal rabbit tibio-fibular bone. *Acta orthopaedica Scandinavica* **49**:497–505.

Packer, C. 1983. Sexual dimorphism: the horns of African antelopes. *Science* **221**:1191–93.

Palmer, A.R. 1994. Fluctuating asymmetry analysis: a primer. In: *Developmental instability: its origins and evolutionary implications*. T.A. Markov (ed.). Dordrecht, Kluwer, pp. 335–64.

Palmqvist, P., Martinez-Navarro, B. and Arribas, A. 1996. Prey selection by terrestrial carnivores in a lower Pleistocene paleocommunity. *Paleobiology* **22**:514–34.

Papadimitriou, H.M., Swartz, S.M. and Kunz, T.H. 1996. Ontogenetic and anatomic variation in mineralization of the wing skeleton of the Mexican free-tailed bat, *Tadarida brasiliensis*. *Journal of zoology, London* **240**:411–26.

Parfitt, A.M. 1994. Osteonal and hemi-osteonal remodeling: the spatial and temporal framework for signal traffic in adult human bone. *Journal of cellular biochemistry* **55**:273–86.

Parkes, E. W. 1974. *Braced frameworks*. Oxford, UK, Pergamon Press.

Pattin, C.A., Caler, W.E. and Carter, D.R. 1996. Cyclic mechanical property degradation during fatigue loading of cortical bone. *Journal of biomechanics* **29**:69–79.

Pauwels, F. 1950. Die Bedeutung der Bauprinzipien der unteren Extremität für die Beanspruchung des Beinskeletes. *Zeitschrift für Anatomie und Entwicklungsgeschichte* **114**:525–38.

———1968. Beitrag zur funktionellen Anpassung der Corticalis der Röhrenknochen. Untersuchung an drei rachitisch deformierten Femora. *Zeitschrift für Anatomie und Entwicklungsgeschichte* **127**:121–37.

———1980. *Biomechanics of the locomotor apparatus*. Berlin, Springer Verlag.

Pead, M.J., Skerry, T.M. and Lanyon, L.E. 1988. Direct transformation from quiescence to bone formation in the adult periosteum following a single brief period of bone loading. *Journal of bone and mineral research* **3**:647–56.

Pennycuick, C.J. 1988. On the reconstruction of pterosaurs and their manner of flight; with notes on vortex wakes. *Biological reviews*. **63**:299–331.

Peterson, R.E. 1974. *Stress concentration factors*. New York, Wiley.

Petit, G. 1955. *Ordre des Sireniens*. In: Traité de zoologie, Vol. 17. P-P. Grassé (ed.). Paris, Masson, p. 926.

Petrtyl, M., Heřt, J. and Fiala, P. 1996. Spatial organization of the haversian bone in man. *Journal of biomechanics* **29**:161–69.

Pfretzschner, H.-U. 1986. Structural reinforcement and crack propagation in enamel. *Mémoires. Museum d'histoire naturelle, serie C. Science de la terre (Paris)* **53**:133–43.

Pidaparti, R.M.V. and Burr, D.B. 1992. Collagen fiber orientation and geometry effects on the mechanical properties of secondary osteons. *Journal of biomechanics*. **25**:869–80.

Pidaparti, R.M.V. and Turner, C.H. 1997. Cancellous bone architecture: advantages of non-orthogonal trabecular alignment under multidirectional joint loading. *Journal of biomechanics* **30**:979–83.

Pidaparti, R.M.V., Chandran, A., Takano, Y. and Turner, C.H. 1996. Bone mineral lies mainly outside collagen fibrils: predictions of a composite model for osteonal bone. *Journal of biomechanics* **29**:909–16.

Piekarski, K. 1973. Analysis of bone as a composite material. *International journal of engineering science* **11**:557–65.

Piney, A. 1922. The anatomy of the bone marrow: with special reference to the distribution of the red marrow. *British medical journal* **2**:792–95.

Polk, C. 1996. Electric and magnetic fields for bone and soft tissue repair. In: Polk, C. and Postow, R. (Editors) *Handbook of biological effects of electromagnetic fields*. C. Polk and R. Postow (eds.). Boca Raton, FL, CRC Press, pp. 231–46.

Pollock, C.M. and Shadwick, R.E. 1994. Relationship between body mass and biomechanical properties of limb tendons in adult mammals. *American journal of physiology* **266** (*Regulatory integrative comparative physiology* **35**): R1016–21.

Pope, M.H. and Outwater, J.O. 1974. Mechanical properties of bone as a function of position and orientation. *Journal of biomechanics* **7**:61–66.

Poplin, C., Poplin, F. and de Riqlès, A. 1976. Quelques particularités anatomiques et histologiques du rostre de l'Espadon (*Xiphias gladius* L.). *Comptes rendues de l' academie des sciences Paris* **282D**:1105–8.

Prange, H.D. Anderson, J.F. and Rahn, H. 1979. Scaling of skeletal mass to body mass in birds and mammals. *American naturalist* **113**:103–22.

Prendergast, P.J. and Taylor, D. 1994. Prediction of bone adaptation using damage accumulation. *Journal of biomechanics* **27**:1067–76.

Preuschoft, H. 1973. Functional anatomy of the upper extremity. In: *The chimpanzee*, Vol. 6. G. H. Bourne (ed.). Basel, Karger, pp. 34–120.

Preuschoft, H., Reif, W-E. and Müller, W.H. 1974. Funktionsanpassungen in Form und Struktur an Haifischzähnen. *Zeitschrift für Anatomie und Entwicklungsgeschichte* **143**:315–344.

Prockop, D.J. and Fertala, A. 1998. The collagen fibril: the almost crystalline structure. *Journal of structural biology* **122**:111–18.

Prockop, D.J. and Kivirikko. K.I. 1995. Collagens: molecular biology, diseases, and potentials for therapy. *Annual review of biochemistry* **64**:403–34.

Puhl, J.J., Piotrowski, G. and Enneking, W.F. 1972. Biomechanical properties of paired canine fibulas. *Journal of biomechanics* **5**:391–97.

Qiu, S-J., Hoshaw, S.J., Gibson, G.J, Lundin-Cannon, K.D and Schaffler, M.B. 1997. Osteocyte apoptosis in reaction to matrix damage in compact bone. *Transactions, Orthopaedic Research Society* **22**:89.

Quinn, T.H. and Baumel, J.J. 1993. Chiropteran tendon locking mechanism *Journal of morphology* **216**:197–208.

Radin, E.L. and Paul, I.L. 1971. Importance of bone in sparing articular cartilage from impact. *Clinical orthopaedics and related research* **78**:342–44.

Ramaekers, J.G. 1977. The dynamic shear modulus of bone in dependence on the form. *Acta morphologica Neerlando-Scandinavica* **15**:185–201.

Randall, F.E. 1944. The skeletal and dental development and variability of the gorilla. *Human Biology* **16**:23–76.

Rasmussen, J., Damsgaard, M. and Voigt, M. 2001. Muscle recruitment by the min/max criterion—a comparative numerical study. *Journal of biomechanics* **34**:409–15.

Rawlinson, S.C.F., El Haj, A.J., Minter, S.L., Tavares, I.A., Bennett, A. and Lanyon, L.E. 1991. Loading-related increases in prostaglandin production in cores of human cancellous bone in vitro: a role for prostacyclin in adaptive bone remodeling? *Journal of bone and mineral research* **6**:1345–51.

Rayfield, E.J., Norman, D.B., Horner, C.C., Horner, J.R., Smith, P.M., Thomason, J.J. and Upchurch, P. 2001. Cranial design and function in a large theropod dinosaur. *Nature* **409**:1033–37.

Reich, F.R., Brenden, B.B. and Porter, N.S. 1967. *Ultrasonic imaging of teeth* Report, Batelle Memorial Institute, Pacific Northwest Laboratory, Richland, Washington.

Reilly, D.T. and Burstein, A.H. 1975. The elastic and ultimate properties of compact bone tissue. *Journal of biomechanics* **8**:393–405.

Reilly, D.T., Burstein, A.H. and Frankel, V.H. 1974. The elastic modulus for bone. *Journal of biomechanics* **7**:271–75.

Reilly, G.C. 2001. Observations of microdamage around osteocyte lacunae in bone *Journal of biomechanics* **33**:1131–34.

Reilly, G.C. and Currey, J.D. 1999. The development of microcracking and failure in bone depends on the loading mode to which it is adapted. *Journal of experimental biology* **202**:543–52.

———2000. The effects of damage and microcracking on the impact strength of bone. *Journal of biomechanics* **33**:337–43.

Rensberger, J.M. 1992. Relationship of chewing stress and enamel microstructure in rhinocerotoid cheek teeth. In: *Structure, function and evolution of teeth*. P. Smith and E. Tchernov (eds.). London, Freund, pp. 163–83.

Rensberger, J.M. 1997. Mechanical adaptation in enamel. In: *Tooth enamel microstructure*. W. v. Koenigswald and P.M. Sander (eds.). Rotterdam, Balkema, pp. 237–57.

Rensberger, J.M. 2000. Pathways to functional differentiation in mammalian enamel. In: Teaford et al. (2000), pp. 252–68.

Rensberger, J.M. and Pfretzschner, H.U. 1992. Enamel structure in astrapotheres and its functional implications. *Scanning microscopy* **6**:495–508.

Rensberger, J.M. and Watabe, M. 2000. Fine structure of bone in dinosaurs, birds and mammals. *Nature* **406**:619–22.

Renson, C.E. and Braden, M. 1975. Experimental determination of the rigidity modulus, Poisson's ratio and elastic limit in shear of human dentine. *Archives of oral biology* **20**:43–47.

Rey, C., Beshah, K., Griffin, R. and Glimcher, M.J. 1991. Structural studies of the mineral of calcifying cartilage. *Journal of bone and mineral research* **6**:515–25.

Rho, J.Y. 2000. Ultrasonic methods for evaluating mechanical properties of bone. In: An and Draughn (2000), pp. 357–69.

Rho, J.Y., Ashman, R.B. and Turner, C.H. 1993. Young's modulus of trabecular and cortical bone material: ultrasonic and microtensile measurements. *Journal of biomechanics* **26**:111–19.

Rho, J.Y., Zioupos, P., Currey, J.D. and Pharr, G.M. 1999. Variations in the individual thick lamellar properties within osteons by nanoindentation. *Bone* **25**:295–300.

Rice, J.C., Cowin S.C. and Bowman, J.A. 1988. On the dependence of elasticity and strength of cancellous bone on apparent density. *Journal of biomechanics* **21**:155–68.

Rice, J.R. 1968. A path independent integral and the approximate analysis of strain concentrations by notches and cracks. *Journal of applied mechanics* **35**:379–86.

Richardson, M.L. and Patten, R.M. 1994. Age-related changes in marrow distribution in the shoulder: MR imaging findings. *Radiology* **192**:209–15.

Ride, W.D.L. 1959. Mastication and taxonomy in the macropodine skull. *Systematics association publications* **3**:33–59.

Riggs, C.M., Lanyon, L.E. and Boyde, A. 1993a. Functional associations between collagen fibre orientation and locomotor strain direction in cortical bone of the equine radius. *Anatomy and embryology* **187**:231–38.

Riggs, C.M., Vaughan, L.C., Evans, G.P., Lanyon, L.E. and Boyde, A. 1993b. Mechanical implications of collagen fibre orientation in cortical bone of the equine radius. *Anatomy and embryology* **187**:239–48.

Rimnac, C.M., Petko, A.A., Santner, T.J. and Wright, T.M. 1993. The effect of temperature, stress and microstructure on the creep of compact bovine bone. *Journal of biomechanics* **26**:219–28.

Rinnerthaler, S., Roschger, P., Jakob, H.F., Nader, A., Klaushofer, K. and Fratzl, P. 1999. Scanning small angle X-ray scattering analysis of human bone sections. *Calcified tissue international* **64**:422–29.

Roach, H.I. 1994. Why does bone matrix contain non-collagenous protein? The possible roles of osteocalcin, osteonectin, osteopontin and bone sialoprotein in bone mineralisation and resorption. *Cell biology international* **18**:617–28.

Robertson, D.M., Robertson D. and Barret, C.R. 1978. Fracture toughness, critical crack length and plastic zone size in bone. *Journal of biomechanics* **11**:359–64.

Robey, P.G. and Boskey, A.L. 1996. The biochemistry of bone. In: Marcus et al. (1996), pp. 95–183.

Robling, A.G. and Stout, S.D. 1999. Morphology of the drifting osteon. *Cells tissues organs* **164**:192–204.

Robling, A.G., Burr, D.B. and Turner, C.H. 2000. Partitioning a daily mechanical stimulus into discrete loading bouts improves the osteogenic response to loading. *Journal of bone and mineral research* **15**:1596–602.

Roesler, H. 1981. Some historical remarks on the theory of cancellous bone structure (Wolff's law). In: Cowin (1981), pp. 27–42.

Rogers, K.D. and Zioupos, P. 1999. The bone tissue of the rostrum of a *Mesoplodon Densirostris* whale: a mammalian biomineral demonstrating extreme texture. *Journal of materials science letters* **18**:651–54.

Rogers, R.R. and LaBarbara, M. 1993. Contribution of internal bony trabeculae to the mechanical properties of the humerus of the pigeon (*Columba livia*). *Journal of zoology, London* **230**:433–41.

Røhl, L. Larsen, E., Linde, F., Odgaard, A. and Jørgensen, J. 1991. Tensile and compressive properties of cancellous bone. *Journal of biomechanics* **24**:1143–49.

Rothenburg, L., Berlin, A.A. and Bathurst, R.J. 1991. Microstructure of isotropic materials with negative Poisson's ratio. *Nature* **354**:470–72.

Ruben, J.A. and Bennett, A.A. 1987. The evolution of bone. *Evolution* **41**: 1187–97.

Rubin, C., Turner, A.S., Bain, S., Mallinckrodt, C. and McLeod, K. 2001. Low mechanical signals strengthen long bones. *Nature* **412**:603–4.

Rubin, C.T. and Lanyon, L. E. 1982. Limb mechanics as a function of speed and gait: a study of functional strains in the radius and tibia of horse and dog. *Journal of experimental biology* **101**:187–211.

———1984. Regulation of bone formation by applied dynamic loads. *Journal of bone and joint surgery* **66-A**:397–402.

———1987. Osteoregulatory nature of mechanical stimuli—function as a determinant for adaptive remodeling in bone. *Journal of orthopaedic research* **5**:300–310.

Rubin, C.T and McLeod, K.J. 1996. Inhibition of osteopenia by biophysical intervention. In: Marcus et al. (1996), pp. 351–71.

Rubin, C.T., Gross, T.S., McLeod, K.J. and Bain, S.D. 1995. Morphologic stages in lamellar bone-formation stimulated by a potent mechanical stimulus. *Journal of bone and mineral research* **10**:488–95.

Runkle, J.C. and Pugh, J.W. 1975. The micromechanics of cancellous bone, II, determination of the elastic modulus of individual trabeculae by buckling analysis. *Bulletin of the Hospital for Joint Diseases*. **36**:2–10.

Russell, A.P. and Dalstra, L.D. 2001. Patagial morphology of *Draco volans* (Reptilia: Agamidae) and the origin of glissant locomotion in flying dragons. *Journal of zoology, London* **253**:457–71.

Russell, A.P. and Thomason, J.J. 1993. Mammalian skull mechanics. In: Hanken and Hall (1993), pp. 345–83.

Ruth, E.B. 1953. Bone studies, II: an experimental study of the Haversian-type vascular channels. *American journal of anatomy* **93**:429–55.

Ryan, S.D. and Williams, J.L. 1989. Tensile testing of rodlike trabeculae excised from bovine femoral bone. *Journal of biomechanics* **22**:351–55.

Sadananda, R. 1991. A probabilistic approach to bone fracture analysis. *Journal of materials research* **6**:202–6.

Saha, S. and Hayes, W.C. 1976. Tensile impact properties of human compact bone. *Journal of biomechanics* **9**:243–51.

Salo, J., Lehenkari, P., Mulari, M., Metsikkö, K. and Väänänen, H.K. 1997. Removal of osteoclast bone resorption products by transcytosis. *Science* **276**:270–73.

Salzstein, R.A. and Pollack, S.R. 1987. Electromechanical potenials in cortical bone, II: experimental analysis. *Journal of biomechanics* 20:271–80.

Salzstein, R.A., Pollack, S.R., Mak, A.F.T. and Petrov, N. 1987. Electromechanical potentials in cortical bone, I: a continuum approach. *Journal of biomechanics* 20:261–70.

Sammarco, G.J., Burstein, A.H., Davis, W.L. and Frankel, V.H. 1971. The biomechanics of torsional fractures: the effect of loading on ultimate properties. *Journal of biomechanics* 4:113–17.

Sano, H., Ciucchi, B., Matthews, W.G. and Pashley, D.H. 1994. Tensile properties of mineralized and demineralized human and bovine dentin. *Journal of dental research* 73:1205–11.

Sano, H., Takatsu, T., Ciucchi, B., Russell, C.M. and Pashley, D.H. 1995. Tensile properties of resin-infiltrated demineralized human dentin. *Journal of dental research* 74:1093–102.

Sasaki, N. 2000. Viscoelastic properties of bone and testing methods. In: An and Draughn (2000), pp. 329–48.

Sasaki, N., Ikawa, T. and Fukuda, A. 1991. Orientation of mineral in bovine bone and the anisotropic mechanical properties of plexiform bone. *Journal of biomechanics* 24:57–61.

Scapino, R. 1981. Morphological investigations into functions of the jaw symphysis in carnivorans. *Journal of morphology* 167:339–75.

Schaffler, M.B. and Burr, D.B. 1988. Stiffness of compact bone: effects of porosity and density. *Journal of biomechanics* 21:13–16.

Schaffler, M.B., Burr, D.B. and Frederickson, R.G. 1987. Morphology of the osteonal cement line in human bone. *Anatomical record* 217:223–28.

Schaffler, M.B., Radin, E.L. and Burr, D.B. 1989. Mechanical and morphological effects of strain rate on fatigue of compact bone. *Bone* 10:207–14

Schaffler, M.B., Radin, E.L. and Burr, D.B. 1990. Long-term fatigue behavior of compact bone at low strain magnitude and rate. *Bone* 11:321–26.

Schaffler, M.B., Pitchford, W.C., Choi, K. and Riddle, J.M. 1994. Examination of compact bone microdamage using back-scattered electron microscopy. *Bone* 15:483–88.

Schechtman, H. and Bader, D.L. 1997. In vitro fatigue of human tendons. *Journal of biomechanics* 30:829–35.

Schirrmacher, K., Lauterbach, S. and Bingmann, D. 1997. Oxygen consumption of calvarial bone cells in vitro. *Journal of orthopaedic research* 15:558–62.

Schultz, A.H. 1937. Proportions, variability and asymmetry of the long bones of the limbs and the clavicles in man and apes. *Human Biology* 9:281–328.

Schultz, A.H. 1939. Notes on diseases and healed fractures of wild apes. *Bulletin of the history of medicine* 7:571–82.

Schultz, A.H. 1941. Growth and development of the Orang-Utan. *Contributions to Embryology* 29:57–110.

Schultz, A.H. 1944. Age changes and variability in Gibbons. *American Journal of physical anthropology* 2:1–129.

Scott, G. and King, J.B. 1994. A prospective, double-blind trial of electrical capacitive coupling in the treatment of non-union of long bones. *Journal of bone and joint surgery* 76-A:820–26.

Sedlin, E.D. and Hirsch, C. 1966. Factors affecting the determination of the physical properties of femoral cortical bone. *Acta orthopaedica Scandinavica* **37**:29–48.

Sedman, A.J. 1993. *Mechanical failure of bone and antler: the accumulation of damage*. Thesis, University of York, UK.

Selker, F. and Carter, D.R. 1989. Scaling of long bone fracture strength with animal mass. *Journal of biomechanics* **22**:1175–83.

Sepulveda, F., Wells, D.M. and Vaughan, C.L. 1993. A neural network representation of electromyography and joint dynamics in human gait. *Journal of biomechanics* **26**:101–9.

Shelton, J.C. 1992. Holographic interferometry. In: Miles and Tanner (1992), pp. 139–55.

Shum, D.K.M. and Hutchinson, J.W. 1990. On toughening by microcracks. *Mechanics of materials* **9**:83–91.

Silva, M.J., Keaveny, T.M. and Hayes, W.C. 1997. Load sharing between the shell and the centrum in the lumbar vertebral body. *Spine* **22**:140–50.

Simkiss, K. 1967. *Calcium in reproductive physiology*. London, Chapman & Hall.

Simkiss, K. and Wilbur, K.M. 1989. *Biomineralization: cell biology and mineral deposition*. New York, Academic Press.

Simone, A.E. and Gibson, L.J. 1998. Effects of solid distribution on the stiffness and strength of metallic foams. *Acta materialia* **46**:2139–50.

Simonian, P.T. and Hanel, D.P. 1996. Traumatic plastic deformity of an adult forearm: case report and literature review *Journal of orthopaedic trauma* **10**:213–15.

Singer, A., Ben-Yehuda, O., Ben-Ezra, Z. and Zaltzman, S. 1990. Multiple identical stress fractures in monozygotic twins. *Journal of bone and joint surgery* **72-A**:444–45.

Singh, I. 1978. The architecture of cancellous bone. *Journal of anatomy* **127**:305–10.

Skerry, T.M. 2000. In: *Development growth and evolution: implications for the study of the human skeleton*. Linnean society symposium series number 20. P. O'Higgins and M.J. Cohn (eds.), pp. 29–39.

Skerry, T.M., Suswillo, R.F.L., ElHaj, A.J., Ali, N.N., Dodds, R.A. and Lanyon, L.E. 1990. Load-induced proteoglycan orientation in bone tissue in vivo and in vitro. *Calcified tissue international* **46**:318–26

Smith, A.J. 2000. Pulpo-dentinal interactions in development and repair of dentine. In: Teaford et al. (2000), pp. 82–91.

Smith, J.W. 1962a. The relationship of epiphysial plates to stress in some bones of the lower limb. *Journal of anatomy* **96**:58–78.

———1962b. The structure and stress relationships of fibrous epiphysial plates. *Journal of anatomy* **96**:209–25.

Smith, M.M. 1985. Petrodentine in extant and fossil dipnoan dentitions: microstructure, histogenesis and growth. *Proceedings of the Linnean Society of New South Wales* **107**:367–407.

Smith, M.M. and Hall, B.K. 1990. Development and evolutionary origins of vertebrate skeletogenic and odontogenic tissues. *Biological reviews* **65**:277–373.

Smith, M.M. and Sansom, I.J. 2000. Evolutionary origins of dentine in the fossil record of early vertebrates: diversity, development and function. In: Teaford et al. (2000), pp. 65–81.

Smith, M.P., Sansom, I.J. and Repetski, J.E. 1996. Histology of the first fish. *Nature* **380**:702–4.

Söderlund, E., Dannelid, E. and Rowcliffe, D.J. 1992. On the hardness of the pigmented and unpigmented enamel in teeth of shrews of the genera *Sorex* and *Crocidura* (Mammalia, Soricidae) *Zeitschrift für Säugetierkunde*. **57**:321–29.

Sørensen, M.S. 1994. Temporal bone dynamics, the hard way. *Acta otolaryngologica (Stockholm)* Suppl. **512**:1–22.

Sørensen, M.S., Bretlau, P. and Jørgensen, M.B. 1992a. Bone repair in the otic capsule of the rabbit. *Acta otolaryngologica (Stockholm)* **112**:968–75.

———1992b. Fatigue microdamage in perilabyrinthine bone. *Acta otolaryngologica (Stockholm)* Suppl **496**:20–27.

Sørensen, M.S., Jørgensen, M.B. and Bretlau, P. 1992c. Drift barriers in the postcartilaginous development of the mammalian otic capsule. *European archives of oto-rhino-laryngology* **249**:56–61.

Spears, I.R. 1997. A three-dimensional finite element model of prismatic enamel: a re-appraisal of the data on the Young's modulus of enamel. *Journal of dental research* **76**:1690–97.

Spengler, D.M., Morey, E.R., Carter, D.R., Turner, R.T. and Baylink, D.J. 1979. Effect of space flight on bone strength. *Physiologist* **22** (Suppl.):75–76.

Stanford, J.W., Weigel, K.V., Paffenberger, G.C. and Sweeney, W.T. 1960. Compressive properties of hard tooth tissues and some restorative materials. *Journal of the American dental association* **60**:746–51.

Starkebaum, W., Pollack, S.R. and Korostoff, E. 1979. Microelectrode studies of stress-generated potentials in four-point bending of bone. *Journal of biomedical materials research* **13**:729–51.

Stefen, C. 1997. Differentiations in Hunter-Schreger bands of carnivores. In: *Teeth enamel structure*. W. v. Koenigswald and P.M. Sander (eds.). Rotterdam, Balkema, pp. 123–36.

Stern, J. T. 1974. Computer modeling of gross muscular dynamics. *Journal of biomechanics* **7**:411–28.

Sterne, L. 1767. *The life and opinions of Tristram Shandy.*

Steudel, K. 1990a. The work and energetic cost of locomotion, I: the effects of limb mass distribution in quadrupeds. *Journal of experimental biology* **154**:273–85.

Steudel, K. 1990b. The work and energetic cost of locomotion, II: partitioning the cost of internal and external work within a species. *Journal of experimental biology* **154**:287–303.

Stokes, I.A.F., Hutton, W.C. and Stott, J.R.R. 1979. Forces acting on the metatarsals during normal walking. *Journal of anatomy* **129**:579–90.

Stover, S.M., Pool, R.R., Martin, R.B. and Morgan, J.P. 1992. Histological features of the dorsal cortex of the third metacarpal bone mid-diaphysis during postnatal growth in thoroughbred horses. *Journal of anatomy* **181**:455–69.

Strömberg, L. and Dalén, N. 1976. Experimental measurements of maximum torque capacity of long bones. *Acta orthopaedica Scandinavica* **47**:257–63.

Sues, H.-D. 1978. Functional morphology of the dome in pachycephalosaurid dinosaurs. *Neues Jahrbuch für Geologie und Paläontologie Monatshefte* **1978**:459–72.

Summers, A.P. 2000. Stiffening the stingray skeleton: an investigation of durophagy in myliobatid stingrays (Chondrichthyes, Batoidea, Myliobatidae). *Journal of morphology* **243**:113–26.

Suresh, S. 1992. *Fatigue of materials*. Cambridge, UK, Cambridge University Press.

Swanson, S.A.V. and Freeman, M.A.R. 1966. Is bone hydraulically strengthened? *Medical and biological engineering* **4**:433–38.

Swanson, S.A.V., Freeman, M.A.R. and Day, W.H. 1971. The fatigue properties of human cortical bone. *Medical and biological engineering* **9**:23–32.

Swartz, S.M., Bennett, M.B. and Carrier, D.R. 1992. Wing bone stresses in free flying bats and the evolution of skeletal design for flight. *Nature* **359**:726–29.

Swartz, S.M., Parker, A. and Huo, C. 1998. Theoretical and empirical scaling patterns and topological homology in bone trabeculae. *Journal of experimental biology* **201**:573–90.

Tague, R.G. 1997. Variability of a vestigial structure: first metacarpal in *Colobus guereza* and *Ateles geoffroyi*. *Evolution* **51**:595–605.

Takano, Y., Turner, C.H., Owan, I., Martin, R.B., Lau, S.T., Forwood, M.R. and Burr, D.B. 1999. Elastic anisotropy and collagen orientation of osteonal bone are dependent on the mechanical strain distribution. *Journal of orthopaedic research* **17**:59–66.

Tappen, N.C. 1953. A functional analysis of the facial skeleton with split-line technique. *American journal of physical anthropology* **11**:503–32.

Tappen, N.C. 1970. Main patterns and individual differences in baboon skull split-lines and theories of causes of split-line orientation in bone. *American journal of physical anthropology* **33**:61–71.

Tattersall, H.G. and Tappin, G. 1966. The work of fracture and its measurement in metals, ceramics and other materials. *Journal of materials science* **1**:296–301.

Tavassoli, M. 1974. Differential response of bone marrow and extramedullary adipose cells to starvation. *Experientia* **30**:424–25.

Taylor, C.A., Shkolnik, A., Dmi'el, A., Baharav, D. and Borut, A. 1974. Running in cheetahs, gazelles, and goats: energy cost and limb configuration. *American journal of physiology* **227**:848–50.

Taylor, C.A., Heglund, N.C. and Maloiy, G.M.O. 1982. Energetics and mechanics of terrestrial locomotion, 1: metabolic energy consumption as a function of speed and body size in birds and mammals. *Journal of experimental biology* **97**:1–21.

Taylor, D. 1998. Fatigue of bone and bones: an analysis based on stressed volume. *Journal of orthopaedic research* **16**:1067–76.

Taylor, D. 2000. Scaling effects in the fatigue strength of bones from different animals. *Journal of theoretical biology* **206**:299–306.

Taylor, D., O'Brien, F., Prina-Mello, A., Ryan, C., O'Reilly, P. and Lee, T.C. 1999. Compression data on bovine bone confirms that a 'stressed volume' principle explains the variability of fatigue strength results. *Journal of biomechanics* **32**:1199–203.

Taylor, M.E. 1971. Bone diseases and fracture in East African viverridae. *Canadian journal of zoology* **49**:1035–42.

Teaford, M.F., Smith, M.M. and Ferguson, M.W.J. 2000. *Development, function and evolution of teeth*. Cambridge, UK, Cambridge University Press.

Ten Cate, A.R. 1989. *Oral Histology: development, structure, and function*. St. Louis, Mosby.

Terai, K., Takano-Yamamoto, T., Ohba, Y., Hiura, K., Sugimoto, M., Sato, M., Kawahata, H., Inaguma, N., Kitamura, Y. and Nomura, S. 1999. Role of osteopontin in bone remodeling caused by mechanical stress. *Journal of bone and mineral research* **14**:839–49.

Thomas, A.A., Yoon, H.S. and Katz, J.L. 1977. Acoustic emission from fresh bovine femora. In: *Ultrasonic symposium proceedings*, cat. 77CH1264. New York, Institute of Electrical and Electronics Engineers, pp. 237–41.

Thompson, A.C. and Ballou, J.E. 1956. Studies of metabolic turnover with tritium as a tracer. V: the predominantly non-dynamic state of body constituents in the rat. *Journal of biological chemistry* **223**:795–809.

Thouzeau, C., Massemin, S. and Handrich, Y. 1997. Bone marrow fat mobilization in relation to lipid and protein catabolism during prolonged fasting in barn owls. *Journal of comparative physiology B—Biochemical, systematic and environmental physiology* **167**:17–24.

Torzilli, P.A., Takebe, K., Burstein, A.H., Zika. J.M. and Heiple, K.G. 1982. The material properties of immature bone. *Journal of biomechanical engineering* **104**:12–20.

Townsend, P.R., Raux P., Rose, R.M., Miegel, R.E. and Radin, E.L. 1975a. The distribution and anisotropy of the stiffness of cancellous bone in the human patella. *Journal of biomechanics* **8**:363–67.

Townsend, P.R., Rose, R.M. and Radin, E.L. 1975b. Buckling studies of single human trabeculae. *Journal of biomechanics* **8**:199–201.

Tschantz, P. and Rutishauser, E. 1967. La surcharge mécanique de l'os vivant: les déformations plastiques initiales et l'hypertrophie d'adaptation. *Annales d'anatomie pathologique* **12**:223–48.

Tsukrov, I. and Kachanov, M. 1997. Stress concentrations and microfracturing patterns in a brittle-elastic solid with interacting pores of diverse shapes. *International journal of solids and structures* **34**:2887–904.

Turner, C.H 1999. Toward a mathematical description of bone biology: the principle of cellular accommodation. *Calcified tissue international* **65**:466–71.

Turner, C.H. 2000. Reply [to Frost]. *Calcified tissue international* **67**:185–87.

Turner, C.H. and Burr, D.B. 1993. Basic biomechanical measurements of bone: a tutorial. *Bone* **14**:595–608.

Turner, C.H., Akhter, M.P., Raab, D.M., Kimmel, D.B. and Recker, R.R. 1991. A noninvasive in vivo model for studying strain adaptive bone modeling. *Bone* **12**:73–79.

Turner, C.H., Forwood, M.R. and Otter, M.W. 1994. Mechanotransduction in bone: do bone cells act as sensors of fluid flow? *FASEB journal* **8**:875–78.

Turner, C.H., Rho, J., Takano, Y., Tsui, T.Y. and Pharr, G.M. 1999. The elastic properties of trabecular and cortical bone tissues are similar: results from two microscopic measurement techniques. *Journal of biomechanics* **32**:437–41.

Tyldesley, W.R. 1959. Mechanical properties of human dental enamel and dentine. *British dental journal* **106**:269–78.

Unsworth, A. 1981. Cartilage and synovial fluid. In: *An introduction to the biomechanics of joints*. D. Dowson and V. Wright (eds.). London, Mechanical Engineering Publications, pp. 107–14.

Vanden Berge, J.C. and Storer, R.W. 1995. Intratendinous ossification in birds: a review. *Journal of morphology* **226**:47–77.

van der Meulen, M.C.H. and Carter, D.R. 1995. Developmental mechanics determine long bone allometry. *Journal of theoretical biology* **172**:323–27.

van der Meulen, M.C.H., Beaupré, G.S. and Carter, D.R. 1993. Mechanobiologic influences in long bone cross-sectional growth. *Bone* **14**:635–42.

van der Velde, J.P., Vermeiden, J.P.W., Touw, J.J.A. and Veldhuzen, J.P. 1984. Changes in activity of chicken medullary bone cell populations in relation to the egg-laying cycle. *Metabolic bone disease and related research* **5**:191–93.

van Lenthe, G.H., van den Bergh, J.P.W., Hermus, A.R.M.M. and Huiskes, R. 2001. The prospects of estimating trabecular bone tissue properties from the combination of ultrasound, dual energy X-Ray absorptiometry, microcomputed tomography and microfinite element analysis. *Journal of bone and mineral research* **16**:550–55.

van Rietbergen, B. and Huiskes, R. 2001. Elastic constants of cancellous bone. In: Cowin (2001a), pp. 15-1 to 15-24.

van Rietbergen, B., Weinans, H., Huiskes, R. and Odgaard, A. 1995. A new method to determine trabecular bone elastic properties and loading using micromechanical finite-element models. *Journal of biomechanics* **28**:69–81.

Van Valkenberg, B. 1988. Incidence of tooth breakage among large, predatory mammals. *American naturalist* **131**:291–302.

Vasciaveo, F. and Bartoli, E. 1961. Vascular channels and resorption cavities in the long bone cortex: the bovine bone. *Acta anatomica* **47**:1–33.

Vashishth, D., Trifonas, J., Behiri, J.C. and Bonfield, W. 1994. Secondary crack propagation in cortical bone. *PD-Vol 64–4 Engineering systems design and analysis* **4**:37–41.

Vaughan, L.C. and Mason, B.J.E. 1975. *A clinico-pathological study of racing accidents in horses*. Dorking, Bartholomew Press.

Verborgt, O., Gibson, G.J. and Schaffler, M.B. 2000. Loss of osteocyte integrity in association with microdamage and bone remodeling after fatigue in vivo. *Journal of bone and mineral research* **15**:60–67.

Villanueva, A.R., Longo, J.A. and Weiner, G. 1994. Staining and histomorphometry of microcracks in the human femoral head. *Biotechnic and histochemistry* **69**:69–88.

Vincent, J.F.V. 1990. *Structural biomaterials*. Princeton, NJ, Princeton University Press.

Vincentelli, R. and Grigorov, M. 1985. The effect of haversian remodeling on the tensile properties of human cortical bone. *Journal of biomechanics* **18**:201–7.

Vogel, K.G. and Koob, T.J. 1989. Structural specialisation in tendons under compression. *International reviews of cytology* **115** 267–293.

von Koenigswald, W. 2000. Two different strategies in enamel differentiation: Marsupialia versus Eutheria. In: Teaford et al. (2000), pp. 107–18.

von Koenigswald, W. and Sander, P.M. (eds.). 1997. *Tooth enamel microstructure*. Rotterdam, Balkema.

von Koenigswald, W., Rensberger, J.M. and Pfretzschner, H.U. 1987. Changes in the tooth enamel of early Paleocene mammals allowing increased diet diversity. *Nature* **328**:150–52.

Wachtel, E. and Weiner, S. 1994. Small-angle X-ray scattering study of dispersed crystals from bone and tendon. *Journal of bone and mineral research* **9**:1651–55.

Wagner, H.D. and Weiner, S. 1992. On the relationship between the microstructure of bone and its mechanical stiffness. *Journal of biomechanics* **25**:1311–20.

Wainwright, S.A., Biggs, W.D. Currey, J.D. and Gosline, J.M. 1982. *Mechanical design in organisms*. Princeton, NJ, Princeton University Press.

Wall, W.P. 1983. The correlation between high limb-bone density and aquatic habits in recent mammals. *Journal of paleontology* **57**:197–207.

Walsh, W.R. and Guzelsu, N. 1993. Mineral-organic interfacial bonding and the mechanical properties of cortical bone. *Biomimetics* **1**:199–217.

———1994. Compressive properties of cortical bone: mineral-organic interfacial bonding. *Biomaterials* **15**:137–45.

Wang, X.T. and Ker, R.F. 1995. Creep rupture of wallaby tail tendons. *Journal of experimental biology* **198**:831–45.

Wang, X.T., Ker, R.F. and Alexander, R.M. 1995. Fatigue rupture of wallaby tail tendons. *Journal of experimental biology* **198**:847–52.

Warren, W.E. and Kraynik, A.M. 1988. The linear elastic properties of open-cell foams. *Transactions of the ASME: journal of applied mechanics* **55**:341–46.

Watts, D.C., El Mowafy, O.M. and Grant, A.A. 1987. Temperature-dependence of compressive properties of human dentin. *Journal of dental research* **66**:29–32.

Waters, N.E. 1980. Some mechanical and physical properties of teeth. In: *The mechanical properties of biological materials*. J.F.V. Vincent and J.D. Currey (eds.). Symposia of the Society for Experimental Biology, no. 34. Cambridge, UK, Cambridge University Press, pp. 99–135.

Wei, X. and Messner, K. 1996. The postnatal development of the insertions of the medial collateral ligament in the rat knee. *Anatomy and embryology* **193**:53–59.

Weibull, W. 1951. A statistical distribution function of wide applicability. *Journal of applied mechanics* **18**:293–305.

Weinbaum, S., Cowin, S.C. and Zeng, Z. 1994. A model for the excitation of osteocytes by mechanical loading-induced bone fluid shear stresses. *Journal of biomechanics* **27**:339–60.

Weiner, S. and Price, P.A. 1986. Disaggregation of bone into crystals. *Calcified tissue research* **39**:365–75.

Weiner, S. and Traub, W. 1986. Organization of hydroxyapatite crystals within collagen fibrils. *FEBS letters* **206**:262–66.

Weiner, S. and Wagner, H.D. 1998. The material bone: structure-mechanical function relations. *Annual review of materials science* **28**:271–98.

Weiner, S., Arad, T. and Traub, W. 1991. Crystal organization in rat bone lamellae. *FEBS letters* **285**:49–54.

Weiner, S., Arad, T., Sabanay, I. and Traub, W. 1997. Rotated plywood structure of primary lamellar bone in the rat: orientations of the collagen fibril array. *Bone* **20**:509–14.

Weiner, S., Traub, W. and Wagner, H.D. 1999. Lamellar bone: structure–function relations. *Journal of structural biology* **126**:241–55.

Weishampel, D.B. 1993. Beams and machines: modeling approaches to the analysis of skull form and function. In: Hanken and Hall (1993), pp. 303–44.

Wellnhofer, P. 1991. *The illustrated encylopedia of pterosaurs.* London, Salamander.

Wenk, H.-R. and Heidelbach, F. 1999. Crystal alignment of carbonated apatite in bone and calcified tendon: results from quantitative texture analysis. *Bone* **24**:361–69.

Wess, T.J., Hammersley, A.P., Wess, L. and Miller, A. 1998a. Molecular packing of type I collagen in tendon. *Journal of molecular biology* **275**:255–67.

———1998b. A consensus model for molecular packing of type I collagen. *Journal of structural biology* **122**:92–100.

White, A.A. III, Panjabi, M.M. and Hardy, R.J. 1974. Analysis of mechanical symmetry in rabbit long bones. *Acta orthopaedica Scandinavica* **45**:328–36.

White, S.N., Paine, M.L., Luo, W., Sarikaya, M., Fong., H., Yu, Z.K., Li, Z.C. and Snead, M.L. 2000. The dentino-enamel junction is a broad transitional zone uniting dissimilar bioceramic composites. *Journal of the american ceramic society* **83**:238–40.

Whitehouse, W.J. 1975. Scanning electron micrographs of cancellous bone from the human sternum. *Journal of pathology* **116**:213–24.

Whitehouse, W.J. and Dyson, E.D. 1974. Scanning electron microscope studies of trabecular bone in the proximal end of the human femur. *Journal of anatomy* **118**:417–44.

Wiffen, J., de Buffrénil, V., de Ricqlès, A. and Mazin, J-M. 1995. Ontogenetic evolution of bone structure in late cretaceous Plesiosauria from New Zealand. *Geobios* **5**:625–40.

Windig, J.J. and Nylin, S. 2000. How to compare fluctuating asymmetry of different traits. *Journal of evolutionary biology* **13**:29–37.

Wink, C.S. and Elsey, R.M. 1986. Changes in femoral morphology during egg-laying in *Alligator mississipiensis*. *Journal of morphology* **189**:183–88.

Witten, P.E. and Villwock, W. 1997. Growth requires bone resorption at particular skeletal elements in a teleost fish with acellular bone (*Oreochromis niloticus*, Teleostei: Cichlidae). *Journal of applied ichthyology* **13**:149–58.

Wolff, J.D. 1892. *Das Gesetz der Transformation der Knochen.* Berlin, A.Hirschwald.

Wong, F.S.L., Elliott, J.C., Anderson, P. and Davis, G.R. 1995. Mineral concentration gradients in rat femoral diaphyses measured by x-ray microtomography. *Calcified tissue international* **56**:62–70.

Woo, S.L-Y. 1981. The relationships of changes in stress levels on long bone remodeling. In: Cowin (1981), pp. 107–29.

Woo, S.L-Y., Kuei, S.C., Amiel, D., Gomez, M.A., Hayes, W.C., White, F.C. and Akeson, W.H. 1981. The effect of prolonged physical training on the properties of long bone: a study of Wolff's law. *Journal of bone and joint surgery.* **63-A**:780–87.

Wright, T.M. and Hayes, W.C. 1976. Tensile testing of bone over a wide range of strain rates: effects of strain rate, microstructure and density. *Medical and biological engineering* **14**:671–80.

———1977. Fracture mechanics parameters for compact bone: effects of density and specimen thickness. *Journal of biomechanics* **10**:419–30.

Yamamoto, T., Domon, T., Takahashi, S., Islam, N. and Suzuki, R. 2000. Twisted plywood structure of an alternating lamellar pattern in cellular cementum of human teeth. *Anatomy and embryology* **202**:25–30.

Yang, G., Kabel, J., Van Rietbergen, B., Odgaard, A., Huiskes, R. and Cowin, S.C. 1999. The anisotropic Hooke's law for cancellous bone and wood. *Journal of elasticity* **53**:125–46.

Yao, J.Q. and Seedhom, B.B. 1993. Mechanical conditioning of articular cartilage to prevalent stresses. *British journal of rheumatology* **32**:956–65.

Yoon, H.D., Caraco, B. and Katz, J.L. 1979. Further studies on the acoustic emission of fresh mammalian bone. In: *Ultrasonic symposium proceedings* cat. 79 CH 1482. New York, Institute of Electrical and Electronics Engineers, pp.399–404.

Yoshikawa, T., Mori, S, Santiesteban, A.J., Sun, T.C., Hafstad, E., Chen, J. and Burr, D.B. 1994. The effects of muscle fatigue on bone strain. *Journal of experimental biology* **188**:217–33.

Young, G.C., Karatajute-Talimaa, V.N. and Smith, M.M. 1996. A possible late Cambrian vertebrate from Australia. *Nature* **383**:810–12.

Young, W.C. 1989. *Roark's formulas for stress and strain.* New York, McGraw-Hill.

Yu, D.L., Ni, H.Y., Gong, Y.K. and Chen, X. 1999. Elastic modulus of human cementum. *Applied mathematics and mechanics—English edition.* **20**:1134–41.

Zaidi, M., Towhidul Alam, A.S.M., Shankar, V.S., Bax, B.E., Bax, C.M.R., Moonga, B.S., Bevis, P.J.R., Stevens, C., Blake, D.R., Pazianas, M. and Huang, C.L.H. 1993. Cellular biology of bone resorption. *Biological Reviews* **68**:197–264.

Zaman, G., Dallas, S.L. and Lanyon, L.E. 1992. Cultured embryonic bone shafts show osteogenic responses to mechanical loading. *Calcified tissue international* **51**:132–36.

Zawin, J.K. and Jaramillo, D. 1993. Conversion of bone-marrow in the humerus, sternum and clavicle: changes with age on MR-images. *Radiology* **188**:159–64.

Zeng, Y., Cowin, S.C. and Weinbaum, S. 1994. A fiber matrix model for fluid flow and streaming potentials in the canaliculi of an osteon. *Annals of biomedical engineering* **22**:280–92.

Zhang, D.J., Cowin, S.C. and Weinbaum, S. 1997. Electrical signal transmission and gap junction regulation in a bone cell network: a cable model for an osteon. *Annals of biomedical engineering* **25**:357–74.

Zhou, H., Chernecky, R. and Davies, J.E. 1994. Deposition of cement at reversal lines in rat femoral bone. *Journal of bone and mineral research* **9**:367–74.

Zihlman, A.L., Morbeck, M.E. and Goodall, J. 1990. Skeletal biology and individual life history of Gombe chimpanzees. *Journal of zoology, London* **221**:37–61.

Zioupos, P. 1998. Recent developments in the study of solid biomaterials and bone: 'fracture' and 'pre-fracture' toughness. *Materials science and engineering C* **6**:33–40.

Zioupos, P. and Currey, J.D. 1994. The extent of microcracking and the morphology of microcracks in damaged bone. *Journal of materials science* **29**:978–76.

Zioupos, P., Currey, J.D. and Sedman, A.J. 1994. An examination of the micromechanics of failure of bone and antler by acoustic emission tests and laser scanning confocal microscopy. *Medical engineering and physics* **16**:203–12.

Zioupos, P., Wang, X-T. and Currey, J.D. 1996a. Experimental and theoretical quantification of the development of damage in fatigue tests of bone and antler. *Journal of biomechanics* **29**:989–1002.

———1996b. The accumulation of fatigue microdamage in human cortical bone of two different ages in vitro. *Clinical biomechanics* **11**:365–375.

Zioupos, P., Currey, J.D., Casinos, A. and De Buffrénil, V. 1997. Mechanical properties of the rostrum of the whale *Mesoplodon densirostris*, a remarkably dense bony tissue. *Journal of zoology, London* **241**:725–37.

Zioupos, P., Currey, J.D. and Casinos, A. 2000. Exploring the effects of hypermineralisation in bone tissue by using an extreme biological example. *Connective tissue research*. **41**:229–48.

Zioupos, P., Currey, J.D. and Casinos, A. 2001. Tensile fatigue in bone: are cycles-, or time to failure, or both, important? *Journal of theoretical biology* **210**:389–99.

Ziv, V. and Weiner, S. 1994. Bone crystal sizes: a comparison of transmission electron microscopic and X-ray diffraction line width broadening techniques. *Connective tissue research* **30**:165–75.

Ziv, V., Sabanay, I., Arad, T., Traub, W. and Weiner, S. 1996. Transitional structures in lamellar bone. *Microscopy research and technique* **33**:203–13.

Zylberberg, L., Traub, W., De Buffrénil, V., Allizard, F., Arad, T. and Weiner, S. 1998. Rostrum of a toothed whale: ultrastructural study of a very dense bone. *Bone* **23**:241–47.

Zysset, P.K. and Curnier, A. 1996. A 3D damage model for trabecular bone based on fabric tensors. *Journal of biomechanics* **29**:1549–58.

Zysset, P.K., Guo, X.E., Hoffler, C.E., Moore, K.E. and Goldstein, S.A. 1999. Elastic modulus and hardness of cortical and trabecular bone lamellae measured by nanoindentation in the human femur. *Journal of biomechanics* **32**:1005–12.

INDEX

The animals indexed do not include bovines, chickens, dogs, horses, humans, mice, pigs, rats, or sheep, which are mentioned too frequently (and often parenthetically) for page numbers to be useful, and rarely include amphibia, fish, mammals, reptiles, or vertebrates, which are not specific enough to be useful. There are relatively few direct references to 'bone' and 'bones' in the index; usually the reference is implicit, unless the context suggests otherwise. 'Strength,' occurring more than 250 times in the text, is also not indexed as an item on its own, except for its displayed numerical values.